136th Annual Meeting of the Florida State Horticultural Society

Proceedings from the Florida State Horticultural Society
Volume 136

Daytona Beach, Florida, USA
11-13 June 2023

ISBN: 979-8-3313-2059-1

Printed from e-media with permission by:

Curran Associates, Inc.
57 Morehouse Lane
Red Hook, NY 12571

CURRAN ASSOCIATES INC.
proceedings.com

Some format issues inherent in the e-media version may also appear in this print version.

Printed by Curran Associates, Inc. (2025)

Florida State Horticultural Society, Inc.
2455 E. Sunrise Boulevard
Suite 816
Fort Lauderdale, FL 33304

www.fshs.org

Additional copies of this publication are available from:

Curran Associates, Inc.
57 Morehouse Lane
Red Hook, NY 12571 USA
Phone: 845-758-0400
Fax: 845-758-2634
Email: curran@proceedings.com
Web: www.proceedings.com

Proceedings of the 136th Annual Meeting

of the

FLORIDA STATE HORTICULTURAL SOCIETY

June 11–13, 2023

Daytona Beach, Florida

Library of Congress Catalogue No. 9-1702. This Library of Congress number is ongoing, having been assigned to these Proceedings in 1909.
ISSN 0886-7283

FLORIDA STATE HORTICULTURAL SOCIETY
2023 Board of Directors

John Griffis, Jr.
Chair of the Board

George Fitzpatrick
President

Chair of the Board	**John Griffis, Jr.,** Fort Myers
President	**George Fitzpatrick,** Miami Springs
President elect	**Ute Albrecht,** Immokalee
Secretary	**Fernando Alferez,** Immokalee
Treasurer	**Kevin Athearn,** Live Oak
Board Member-at-Large	**Richard Campbell,** Homestead
Board Member-at-Large	**Craig Campbell,** Winter Garden
Editor	**Mary Lamberts,** Miami Springs
Program Coordinator	**Germán V. Sandoya Miranda,** Belle Glade
Newsletter Editor	**Alicia Lamborn,** Macclenny
Poster Coordinator	**Anna Mészáros,** West Palm Beach
Student Best Paper Awards	**Xin Zhao,** Gainesville

Sectional Vice Presidents

Agroecology & Natural Resources	**Ashely Stonecipher,** DeLand
Citrus	**Lorenzo Rossi,** Fort Pierce
Handling and Processing	**Angelos Deltsidis,** Tifton, GA
Krome Memorial Institute	**Kimberly Morgan,** Immokalee
Ornamental, Garden & Landscape	**Bonnie Wells,** Cocoa
Vegetable	**Lisa Hickey,** Palmetto

Proceedings of the Florida State Horticultural Society 2023

CONTENTS

VOLUME CXXXVI PRINTED JUNE 2024

Published Papers

KROME MEMORIAL INSTITUTE (TROPICALS)

—Mango Summit—

CITRUS SECTION

Vegetable Section

Handling & Processing Section

Ornamental, Garden & Landscape Section

AGROECOLOGY AND NATURAL RESOURCES SECTION

2023 PRESIDENTIAL ADDRESS

George FitzPatrick

President of the Florida State Horticultural Society

Welcome to the 136th Annual Conference of the Florida State Horticultural Society. When we met last year in Sarasota, the very last thing on my mind, and very probably your minds also, was that I would be presenting the Presidential Address at this year's Conference. As most of you know, our president, Alan Chambers, resigned both his faculty position at the University of Florida and his presidency of our society. Alan has accepted a position in industry and has relocated from Florida to the State of Maryland. I am sure you all join me in wishing Alan all the best as he embarks on this new phase of his career. I think that it is safe to say that Florida's loss is Maryland's gain.

The bylaws of our society provide that the Board of Directors select a member to complete the unexpired term of an officer who resigns or who otherwise leaves office. I was very surprised to be asked if I would allow my name to be entered into consideration and even more surprised to be chosen.

My selection to serve as the Society's president has created an interesting precedent. You see, my wife, Mary Lamberts, who is currently serving as the Editor of the Society's Proceedings, was the Society's president back in 2009. I would have to check with our Society's history expert, Ed Etxeberria, for absolute accuracy, but I believe that this is the first time that a husband and wife have served as president of the Florida State Horticultural Society. There have been situations in the past where members of the same family have been presidents, such as Carl Campbell, who served as our president in 1984 and his son, Craig Campbell, who served as president in 2004.

Naturally, when I was selected to serve in this capacity, I had many questions about what a good president should do. Fortunately, I could ask Mary, and she was the source of very useful information. Also, I am very grateful to John Griffis, Jr. for his wise counsel. I also checked the back issues of the Society's Proceedings, to see if any of the previous presidents had the same sense of wonderment that I had. One of the more interesting sources of what a good president should do is published in the Presidential Address given by S. John Lynch in 1958.

When Mr. Lynch became the Society's president, in 1958, he asked the previous president, A.F. Camp, for advice. Dr. Camp said "The President, if he is wise, delegates all work, offers an occasional bit of advice, which no one utilizes, presides at meetings, but, in general, keeps his mouth shut."

In addition to thanking Mary, John and previous Society presidents who gave such helpful advice, there are many other people who deserve my gratitude. The Society Board of Directors, including President-Elect Ute Albrecht, Secretary Fernando Alverez, Treasurer Kevin Athearn, Program Coordinator Germán Sandoya-Miranda, Communications Coordinator Mark Ritenour, our two Membership Coordinators, Ali Sarkosh and Davie Kadyampakeni and our two Board Members at Large, brothers Craig Campbell and Richard Campbell. Also, I am very grateful to our 6 Sectional Vice Presidents: Ashley Stonecipher, Lorenzo Rossi, Angelos Deltsidis, Kimberly Morgan, Bonnie Wells and Lisa Hickey and to our Poster Session Coordinator, Anna Meszaros, who is not yet an official Board member, but who attends Board meetings and offers her words of wisdom.

In addition to those Board members, I would like to extend special thanks to Xin Zhao, David Austin and Hayk Khachatryan for their important work with the Student Program. Those of you who read the brief biographical note about me published in this past February's issue of our Society's Newsletter know that I have had many different jobs over the years with our Society, including Local Arrangements Chair, Presidential Gold Medal Award Selection Committee, Proceedings Associate Editor, Co-Editor, Editor-Publisher, and several others. They were all interesting and rewarding assignments, but none were as rewarding as my serving, under Xin Zhao's direction, as a judge in the Student Oral Paper competitions. You see, I tend to be a worrier, and I often feel anxiety about a number of issues, including the future of horticulture in Florida as well as the future of our Society. The best antidote for these kinds of feelings is to be in the audience when our current students present their work. I have been, and continue to be, extremely impressed with their intelligence, dedication, resourcefulness and persistence. Listening to these students, as I have done for many years now, convinces me that our horticultural future is in very good hands. Our Society's by laws provide that when I complete my term as president, I will serve as the Chair of the Society's Board of Directors for one year. Because of these responsibilities, I am unable to assist Xin in student paper judging, but as soon as I am able, I intend to ask Xin if she will let me come back into the fold of the judges who assist her in this very important, and most pleasant, task.

I also thank Gene McAvoy his leadership of the Sponsorship Committee, our Newsletter Editor Alicia Lamborn, our Social Media Coordinator Carol De Favari Tardivo and Karen Stauderman for her leadership on the Local Arrangements Committee. If I accidentally left anyone out, I apologize, but you can clearly see that our Society's success is due to the hard work of a large number of highly capable and dedicated people.

When John Griffis, Jr. spoke at last year's Annual Conference, he told us that the management agreement between the American Society for Horticultural Science (ASHS) and our Society would not be renewed past June 30, 2022. During the previous 20 years, ASHS staff members had managed our website, handled accounts receivable and accounts payable, managed our Annual Conferences, negotiated with the Conference Hotels, served as the publisher of our Proceedings and other associated functions. During the mid-part of last year, the Society's Board of Directors conducted a search of association management firms that deal with scientific societies, requested proposals from several of them, and approved a contract with AMCSource of Fort Lauderdale to manage FSHS business affairs, our website, and also serve as the site coordinators for the FSHS Annual Conferences. It is doubtful that the FSHS search considered the neighborhood of the physical office, but AMCSource's office address, 2455 East Sunrise Boulevard, in Fort Lauderdale, is only about 6 blocks west of the Glenn F. Bates Garden Center, at 3109 East Sunrise Boulevard. Headquarters of the Fort Lauderdale Garden Club (founded in 1927), it was named in honor of local nursery grower Glenn F. Bates (1873–1944), who, for many years operated Sandy Acres, a local plant nursery. . Mr. Bates was interesting for many reasons, one of which was that his daughter-in-law was Nancy Fairchild, daughter of Dr. David Fairchild. Mr. Bates and Dr. Fairchild co-authored a paper in the 1944 Proceedings (Bates, Glenn and David Fairchild. 1944. Protecting papaya plants from nematodes by the planting of Crotalaria spectbilis. [Proceedings of the Florida State Horticultural Society 57:181-182]); this was Mr. Bates's final contribution to our Society, as he died later that year. As a tribute to his memory, the Fort Lauderdale Garden Club members named their new building, opened in 1949, in his honor. So, among their other stellar attributes, AMCSource is located in a very accomplished horticultural neighborhood! Although AMCSource has a sizeable staff of professionals, I would like to highlight, and thank, three of them, Lindy Rome, Leslie Wysowski and Lisa Marburger, who have met with me and other Board members on a weekly basis for the past several months and have done a great job with the various transition issues.

In this period of transition, we have had some website issues, particularly with links to back issues of the Proceedings, and abstract submissions, dues payments and Conference registration issues. We are working hard to resolve these issues and I am happy to report significant progress.

One issue that was significantly affected during this transition from ASHS to AMC Source has been the Proceedings of the FSHS. Speaking as one of the former editors, I can attest that the editing and publishing of the Proceedings can be a very time-consuming, and quite expensive process. At Tuesday's Business Meeting, our Editor will explain several things which have been updated since our last Annual Meeting as she has been working closely with our Publisher to make it easier to submit a variety of publications.

One important Proceedings issue that the FSHS Board will be dealing with during the next year is the increasing tardiness of manuscript submission by some authors. As you all know, manuscripts are due at the time of the Annual Conference. While all authors agree to this provision when they submit their author's abstract, some authors nevertheless submit their manuscripts later. Editorially, we have always allowed some slack in submission timing, since some authors may face time challenges that are beyond their immediate control. However, this past year, a substantial number of manuscripts were submitted very late, including several submitted to the Editor in April of 2023 for papers that had been presented in June of 2022, a 10-month delay! Since the Editor cannot edit a manuscript until it is submitted, and the Publisher cannot do the layout work until the Editor sends her the edited manuscripts, the extreme tardiness in submitting manuscripts causes preventable delays in the publication of our Proceedings. I believe that some compromise will be necessary, but I also believe that a 10-month delay in submitting manuscripts is unreasonable as this affects more than the individual author. If and when the Board takes action on this issue, the members will be informed in the Society's Newsletter and under the Instruction to Authors tab on the FSHS website.

In reading through the previous issues of our Proceedings, I have noted in the Presidential Addresses several themes, such as the advice that Dr. Camp gave to Mr. Lynch in 1958. Another very common theme in Presidential Addresses is our previous presidents telling the membership that there are very serious challenges facing horticulture in Florida. There are many specific examples as one reads these addresses over the years: cold damage to horticultural crops, water management issues, devasting introduced peats and many more. I have been attending FSHS Annual Conferences since the late 1970s. My first published paper in the 1977 Proceedings, was on citrus blackfly control. dealt with the problem of controlling a newly introduced, exotic insect pest.

The establishment of sustainable control measures for introduced, exotic pests is an old story, but it is one of many. Yes, problems, threats, challenges and emergencies affecting Florida's horticulture have occurred in the past, are occurring at the present time, and can be expected to happen in the future. If the past, as reported in our Proceedings, can be seen as indicative, we will solve these present issues, just as our predecessors have solved the problems they encountered . A substantial number of research papers in this year's Annual Conference will describe the progress in solving many of these current problems.

I always come away from our Annual Conferences feeling optimism on the state of horticulture and I hope you all do, too.

2023 MERITORIOUS SERVICE AWARDS

For significant contributions to Florida horticulture

Presidential Gold Medal

The presidential gold medal is awarded to the individual who has contributed most to Florida horticulture through works published in the Proceedings of the Florida State Horticultural Society over the preceding six-year period in the Citrus Section.

Davie M. Kadyampakeni

Outstanding Commercial Horticulturist Award

This award shall be presented annually to an individual who has made significant contribution to the commercial Florida horticultural industry and to the FSHS.

Craig Campbell

President's Industry Award

The award will be presented to the senior author of a single best paper given at the Florida State Horticultural Society annual meeting by an industry author.

Not awarded

2023 Social Media Award

Caroline De Favari Tarvido

Florida Tomato Research Award

This award shall be given to the individual or group who have done work with the most potential to further the fresh market tomato industry in Florida through advances reported in any single publication during the two (2) previous calendar years

Rafia Khan, Dakshina Seal, Catherine Sabines, and Shawbeta Seal

University of Florida, IFAS, Tropical Research and Education Center, Homestead, FL

Friends of the Florida State Horticultural Society Award

To recognize individuals for their longtime commitment and support of the Florida State Horticulture Society

Not awarded

2023 Patron Members

Members who annually pledge additional support for the student programs of the Society

Ed Etxeberria

Gene McAvoy

Karen Stauderman

Erika Wallace

SECTIONAL BEST PAPER AWARDS

From the Proceedings of the Florida State Horticultural Society Volume 135

Awarded to the senior author of the best and most meritorious paper as printed in the previous year's Proceedings in each of the six sections

Agroecology and Natural Resources Section

Karen Stauderman

Citrus Section

Lamiaa Mahmoud and Manjul Dutt

Handling & Processing Section

Robert Madden, Joonhyuk Suh, Jonathan Crane, Alan Chambers, Yu Wang

Krome Memorial Institute

Cecilia Rubert Heller (*pictured*), Gerardo H. Nunez, and Jeffrey Williamson

Ornamental, Garden & Landscape Section

John L Griffis Jr., Malcolm M. Manners, Mollie M. Tew, Kazana Bennett (*pictured*), and Özlem Tuncay

Vegetable Section

Rafia Khan (*pictured*), Dakshina Seal, Catherine Sabines and Shawbeta Seal

STUDENT AWARDS

Student Scholarship Recipients

Ramsey Corn

Kara Fischer

Orestis Giannopoulos

Amit Godara

Shradd Hasharma

Nagamani Kanchupati

Byron Manzanero

Jean-Maudel Louizias

Susan Mostert

Dominick Padilla

Ankit Pokhrel

Jose Prieto

Zachary Ray

Josue St. Fort

Yasmeen Saleemdk

Duplicate Sambani

Mari Schroeder

Joseph Volpe

Zimmy Wang

Emily Worbington

STUDENT AWARDS

Student Best Oral Presentation Awards – 2023

First Place—Interactions of Four Citrus Scion/Rootstock Combinations in Response to Salinity Stress

Neus Alcon Bou[1], Sandra Guzman[1], Edzard van Santen[2], and Lorenzo Rossi[1]

[1]Indian River Research and Education Center, University of Florida, IFAS, Fort Pierce, FL;
[2]Agronomy Department, University of Florida, Gainesville, FL

Second Place—Profiling of HLB-Symptomatic Sweet Orange Fruit to Understand the Cause and Mechanism of Poor Fruit Attributes

Mary Sutton and Tripti Vashisth

Citrus Research and Education Center, University of Florida, IFAS, Lake Alfred, FL

Third Place—Evaluation of Tropical Pumpkin for Yield and Horticultural Traits in Florida

Prerna Sabharwal, Yuqing Fu, Swati Shrestha, Geoffrey Meru

Tropical Research and Education Center, University of Florida. IFAS, Homestead, FL

From L: George Fitzpatrick (President), Neus Bou, Mary Sutton, Zimmy Wang, Prerna Sabharwal, Xin Zhao (Student Coordinator)

Best Written Student Paper Awards—2023

not awarded

Best Poster Awards – 2023

First Place—Hedging and Thinning to Mitigate Fruit Disorder Incidence in HLB-affected 'Sugar Belle' Mandarin

Joseph H. Volpe and John M. Chater
Citrus Research and Education Center – University of Florida, IFAS, Lake Alfred, FL

Second Place—High-throughput Phenotyping and Fruit Juice Quality of 'Oll-8' Sweet Orange on Different Rootstocks

Emily Worbington, Heaven Heflin, Joe H. Volpe, and John M. Chater
University of Florida, IFAS, Citrus Research and Education Center, Lake Alfred, FL

Third Place—Water Use Efficiency and Soil C And N Dynamics for Tomatoes Produced in Conventional and Compact Bed Geometries

Shahd Alrabia and Rao Mylavarapu, Soil, Water and Ecosystem Sciences Department, University of Florida, IFAS, Gainesville, FL

Lisa Hickey (left) and Dominick Padilla

Best Poster Awards—2022

First Place—Toward Developing a Specialty Pepper (Capsicum annuum) Breeding Line with Bacterial Spot Resistance and Improved Fruit Quality Traits

Dominick Padilla[1], Lucas Manko[1], Jeff Jones[2], and Bala Rathinasabapathi[1]

[1]Horticultural Sciences Department, University of Florida, Gainesville, FL; [2]Plant Pathology Department, University of Florida, Gainesville, FL

Second Place—Success Rate of Propagation by Cuttings in American Beautyberry (*Callicarp americana*)

Anna Goebel[1*], Brian Bovard[2], and Sean Wortman-Goebel[3]

[1]Department of Biological Sciences, Florida Gulf Coast University, Fort Myers, FL; [2]Department of Ecology and Environmental Studies, Florida Gulf Coast University, Fort Myers, FL; [3]Florida State University, Tallahassee, FL

Third Place—Evaluation of Micro-sprinkler Irrigation for Reducing Water Use During Strawberry Establishment and Frost Protection in Florida

Josue St Fort[1], Carlene A. Chase[2*], and Vivek Sharma[3]

[1]School of Natural Resources and Environment, University of Florida IFAS, Gainesville, FL; [2]Horticultural Sciences Department, University of Florida IFAS, Gainesville, FL; [3]Agricultural and Biological Engineering, University of Florida IFAS, Gainesville, FL

Honorary Members (1988 to Present)*

Albrigo, L. Gene	2012
Anderson, J.B.	1922
Anderson, Shirley F.	2002
Beckenbach, J.R.	1967
Berckmens, P.J.	1893
Berger, E.W.	1940
Berry, Robert E.	1987
Blackmon, G.H.	1964
Bosanquet, L.P.	1924
Brecht, Jeffrey K.	2014
Brown, Arthur C.	1952
Bryan, Herbert H.	2003
Bullock, Robert C.	2008
Burgis, Donald S.	1980
Calvert, David V.	1997
Camp, A.F.	1956
Campbell, Beverly	2010
Campbell, C.W.	1988
Cantliffe, Daniel J.	2006
Carlton, R.A.	1962
Castle, William S.	2011
Chase, J.C.	1939
Chase, S.O.	1939
Childers, N.F.	1993
Clayton, H.G.	1956
Colburn, Burt	1970
Commander, C.C.	1952
Cooper, W.C.	1981
Dickey, R.D.	1968
Edsall, R.S.	1967
Etxeberria, Ed	2016
Everett, Paul H.	1986
Fairchild, David	1922
Fifield, Willard M.	1955
Flagler, H.M.	1903
Floyd, Bayard F.	1944
Floyd, W.L.	1939
Ford, Harry	1985
Forsee, W.T., Jr.	1973
Gaitskill, S.H.	1909
Gardner, Frank E.	1967
Garrett, Charles A.	1951
Goldweber, Seymour	1984
Grierson, William	1979
Guzman, Victor L.	1987
Haden, Mrs. Florence P.	1934
Hall, David J.	2011
Harding, Paul L.	1968
Hart, W.S.	1909
Hastings, H.G.	1939
Hatton, Thurman T.	1987
Hayslip, Norman C.	1981
Henricksen, H.C.	1939
Holland, Frank L.	1962
Holland, Spessard L.	1945
Hoyt, Avery S.	1950
Hoyt, R.D.	1914
Hubbard, E.S.	1922
Hume, H. Harold	1927
Ismail, Mohamed A.	1996
Jackson, Larry K.	2000
Jamison, F.S.	1962
Johnson, Warren O.	1965
Jones, John Paul	1997
Kender, Walter J.	2000
Knight, Robert J., Jr.	2006
Koo, R.C.J.	1978
Krezdorn, A.H.	1979
Krome, William H.	1973
Krome, William J.	1927
Krome, Mrs. Isabelle B.	1960
Lawrence, Fred P.	1973
Lipsey, L.W.	1924
Locascio, Salvadore J.	1996
Logan, J.H.	1965
Lynch, S. John	1975
MacDowell, Louis G.	1968
Magie, Robert O.	1977
Mathias, A.F.	1972
Matthews, Richard F.	1992
Mayo, Nathan	1940
McCornack, A.A.	1986
Menninger, Edwin A.	1964
Miller, Leon W.	1972
Miller, Ralph L.	1972
Montelaro, James	1985
Morton, Julia F.	1989
Mounts, M.U.	1958
Mowry, Harold	1950
Murdock, Del I.	1984
Newell, Wilmon	1940
Norman, Gerald G.	1967
Norris, Robert, E.	1962
O'Byrne, Frank M.	1962
Overman, A.J.	1988
Painter, E.O.	1909
Peterson, J. Hardin	1950
Pratt, J.B.	1980
Redmond, D.	1893
Reed, R.R.	1970
Reitz, Herman J.	1970
Reitz, J. Wayne	1955
Robinson, T. Ralph	1942
Rolfs, P.H.	1921
Rolfs, Mrs. P.H.	1921
Ruehle, George D.	1958
Sargent, Steven A.	2018
Saunders, Fred	1999
Sharpe, Ralph H.	1974
Shaw, Miss Eleanor G.	1927
Sherman, Wayne B.	2003
Showalter, Robert K.	1984
Singleton, Gray	1962
Skinner, L.B.	1931
Sloan, G. Dexter	1964
Smith, Paul F.	1972
Smoot, John J.	1986
Spalding, Donald H.	1987
Spencer, E.L.	1962
Stamps, Robert	2010
Steffani, C.H.	1958
Stephens, James M.	1995
Stevens, H.B.	1934
Swingle, W.T.	1941
Syvertsen, James P	2013
Taber, George L.	1914
Tait, W.L.	1941
Talbott, George M.	1980
Tenny, Lloyd S.	1956
Thompson, Ralph P.	1962
Thompson, W.L.	1962
Thullbery, Howard A.	1962
Todd, Norman	1991
Tucker, David P.	1999
Veldhuis, M.K.	1972
Ward, W.F.	1962
Waters, Will E.	1997
Webber, H.J.	1941
Wedgworth, Ruth S.	1965
Wenzel, F.W.	1973
Wheaton, T. Adair	1994
Wilfret, Gary J.	1998
Wilson, Lorenzo A.	1934
Wiltbank, William J.	1987
Winsberg, Theodore W.	2007
Winston, J.R.	1960
Wolfe, H.S.	1964
Young, T.W.	1978
Yothers, W.W.	1976
Ziegler, L.W.	1976

**Date listed is the year in which the award was received.*

Presidents of the Florida State Horticultural Society, 1888–Present

1888–96	Dudley W. Adams	1965	Ernest L. Spencer	1995	Mohamed A. Ismail
1897–1904	George L. Taber	1966	Arthur F. Mathias	1996	Walter J. Kender
1905–06	C.T. McCarty	1967	Ed H. Price, Jr.	1997	Fred Saunders
1907–08	P.H. Rolfs	1968	J.R. Beckenbach	1998	Larry E. Beasley
1909	William C. Richardson	1969	G.M. Talbott	1999	David W. Buchanan
1910–22	H.H. Hume	1970	F.E. Gardner	2000	Will E. Waters
1923–29	L.B. Skinner	1971	O.R. Minton	2001	Frederick S. Davies
1930–36	John S. Taylor	1972	R.A. Dennison	2002	William S. Castle
1937	C.W. Lyons	1973	B.E. Colburn	2003	Jonathan H. Crane
1938–40	Charles I. Brooks	1974	G.G. Norman	2004	Craig Campbell
1941	T. Ralph Robinson	1975	Leon Miller	2005	George J. Hochmuth III
1942	Henry C. Henricksen	1976	John W. Sites	2006	Jacqueline K. Burns
1943–47	Frank M. O'Byrne	1977	J.B. Pratt	2007	Peter McClure
1948	William F. Ward	1978	R.R. Reed	2008	James P. Syvertsen
1949	Frank Stirling	1979	J.F. Morton	2009	Mary Lamberts
1950	Leo H. Wilson	1980	C. Wayne Hawkins	2010	Jeffrey K. Brecht
1951	G. Dexter Sloan	1981	W. Grierson	2011	Richard Tyson
1952	Frank L. Holland	1982	Roger Young	2012	Juanita Popenoe
1953	R.S. Edsall	1983	Charles A. Conover	2013	Nancy Roe
1954	M.U. Mounts	1984	Carl W. Campbell	2014	Ed Etxeberria
1955	H.A. Thullbery	1985	Fred Bistline	2015	Steve Sargent
1956	R.A. Carlton	1986	Al H. Krezdorn	2016	Chris Oswalt
1957	R.E. Norris	1987	Richard F. Matthews	2017	Mark Ritenour
1958	A.F. Camp	1988	T.T. Hatton	2018	Eric Simonne
1959	S. John Lynch	1989	W.H. Krome	2019	Gene McAvoy
1960	W.L. Thompson	1990	Tom J. Sheehan	2020	Jeffrey Williamson
1961	Ruth S. Wedgworth	1991	Larry K. Jackson	2021	Noris Ledesma
1962	John H. Logan	1992	Daniel J. Cantliffe	2022	John Griffis, Jr.
1963	Herman J. Reitz	1993	Michael O. Taylor	2023	George Fitzpatrick
1964	Willard M. Fifield	1994	Salvadore J. Locascio		

Outstanding Growers or Commercial Horticulturists

1997	Norman Todd	Grove Crafters, Labelle, FL
1998	Ted Winsberg	Green Cay Farms, Boynton Beach, FL
1999	Larry K. Jackson	Horticultural Consultant, Auburndale, FL
2000	Gary E. Zill	Zill High Performance Plants, Boynton Beach, FL
2001	Murray J. Corman	Garden of Delights, Davie, FL
2002	Craig Campbell	Valent Biosciences, Orlando, FL
2003	Derek Burch	Masterworks, Plantation, FL
2004	Scott Emerson	Citrus & Vegetable Magazine, Tampa, FL
2005	James "Buster" Pratt	Hanes City, FL
2008	David J. Hall	HDH Agri Products, Tavares, FL
2009	Kenneth Shuler	Stephen's Produce, Jupiter, FL
2010	Peter McClure	Evans Properties, Okeechobee, FL
2011	Michael Edenfield	Bayer CropScience, Windemere, FL
2012	Robert T. McMillan, Jr.	Kerry's Bromeliads Nursery Inc., Homestead, FL
2013	Alvin Cheng	JBT FoodTech, Lakeland, FL
2014	*not awarded*	
2015	*not awarded*	
2016	Steve Rogers	Ecostat Inc., Highland City, FL
2017	Noris Ledsesma	Fairchild Tropical Bodanic Garden, Homestead, FL
2018	Erin Harlow	Dubal County Extension, Jacksonville, FL
2019	*not awarded*	
2020	Noris Ledesma	Miami, FL
2021	Noris Ledesma	Miami, FL
2022	*not awarded*	
2023	Craig Campbell	Valent Biosciences, Orlando, FL

FSHS Annual Meeting Host Cities

1888	Ocala
1889	Orlando
1890	De Land
1891	Interlachen
1892	Ormond Beach
1893	Pensacola
1894	Jacksonville
1895	Jacksonville
1896	Jacksonville
1897	Orlando
1898	Orlando
1899	Jacksonville
1900	Jacksonville
1901	St. Augustine
1902	Tampa
1903	Miami
1904	Jacksonville
1905	Jacksonville
1906	Jacksonville
1907	St. Petersburg
1908	Gainesville
1909	Daytona
1910	Orlando
1911	Jacksonville
1912	Miami and Havana, Cuba
1913	DeLand
1914	Palatka
1915	Tampa
1916	Arcadia
1917	Arcadia
1918	Ft. Myers
1919	Orlando
1920	Ocala
1921	Miami
1922	Lakeland
1923	Orlando
1924	Tampa
1925	Eustis
1926	Cocoa Beach
1927	Bradenton
1928	Winter Haven
1929	Clearwater
1930	Sebring
1931	Miami and Homestead (spring meeting) Mariana (fall meeting)
1932	Gainesville
1933	Lake Wales
1924	Orlando
1935	Vero Beach
1936	DeLand
1937	Ocala
1939	Winter Haven
1939	Hollywood
1940	Tampa
1941	Orlando
1942	West Palm Beach
1943	Winter Haven
1944	Winter Haven
1945	Orlando
1946	Miami
1947	St. Petersburg
1948	West Palm Beach
1949	Tampa
1950	Winter Haven
1951	West Palm Beach
1952	St. Petersburg
1953	Daytona Beach
1954	Miami Beach
1955	Clearwater
1956	Orlando
1957	Miami Beach
1958	Clearwater
1959	Miami
1960	Tampa
1961	Miami Beach
1962	Miami Beach
1963	Miami Beach
1964	Miami Beach
1965	Miami Beach
1966	Miami Beach
1967	Miami Beach
1968	Miami Beach
1969	Miami Beach
1970	Miami Beach
1971	Miami Beach
1972	Miami Beach
1973	Miami Beach
1974	Miami Beach
1975	Lake Buena Vista
1976	Miami Beach
1977	Lake Buena Vista
1978	Miami Beach
1979	Lake Buena Vista
1980	Miami Beach
1981	Lake Buena Vista
1982	Miami Beach
1983	Daytona Beach
1984	Miami Beach
1985	Tampa
1986	Miami Beach
1987	Orlando
1988	Miami Beach
1989	Tampa
1990	Lake Buena Vista
1991	Miami Beach
1992	Tampa
1993	Miami Beach
1994	Orlando
1995	Orlando
1996	Orlando
1997	Orlando
1998	St. Petersburg
1999	Stuart
2000	Lake Buena Vista
2001	Stuart
2002	Marco Island
2003	Orlando
2004	Orlando
2005	Tampa
2006	Tampa
2007	Palm Beach Gardens
2008	Ft. Lauderdale
2009	Jacksonville
2010	Crystal River
2011	St. Petersburg
2012	Delray Beach
2013	Sarasota
2014	Clearwater
2015	St. Augustine
2016	Stuart
2017	Tampa
2018	Fort Lauderdale
2019	Maitland
2020	Online via UF TEAMS
2021	Daytona Beach
2022	Sarasota
2023	Daytona Beach

Proc. Fla. State Hort. Soc. 136:1. 2023.

—Scientific Note—

Identifying and Overcoming Barriers to Expanding Passion Fruit Production in Florida

Mark Bailey*

Marion County Cooperative Extension, University of Florida, IFAS, Ocala, FL

Keywords. *Passiflora edulis*, marketing, sustainable

Passion fruit (*Passiflora edulis*) is an emerging alternative crop that has significant growth potential to generate profitability for farms of any size, and benefit consumers with high-quality Florida-grown produce. Currently, passion fruit production in Florida is mainly limited to South Florida. First-time commercial production in Central and North-Central Florida began in 2020. As of 2023 production in these regions has expanded to 11 small farms. Presently, most of these farms are producing passion fruit on acreage ranging from ¼ - 2 acres. In order for sustainable passion fruit production and expansion to occur, there are several significant barriers to overcome. The first barrier is the grower adoption of passion fruit production. The second barrier is establishing effective and sustainable crop production methods. The third barrier is evaluating consumer preferences and effective marketing of passion fruit. Funding for this ongoing research has been provided by a SEEDIT and SARE grant.

Surveys of current growers and University of Florida faculty who have experience with passion fruit production identified key barriers to expanding passion fruit production. Freeze damage of passion fruit is the primary threat to expanding production in North-Central Florida. Effective freeze damage mitigation strategies have been successfully demonstrated across several farms from 2021–2023. Pests and pathogens are additional significant threats to passion fruit production throughout Florida. In North-Central Florida significant fungal disease damage has been observed in locations with poor soil drainage, most notably *Fusarium* sp. and *Colletotrichum* sp. (Anthracnose) have been widely observed across multiple farm locations and cause rapid and significant damage to new vegetative growth as well as all stages of flower and fruit development. Integrated pest management strategies and sustainable production practices are currently being explored. Grower adoption of passion fruit as a commercial crop is in part dependent upon potential consumer awareness and interest in purchasing passion fruit. To determine potential consumer interest in passion fruit, a sensory evaluation of passion fruit was conducted in Marion County, FL at the Ocala Downtown Market. A total of 111 participants took the anonymous survey. The gender demographics of the participants included 79 females, 29 males, 2 preferred not to say, and 2 preferred a different term. The ethnic and racial demographics included 79 non-Hispanic and 21 Hispanic, 78 White, 9 Asian/Pacific Islander, 5 Black/African-American, and 1 Native American.

Participants were asked if they had previously consumed fresh passion fruit. They responded 54% (n = 60) "no" they had not, while 46% (n=51) responded "yes" they had. Participants were then asked if they had consumed *any* kind of passion fruit product. They responded 34% (n = 38) "no" they had not, while 66% (n = 73) "yes" they had. Next, participants were provided with a passion fruit sample consisting of freshly cut passion fruit and passion fruit juice that consisted of 90% pear juice and 10% passion fruit juice.

*Corresponding author. Email: ironhill@ufl.edu

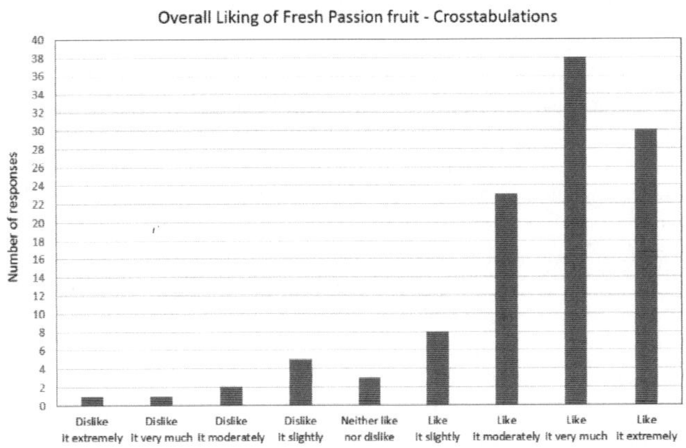

Fig. 1. Overall liking of fresh passion fruit with the number of responses.

A likability rating scale of 1–9 was used for the survey. The likability scale ranged from a rating of 1 ("dislike it extremely") to a 9 ("like it extremely"), with seven rating choices in between. Participants provided the following ratings for fresh passion fruit: overall appearance of 6.39, aroma of 7.86; overall texture of the fresh fruit pulp and seeds of 7.1; flavor of 7.46; and overall liking of 7.45 (Fig. 1). Participants next evaluated the juice product. They made the following ratings: aroma–7.43; flavor–7.92; and overall liking–8.03. The survey concluded with a question about whether or not they would buy fresh passion fruit grown in Florida if it were available. Participants responded with 93% (n = 102) "yes" and 7% (n = 8) "no".

Suggestions for Future Direction

If prospective growers are provided with an effective marketing plan based upon evidence that consumers are interested in purchasing passion fruit, this could provide an incentive to plant more commercial acreage. Combined with effective sustainable production, passion fruit has the potential to become a significant commercial crop in Florida.

Proc. Fla. State Hort. Soc. 136:2–5. 2023.

Evaluation of Papaya Cultivars in the Valley of Les Cayes, Haiti

LUDGER JEAN SIMON*[1], JUNIOR ARISTIL[1,2], CLÉNOLD RAYMOND[1], SERGIO DUCLONA[1], AND PIERRE DUCLONA[1]

[1]*American University of the Caribbean, Pierre Toussaint Campus, Charpentier, Les Cayes, Haïti*

[2]*Centre de Recherche Interdisciplinaire pour la Vulgarisation Agricole et le Développement Local (CREIVADEL), Université Notre Dame d'Haïti, Faculté d'Agronomie, Redon, Torbeck, Haïti*

Keywords. fruit yield, disease index, papaya ringspot virus

Papaya is one of the most nutritionally and economically important fruits grown in Haiti. A field experiment was conducted at Ducis, in the Valley of Les Cayes, South Haiti, to evaluate six commercial papaya cultivars (Red Lady# 786, Gold Maradol, Red Maradol, Hawaiian Solo Sunrise, Tainung#1, and Known You #1) and two local accessions (CA 1208-1, CA 1310-1) for plant vigor, fruit yield, and tolerance to papaya ringspot virus (PRSV). The experimental layout was a randomized complete block design of eight papaya cultivars in three replications. The experimental unit comprised a row of 5 papaya plants spaced 2.5 m within and between the rows. The variables measured were stem diameter, plat height, disease index, and fruit yield. There was no significant difference between the papaya cultivars for stem diameter (*P*-value = 0.426818), but there were significant differences for plant height (*P*-value = 0.00559), disease index (*P*-value = 0.0464), and fruit yield (*P*-value = 0.000101). The papaya cultivar Tainung #1 (33124.27 kg·ha^{-1}), followed by Known You #1 (22813.33 kg·ha^{-1}), had the highest fruit yield while the cultivar Gold Maradol (Disease Index = 1.80) had the highest tolerance to PRSV at 6 months after planting.

Papaya (*Carica papaya* L.) is one of the most nutritionally and economically important fruits grown in Haiti. It is a rich source of antioxidant nutrients such as carotenes, vitamin C, and flavonoids (Le Bellec and Renard, 2002; Purseglove, 1991; Rice et al., 1990). Papaya is also the source of the digestive enzyme papain, which is an industrial ingredient used in brewing, meat tenderizing, pharmaceuticals, and cosmetics (Samson, 1989; Morton, 1987). In 2011, the national production estimate was 98 MT for an area of 9 ha (FAO, 2014). The production in Jamaica and the Dominican Republic was 5846 and 891,731 MT respectively in 2011 (FAO, 2014).

Papaya ringspot virus (PRSV) is the most severe papaya disease and is often the limiting factor in papaya production throughout the world (Nishina et al., 2000). PRSV is widespread in Haiti and the important papaya production countries of the Caribbean islands. The symptoms, biology and epidemiology of the Potyvirus causing PSRV in papaya are well documented (Tripathi et al., 2008; Gonsalves et al., 2010). The earliest symptoms are yellowing and vein-clearing of young leaves (Crane, 2013). Symptoms on leaves begin with a yellow mottling sometimes followed by severe blistering and distortion. Dark-green streaks and rings also appear on leaf stalks and stems. Symptoms on ripe fruits are dark orange-brown rings. The age of the plant when infected influences the reduction of tree vigor and fruit set. The disease adversely affects fruit flavor (Gonsalves, 1993).

Successful commercial production of papaya in Haiti is dependent on a well-defined PRSV management strategy. PRSV tolerant papaya cultivars have been developed from breeding programs in countries such as the United States and Taiwan. However, these materials have not yet been sufficiently evaluated in Haiti. Hence, the aim of this study was to evaluate six introduced papaya cultivars and two local selections for plant vigor, fruit yield, and tolerance to PRSV.

Materials and Methods

A field experiment was conducted from February 2014 to March 2015 at the Experimental Farm of the Club d'Agriculture pour la Recherche et l'Encadrement Technique (CARET) at Ducis, South Haiti (18.23590°N latitude; 73.88023°W longitude), and an altitude of 94 m above sea level. The location is classified as Af by Köppen and Geiger. The average temperature is 26.2 °C and the average annual rainfall is 2344 mm (Climate data, 2016). The soil of the experiment site was classed as a sandy clay, containing 37% clay, 13% silt, and 50 sand. The soil pH was 7.2.

Two local papaya cultivars (CA1208-1 and CA1310-1) and six commercial cultivars (Red Lady#786 F-1 Hybrid, Gold Maradol, Red Maradol, Hawaiian Solo Sunrise, Tainung#1 F-1 Hybrid, and Known You #1 F-1 Hybrid) were evaluated for tolerance to PRSV, plant vigor, and fruit yield (Table 1). Seeds were sourced from Aloha Seed & Herb for the commercial cultivars and from two different papaya orchards in Les Cayes for the local cultivars.

The experimental design was a randomized complete block of 8 papaya cultivars (CA 1208-1, CA 1310-1, Red Lady# 786, Gold Maradol, Red Maradol, Hawaiian Solo Sunrise, Tainung#1, and Known You #1) and three replications. The local cultivars CA-1208-1 and CA1310-1 were used as controls to test the performance of the commercial cultivars. The experimental unit comprised one row of 5 plants, with a spacing of 2.5 m between plants within and between the rows. This translates into an area of 31.25 m^2 per experimental unit, and a planning density of 1600 plants/hectare.

This research was supported by Catholic Relief Services

*Corresponding author. Email: ljeansimon@auc.edu.ht

Table 1. Features of the papaya cultivars evaluated in the Valley of Les Cayes, Haiti.

Cultivar[z]	Fruit size (kg)	Flesh color	Sweetness	Sex ratio	Other features
CA 1208-1	1.0–1.5	Yellow	Medium	50% Hermaphrodite 50% Female	Vigorous
CA1310-1	1.4–2.0	Yellow	Medium	50% Female 50% Male	Vigorous
Hawaiian Solo Sunrise	0.465	Red/Pink	Very sweet	66% Hermaphrodite 33% Female	No virus tolerance
Red Lady #786	1.5–2.0	Red	Medium	50% Hermaphrodite 50% Female	PRSV tolerant; heat tolerant
Known You #1	1.6–3.0	Yellow	Sweet	50% Hermaphrodite 50% Female	Early producer; sturdy plants
Tainung No.1	1.1	Red	Medium	50% Hermaphrodite 50% Female	Tolerant to PRSV; vigorous; very high yielding
Gold Maradol	2.8–3.7	Yellow	Low	50% Hermaphrodite 50% Female	Tolerant to *Cercospora* spp.
Red Maradol	1.5–2.6	Red	Medium	66% Hermaphrodite 33% Female	No virus tolerance

[z]Information for the commercial cultivars are from the seed supplier (Aloha Seed, El Centro, CA (https://www.alohaseed.com/). Information for the local cultivars are from the authors.

Papaya seedlings were planted in a nursery on 18 December 2013 at the Experimental Farm of the American University of the Caribbean (AUC) at Charpentier, Les Cayes, and were transplanted in the field on 3 February 2014 when they were 10–12 cm high. The fertilizer program consisted of two applications of 12N–12P–20K fertilizer at the rate of ¼ lb per papaya plant, one month and four months after planting (MAP). This is the equivalent of approximately 364 $kg \cdot ha^{-1}$ of the 12N–12P–20K fertilizer during the crop cycle.

Irrigation water was scheduled at weekly intervals during the first month after transplanting. Thereafter, irrigation was scheduled based on an estimated water requirement of 50–74 mm every three to four weeks (Samson, 1989). Irrigation decisions were based on weekly assessment of soil moisture level, using the feel method (USDA, 2001). The trial plots were watered when the soil moisture level reached 50% field capacity. Weeds were controlled using garden hoes. There were no intentional control measures for pests and diseases.

Stem diameter and plant height were measured at 3 months after transplanting (MAT), when at least 50% of the plants were flowering. Disease index (DI) was calculated for PRSV three times, at approximately one-month intervals, during initial fruit setting stage. DI was obtained by multiplying the incidence of PRSV (number of diseased plants divided by number of observed plants) by the severity of PRSV, using a scale of 1–9, with 1 referring to absence of disease and 9 to plants dying or completely died (Rohrmoser, 1985).

Papaya fruits were harvested, and all marketable fruits per experimental unit were counted, weighed, and yield data registered, when traces of yellow or orange color appeared on the skin (Purseglove, 1991; Crane, 2013), the first harvest occurred 7 MAT, and the last harvest approximately 14 MAT. All fruit harvested in each experimental unit were weighted in the field using a hanging scale. Fruits harvested at 7 MAT, 8 MAT and 9 MAT were considered as early yield, and fruits harvested from 10 MAT through 14 MAT as late yield.

The measured data were analyzed with R Studio version 4.2.2 (RStudio Team, 2020) and the statistical package Agricolae (de Mendiburu and Yaseen, 2020). Data exploration was performed using descriptive statistics procedures, followed by analysis of variance (ANOVA) at the 5% significance level, and mean separation using the Tukey procedure.

Results and Discussions

There was no significant difference between treatment means for stem diameter (P-value = 0.426818). In contrast, there was a highly significant difference for plant height (P-value = 0.00559). The local selection CA 1310-1 was the tallest while Red Maradol, Known You #1, and CA 1208-1 were the shortest (Table 2). Plant heights for the other cultivars were intermediate. The stem diameters and plant heights obtained in this experiment at flowering are consistent with those reported by Silva et al. (2018). A taller papaya plant means more effort to harvest the fruits which translates into increased cost of labor for the commercial grower. Hence, the cultivars Known You #1 and CA 120801 could be preferable for this growth trait. Phenological differences are attributed to genetic dissimilarities in growth responses among different cultivars (Jeyakumar et al., 2007; Galan Sauco and Rodriguez Pastor, 2007).

Table 2. Papaya stem diameter and plant height at flowering, and disease index at 6 months after planting.

Cultivar	Stem diameter (cm)	Plant height (cm)	Disease index
CA1208-1	3.14 a	80.33 b	2.30 ab
CA1310-1	3.92 a	127.50 a	5.00 ab
Red Lady #786	3.74 a	84.83 ab	3.33 ab
Gold Maradol	4.06 a	106.33 ab	1.80 b
Red Maradol	3.11 a	64.87b	4.50 ab
Hawaiian Solo Sunrise	3.66 a	107.93 ab	7.00 a
Tainung #1	4.15 a	107.37 ab	4.13 ab
Known You #1	3.52 a	88.00 b	3.67 ab
MSD5%	1.873786	44.85512	4.829323
P-value	0.426818	0.00559*	0.0464**

Cultivars with the same letter are not significantly different. Signif. codes: 0 '***' 0.001 '**' 0.01 '*' 0.05 '.' 0.1 ' ' 1.

Table 3. Marketable fruit yield for papaya cultivars evaluated in the valley of Les Cayes from February 2014 through March 2015.

Cultivar	Fruits harvested (fruits/ha)	Total yield (kg·ha^{-1})	Early yield (kg·ha^{-1})	Late yield (kg·ha^{-1})
CA1208-1	8,746.67 c	8824.00 bc	2320.00 b	6504.00 b
CA1310-1	6,720.00 c	9690.63 bc	1960.00 b	7730.63 b
Red Lady #786	11,413.33 bc	20226.67 abc	15213.33 a	5013.33 b
Gold Maradol	9,808.93 c	12925.60 bc	2813.33 b	10112.27 b
Red Maradol	7,360.00 c	7833.33 c	4213.33 b	3620.00 b
Hawaiian Solo Sunrise	20,800.00 ab	8822.27 bc	2293.33 b	6528.93 b
Tainung #1	31,253.33 a	33124.27 a	4973.33 b	28150.93 a
Known You #1	12,800.00 bc	22813.33 ab	15160.00 a	7653.33 b
MSD5%	10701.09	14011.52	9129.39	10315.9
P-value	1.42e-05 *	0.000101 **	0.000155***	1.51e-05 ***

Cultivars with the same letter are not significantly different. Signif. codes: 0 '***' 0.001 '**' 0.01 '*' 0.05 '.' 0.1 ' ' 1

There was no significant difference between treatment means for disease index at 4 MAT and 5 MAT (*P*-value = 0.2666 and *P*-value = 0.0482, respectively), but a significant difference was detected among the cultivars at 6 MAT (*P*-value = 0.0464). The papaya cultivar Hawaiian Solo Sunrise had the highest disease index, indicating its great susceptibility to PRSV compared to the other varieties tested (Table 2). The papaya cultivar Gold Maradol had the highest tolerance to PRSV at 6 MAT. Several authors reported the high susceptibility of solo papaya to PRSV (Davis et al., 2004; Escudero et al., 1994).

A significant difference was found among cultivar means for both the number of fruits harvested (*P*-value = 1.42e-05) and total fruit yield expressed in tons per hectare (*P*-value = 0.000101). "Tainung #1", followed by "Hawaiian Solo Sunrise" had the highest number of fruits per hectare, but there were no significant differences among the other six cultivars (Table 3). The papaya cultivar Tainung #1, followed by "Known You #1" had the highest total marketable fruit yield in kg·ha^{-1}. There were no significant differences among the remaining cultivars. The papaya cultivars Red Lady #786 and Known You #1 had the highest early yield (*P*-value = 0.000155***) while "Tainung #1" had the highest late yield (*P*-value = 1.51e–05 ***).

Although 'Hawaiian Solo Sunrise' was among the two cultivars with the highest number of fruits per hectare, it was among the group with the lowest total yield in kg ha^{-1} due to the small size of its fruits. Geonaga et al (2001) did not find any significant differences in weight of marketable fruits per hectare among the cultivars Tainung #1, Known You #1 and Red Lady under intensive management in Puerto Rico. Silva et al. (20018) reported higher fruit yields for the commercial cultivars Tainung #1 and Solo Sunrise compared to the local control in Botucatu, São Paulo, but no significant difference between Tainung #1 and Solo Sunrise.

The results of this experiment show that papaya can be successfully produced in the Valley of Les Cayes if PRSV tolerant cultivars are used by the producers. The cultivars Tainung #1 and Known You # 1 can be recommended for their high yield potentials. The cultivar Red Lady #786 can also be recommended for its high early yield. Further studies should assess under different environments and/or management settings the expression of disease tolerance, fruit yield, and fruit quality traits of papaya cultivars selected for PRSV resistance.

Literature Cited

Climate data. 2016. Accessed 18 April 2023. https://en.climate-data.org/location/987588/

Crane, J.H. 2013. Papaya growing in the Florida home landscape. Fact Sheet HS11. Florida Cooperative Extension Service, IFAS. University of Florida, Gainesville, FL. 7 pp.

Davis M.J, White T.L., & Crane J.H. 2004. Resistance to papaya ringspot virus in transgenic papaya breeding lines. *Proc. Fla. State Hort. Soc.* 117:241–245

de Mendiburu F. & Muhammad Yaseen M. 2020. Agricolae: Statistical procedures for agricultural research. R package version 1.4.0. https://myaseen208.github.io/agricolae/https://cran.r-project.org/package=agricolae.

Escudero J., Acosta A., Ramírez L.V., Caloni I.B., & Sifre G.R. 1994. Yield of three papaya genotypes and their tolerance to papaya ringspot virus in Puerto Rico. *J. Agric. Univ. P.R.* 78:111—121.

Food and Agriculture Organisation (FAO). 2014. FAO Stat. Accessed 18 April 2023. http://faostat3.fao.org/download/

Galan Sauco V. & Rodriguez Pastor M.C. 2007. Greenhouse cultivation of papaya. *Acta Hortic.* 740:191–195. https://doi.org/10.17660/ActaHortic.2007.740.22

Geonaga R., Irizarry H., & Rivera-Amador E. 2001. Yield and fruit quality of papaya cultivars grown at two locations in Puert Rico. *J. Agric. Univ. P.R.* 85(3–4):127–134

Gonsalves, D. 1993. Papaya Ringspot Virus (P-strain). *Crop Knowledge Master.* http://www.extento.hawaii.edu/kbase/crop/type/papring.htm

Gonsalves D., Tripathi S., Carr JB, & Suzuki J.Y. 2010. Papaya Ringspot virus. *The Plant Health Instructor.* https://doi.org/10.1094/PHI-I-2010-1004-01

Jeyakumar P., Kavino M, Kumar N., & Soorianthasundaram K. 2007. Physiological performance of papaya cultivars under abiotic stress conditions. *Acta Hortic.* 740:209–215. https://doi.org/10.17660/ActaHortic.2007.740.25

Le Bellec F. & Renard V. 2002. Le Grand Livre des Fruits Tropicaux. Editions Orphie. CEE. 189 pp.

Morton, J.F. 1987. Papaya. p. 336–346. In: Fruits of warm climates. Florida Flair Books, Miami.

Nishina M., Zee F., Ebesu R., Arakaki A., Hamasaki R., Fukuda S., Nagata N, Chia C.L., Nishijima W., Mau R., & Uchida R. 2000. Papaya production in Hawaii. F&N-3. College of Tropical Agriculture & Human Resources. University of Hawaii at Manoa. 8 pp.

Purseglove J.W. 1991. Tropical crops: Dicotyledons. Longman Scientific & Technical. Co-published with John Wiley & Sons Inc. New York. 719 pp.

Rice R.P., Rice L.W., & Tindall H.D. 1990. Fruit and vegetable production in warm climates. MacMillan Education Ltd. London and Basingstoke. 486 pp.

Rohrmoser, K. 1985. Handbook for field trials in technical cooperation. DeutscheGesellschaft für Technische Zusammenarbeit (GTZ) GmbH, Eschborn, Germany/Technical Centre for Agricultural and Rural Cooperation (CTA) Wageningen Netherlands. 300 pp.

R Studio Team 2020. RStudio: Integrated Development for R. RStudio,

PBC, Boston, MA. Accessed 18 April 2023. http://www.rstudio.com/
Samson J.A. 1989. Tropical fruits (Second ed.). Longman Scientific & Technical. Co-published with John Wiley & Sons, Inc. New York. p. 256–269.
Silva M.S., Leonel S., Souza J.M.A., Modesto J.H., Ferreira R.B., & Bolfarini A.C.B. 2018. Evaluation of papaya genotypes using agronomic descriptors and estimation of genetic parameters. *Biosci. J.*, Uberlândia. 34(4):943–951, ref.37.
Tripathi S., Suzuki J.Y., Ferreira S.A., & Gonsalves D. 2008. Papaya ringspot virus-P: Characteristics, pathogenicity, sequence variability and control. *Molecular Plant Pathology*. 9:269–280. https://doi.org/10.1111/j.1364-3703.2008.00467.x
United States Department of Agriculture (USDA). 2001. Estimating soil moisture by feel and appearance. USDA Natural Resources Conservation Service. 7 pp. Accessed 18 April 2023. https://nutrientmanagement.tamu.edu/content/tools/estimatingsoilmoisture.pdf

Krome Memorial Section

Proc. Fla. State Hort. Soc. 136:6. 2023.

Integration of Volatilomics and Metabolomics Unveils Key Flavor-related Biological Pathways in Different Carambola Cultivars

JINGWEN LI[1], ZHIXIN WANG[1], JONATHAN CRANE[2], AND YU WANG[1]*

[1]*Citrus Research and Education Center, University of Florida, IFAS, Lake Alfred, FL*

[2]*Tropical Research and Education Center, University of Florida, IFAS, Homestead, FL*

Keywords. carambola, flavor, metabolomics, non-volatile organic compounds, volatile organic compounds, volatilomics

—abstract—

Carambola is a tropical fruit that is highly sought after by consumers due to its unique flavor, star shape, and nutritional value. Enhancing the flavor quality of this fruit can increase the consumer acceptance and market demand. However, flavor is an intrinsic characteristic of fruits. Its decoding requires in-depth knowledge based on recognizing key biological pathways relevant to flavor formation and development. In this study, the volatile and non-volatile metabolites contributing to the flavor variation of five carambola cultivars were investigated by a novel strategy combining GC–MS/O-based volatilomics with LC–MS-based metabolomics. Several significant flavor-related pathways, involving biosynthesis or metabolism of amino acids, terpenoids, fatty acids, sugar and organic acid, and flavonoids were identified based on the enrichment analysis of important volatile and non-volatile metabolites. The results indicated that there were metabolites in the flavor-related pathways being up- or downregulated, leading to the differences in flavor traits of different carambola cultivars. This study could provide a valuable reference for breeders and researchers of interest in the mechanisms underlying the regulation of flavor, which would ultimately lead to the creation of carambola cultivars with more attractive flavor profiles and pleasurable consuming experiences.

The full paper can be found at: https://doi.org/10.1021/acs.jafc.3c02015

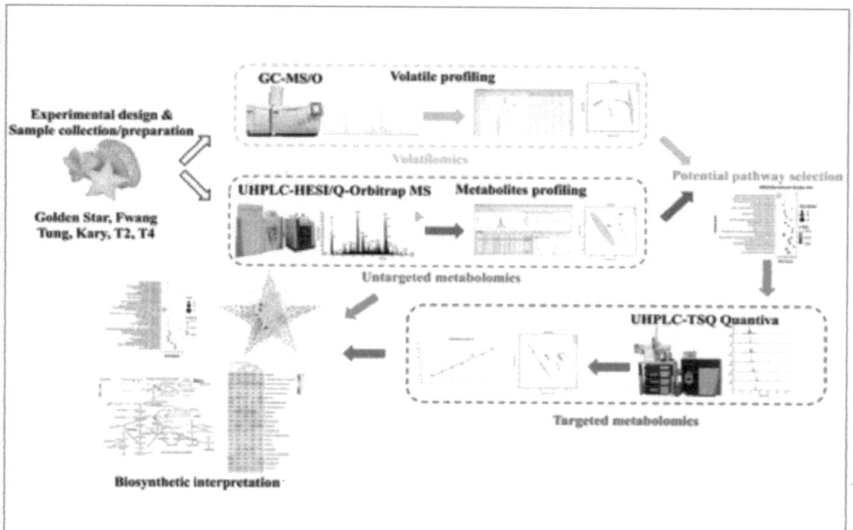

*Corresponding author. Email: yuwang@ufl.edu

Proc. Fla. State Hort. Soc. 136:7. 2023.

Analysis of Metabolomic Profiles and Evaluation of Biological Activities of Six Blackberry Cultivars

Hana Lee[1], Zhanao Deng[2], and Yu Wang[1]*

[1]*Citrus Research and Education Center, University of Florida, IFAS, Lake Alfred, FL*

[2]*Gulf Coast Research and Education Center, University of Florida, IFAS, Wimauma, FL*

—abstract—

Blackberries (*Rubus* sp.) are widely consumed fruits and are used in many processing products. Blackberry is known to have high antioxidant activity and prevent various diseases by its bioactive compounds including flavonoids and phenolic acids. Particularly, cyanidin 3-*O*-glucoside is a consistently predominant anthocyanin, which is responsible for the dark color in blackberry and has been reported to exert prominent bioactivity including anti-inflammatory, antioxidant, anti-cancer, and anti-neurodegenerative activities. Considering the rapid expansion of blackberry applications and the importance of their efficient use, it is essential to comprehend their nutritional profiles. This study aimed to compare the metabolomic profiles and biological activities of six different blackberry cultivars to better understand their nutritional and therapeutic potential. The targeted and untargeted analysis was performed using liquid chromatography-mass spectrometry to identify and quantify the metabolites present in blackberries. Partial least squares-discriminant analysis revealed a distinct clustering pattern of metabolites for each blackberry. The BL cultivars indicated high amounts of sugar, organic acid, and amino acid. The F cultivars showed high amounts of anthocyanin. We also measured the antioxidant contents and activities and the anti-inflammatory properties of six blackberry cultivars. Among six blackberries, the F6 cultivar had the highest antioxidant content and activities. In addition, the F6 cultivar protected the lipopolysaccharide-induced macrophages by decreasing the generation of nitric oxide, tumor necrosis factor-α, and reactive oxygen species. These cytoprotective effects of the F6 cultivar in RAW 264.7 cells may be because of its high antioxidant content and activities by anthocyanin, thus exerting anti-inflammatory properties. The metabolites composition of each blackberry cultivar through different biosynthetic pathways might affect nutritional and functional benefits. Moreover, our results provide information for potential applications of blackberry cultivars in the development of functional foods or processed products. However, further studies are required to fully understand the mechanism of blackberries in the prevention of inflammation-related diseases.

*Corresponding author. Email: yu.wang@ufl.edu

Proc. Fla. State Hort. Soc. 136:8. 2023.

—Scientific Note—

Effects of Foliar Application of Selenium and Iodine on Chemical and Sensory Characteristics of Blackberry and Raspberry Fruits

Carlos Henrique Milagres[1], Francielle Roberta Dias de Lima[2], Everton G. Morais[1], Gilson Gustavo Lucinda Machado[1], Elano Pinheiro Pereira[1], Shinsuke Agehara[2], Luiz Roberto Guimarães Guilherme[1], and Rafael Pio[1]*

[1]*University of Lavras, Lavras, Minas Gerais, Brazil*

[2]*Gulf Coast Research and Education Center, University of Florida, Wimauma, FL*

Keywords. biofortification, food security, phenolic compounds, small fruits

Selenium (Se) and iodine (I) are micronutrients that are essential for proper thyroid function in humans. Approximately 15% and 30% of the global population are affected by deficiencies of Se and I, respectively (White and Broadley 2009). For plants, Se and I are considered beneficial elements (Brown et al. 2022). It was found that applications of Se and I increase the accumulation of bioactive compounds and antioxidant activity in several fruit and vegetable crops (Medrano-Macías et al. 2016; Groth et al. 2020). Small fruit crops such as blackberry and raspberry (Rubus spp.) have significant amounts of bioactive compounds like phenolics (De Souza et al. 2014). Phenolic compounds in plant tissues can be divided into two groups: flavonoids (anthocyanins, flavonols, and isoflavones) and non-flavonoids (phenolic acids). This study aimed to examine the effects of foliar Se and I applications on chemical and sensory characteristics of blackberry (cv. BRS-Tupy) and raspberry (cv. Heritage) fruits.

The field experiment was carried out in Aiuruoca, Minas Gerais, Brazil (Subtropical highland climate - Cwb), in clay soil (39% clay), with adequate fertility attributes after addition of limestone and fertilizers. Seven treatments were evaluated: control, sodium selenate (SS) (inorganic Se), commercial organic Se (COS), commercial inorganic Se (CIS), SS and potassium iodide (PI) (SS+PI), COS+PI, and CIS+PI. Se and I application rates were 30 and 200 $g \cdot ha^{-1}$, respectively. The fruits were harvested at physiological maturity at 40, 54, and 73 days after treatment (DAT). Moisture, pH, total anthocyanin (cyanidin-3-glucoside), and color indices (L, C, and h) were measured. The treatments were arranged in a split-plot design with four replications, with harvest time as the main plot factor and fertilizer treatments as the sub-plot factor. Each blackberry and raspberry plots consisted of 6 and 10 plants, respectively. When the significance of the F-test was reached ($P \leq 0.05$), means were compared using the Duncan multiple range test ($P \leq 0.05$).

In both crops, lower moisture was found for COS and COS+PI compared with the other treatments. Lower pH values were found at 40 DAT than at 54 and 73 DAT, but they were not affected by Se and I treatments. At 54 DAT, COS and CIS increased total anthocyanin concentration in blackberry fruits by (83%) and (71%) respectively. CIS+PI increased total anthocyanin concentration in raspberry fruits (37%) compared with the control. Color indices increased in all Se and I treatments compared with the control, with the greatest increases induced by the I-containing treatments. The results demonstrate improved chemical and sensory characteristics of blackberry and raspberry fruits using foliar application of Se and I. To fully characterize the biofortification effects of Se and I, we are currently analyzing fruit samples to quantify additional quality variables, including Se, I, ascorbic acid, carotenoids, phenolics, and total antioxidant activity.

Literature Cited

Brown P.H., Zhao F.Z. & Dobermann A. 2022. What is a plant nutrient? Changing definitions to advance science and innovation in plant nutrition. *Plant and Soil*. 476(1–2):11–23. https://doi.org/10.1007/s11104-021-05171-w.

Groth S., Budke C., Neugart S., Ackermann S., Kappenstein F.S., Daum D., & Rohn S. 2020. Influence of a selenium biofortification on antioxidant properties and phenolic compounds of apples (*Malus domestica*). *Antioxidants*. 9(2):1–22. https://doi.org/10.3390/antiox9020187.

Medrano-Macías J., Leija-Martínez P., González-Morales S., Juárez-Maldonado A. & Benavides-Mendoza A. 2016. Use of iodine to biofortify and promote growth and stress tolerance in crops. *Front. Plant Sci*. 7. https://doi.org/10.3389/fpls.2016.01146.

De Souza V.R., Pereira P.A.P, Da Silva T.L.T., De Oliveira Lima L.C., R Pio R. & Queiroz F. 2014. Determination of the bioactive compounds, antioxidant activity and chemical composition of Brazilian blackberry, red raspberry, strawberry, blueberry and sweet cherry fruits. *Food Chemistry*. 156:362–368. https://doi.org/10.1016/j.foodchem.2014.01.125.

White P.J. & Broadley M.R.. 2009. Biofortification of crops with seven mineral elements often lacking in human diets - Iron, zinc, copper, calcium, magnesium, selenium and iodine. *New Phytologist*. 182(1):49–84. https://doi.org/10.1111/j.1469-8137.2008.02738.x.

The authors would like to thank the National Council of Technology and Scientific Development (CNPq), the Coordination for the Improvement of Higher Education Personnel (CAPES), and the Foundation for Research Support of the State of Minas Gerais (FAPEMIG) for financial support for this study.

*Corresponding author email: rafaelpio@ufla.br

Proc. Fla. State Hort. Soc. 136:9. 2023.

—Scientific Note—

Exploring Thornless Blackberry Cultivars while Comparing Trellising Styles in Containerized Production in North Florida

EVELYN "PRISSY" FLETCHER*

St. Johns County Extension, University of Florida, IFAS, St. Augustine, FL

Keywords. blackberry, cultivar, thornless, trellising

While wild blackberries grow commonly in pastures and natural areas of north Florida, they are commercially problematic, especially in a u-pick operation with their thorny growth habit. Fortunately, with breeding efforts from the University of Arkansas, thornless cultivars have been released with chill requirements low enough to produce fruit in Florida. While blackberry production has been in existence on small farms in neighboring Putnam County, FL, for many years, the coastal impacts of saltwater intrusion has been the limiting factor for St. Johns County, FL. To mitigate this environmental impact, the study included container-grown plants with soilless media and drip irrigation. Using 20-gallon squat pots, 'Caddo,' 'Ouachita,' 'Ponca,' and 'Prime Ark Freedom' were trialed at the University of Florida/IFAS (UF/IFAS) Hastings Agricultural Extension Center—Downtown Demonstration Unit, under 30% shade cloth. It consisted of 80 total plants divided by 4 cultivars, with each row representing one cultivar with 10 pots each. Rows 1–4 were designed with the T-shape trellis, and rows 4–8 were designed with the V-shaped trellis. Each pot contained 40% perlite, 30% peat moss and 30% coconut coir, in addition to a coconut coir weed mat on the surface after establishment. Drip irrigation was installed with one dripper/emitter per pot, using daily irrigation and weekly fertigation injections via Dosatrons (except during the dormant season). Plugs were transplanted in December 2020 from trays to 1-gal pots until the trellising design was completed in January 2021. They were then transplanted to 20-gal squat pots. During this winter season, Hastings received 283.75 hours of chill according to the UF/IFAS FAWN (weather) report. The trellising styles were meant to compare the V-shape to the T-shape methods to determine if the impacted the marketable yield by cultivar. The hypothesis was that the V-shaped trellising would allow for more air flow, which would decrease the chances of disease incidence, and increase marketable yields. Crops received integrated pest management with weekly scouting and "caution" labeled products throughout the growing season. Rust and Septoria leaf spots were the most common pathogens, and spider mites being the most problematic into harvest time.

Findings

Ripe and marketable berries were hand-harvested starting on April 18, 2022 for 'Prime Ark Freedom,' May 16 for 'Ponca,' and May 18 for 'Ouachita' and 'Caddo.' All cultivars completed their harvest by June 28 except for 'Prime Ark Freedom', which concluded earlier on June 10. 'Prime Ark Freedom' had the largest berry by weight. The others in descending order, were with 'Caddo', then 'Ouachita', and finally 'Ponca' with the smallest berry. Averages of individual berries by cultivar and total yield are in Table 1. Each total represents the yield of 10 containerized blackberry plants.

Table 1. Average individual berry weight and total yield by cultivar and trellising style.

Cultivar	Avg. berry wt (g)	Total yield (g)
Caddo – V	7	4358
Caddo – T	7.2	6092
Ouachita – V	6.6	12,284
Ouachita – T	6.2	14,102
Ponca – V	5.9	7367
Ponca – T	5.2	7730
Prime Ark Freedom – V	10.9	16,628
Prime Ark Freedom – T	7.4	11,516

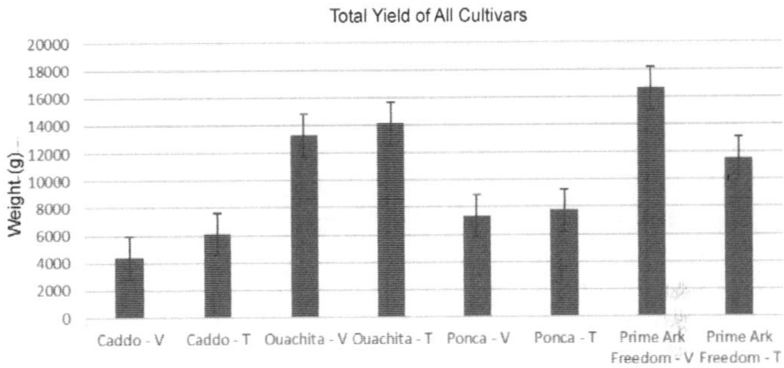

Fig. 1. Total yield by cultivar for both trellis types: "V" compared to "T" shaped design.

In every cultivar except 'Prime Ark Freedom,' the T-trellis method produced a higher yield than the 'V' style trellis, however, it was not statistically significant. The only cultivar where there was a statistical difference in trellising methods was with 'Prime Ark Freedom', where the V-shaped trellising resulted in a higher yield. From this one-season snapshot, either trellising style will be beneficial for thornless blackberry production, but 'Prime Ark Freedom' did respond better to the V-shape design. 'Caddo–V' was the lowest yielding cultivar in this demonstration, as seen in Fig. 1.

Suggestions for Future Direction

This short study was based on the first year of establishment and fruit production of the thornless blackberry plugs. It would be beneficial to continue this study with the plants being more mature to compare yields while assessing the longevity of containerized blackberry plants. We also did not apply any plant growth regulators, and that production practice could be impactful, especially in seasons without adequate chill.

*Corresponding author. Email: pfletch@ufl.edu

Proc. Fla. State Hort. Soc. 136:10. 2023.

—Scientific Note—

Utilizing Foliar Fertilization to Overcome Poor Soil Quality on a Blueberry Farm—A Case Study from Duval County

Stephen Jennewein* and George Richardson

University of Florida, IFAS Extension, Duval County, Jacksonville, FL

Keywords. soil quality, high pH soils, foliar fertilization, nutrient deficiencies, blueberry cultivation

This paper aims to showcase a successful foliar fertilization strategy employed by a Duval County blueberry grower for managing a high pH soil (> 7.5). The efficacy of this approach in mitigating nutrient deficiencies along with the current knowledge gap in foliar fertilization techniques among growers, illustrates the potential for impactful extension programming.

Findings

Pasture conversion in Duval County commonly leads to negative implications for soil quality, presenting challenges that may persist for years. The prevalence of calcium in Florida soils contributes to strong buffering capacity, hindering efforts to lower pH effectively. Irrigation water in the region is typically hard, providing a liming action on the soil and confounding the effectiveness of amendments. Water testing concluded this to be the case at the farm utilized in this case study. Conventional wisdom suggests cultivating tolerant plant species on high pH soils and not focusing efforts on acidifying soil through elemental sulfur applications. However, certain growers through persistence or circumstance intend to cultivate low soil pH-thriving plants on high pH soils, necessitating effective fertilization strategies. There is the potential for additional mitigation methods such as irrigation acidification but the knowledge gap and concern of damage for smaller scale operations is a hindrance.

In this case study, a Duval County blueberry grower successfully addressed the challenges posed by a high pH soil through the implementation of a foliar fertilization strategy. Before adopting this approach, the grower faced extensive nutrient deficiency symptoms, poor vegetative growth, suboptimal blooming, and increased pest pressure. Pests and pathogens observed included flea beetles (*Altica* spp.), rust (*Naohidemyces vaccinii*), target spot (*Corynespora cassiicola*), and algal stem blotch (*Cephaleuros virescens Kunze*; Phillips et al., 2022).

Mitigation Strategy

The grower utilized a commercially available chelated nutrient liquid fertilizer as a foliar spray, complemented by continued use of pine bark soil amendments. The use of foliar sprays enhanced absorption efficiency by bypassing the alkaline soil conditions, effectively mitigating nutrient deficiencies. Chelated metal compounds resist oxidation in alkaline soil conditions and may have also enhanced nutrient uptake from the roots (Liu et al., 2022). As a result, the blueberry grower witnessed a transformation with improved vegetative growth, blooming, and reduced pest pressure within 1 year of beginning applications. Once amendments have lowered soil pH to within the appropriate range and nutrient deficiency symptoms abate, foliar fertilizer application rate can be decreased or ceased. This strategy effectively deals with providing sufficient short-term nutrient needs while long-term issues can be addressed through additional mitigation strategies.

Suggestions for Future Direction

This case study provides a valuable plan of action for growers encountering similar challenges with high pH soils. Foliar fertilization can be a successful strategy for managing nutrient deficiencies and optimizing blueberry cultivation in locations with soil pH higher than the optimum range of 4.0–5.5 (Williamson et al., 2018). Future research should focus on refining and expanding this approach to other crops and regions with similar soil challenges. If ligands could be identified, or developed, for use with specific combinations of soil characteristics and crop, the impact of this method would be substantially enhanced. An extension program aimed at educating growers on matching ligands to soil conditions, implementing a fertilization plan based on foliar products, and identifying nutrient deficiencies resulting from poor soil quality would prove impactful. Programming that includes coupling this strategy with techniques such as acid injection in irrigation water and application of soil amendments will take care of short-term nutrient deficiencies while addressing long-term soil quality.

Literature Cited

Phillips D.A., Flor N.C., & Harmon P.F. 2022. Florida blueberry leaf disease guide. Florida Cooperative Extension, Gainesville, FL. Publication #PP348. https://doi.org/10.32473/edis-pp348-2019

Liu G., Hanlon E., & Li Y. 2022. Understanding and applying chelated fertilizers effectively based on soil pH. Florida Cooperative Extension, Gainesville, FL. Publication #HS1208.

Williamson J.W., Lyrene P.W. & Olmsted J.W. 2018. Blueberry Gardener's Guide. Florida Cooperative Extension, Gainesville, FL. Publication #CIR1192.

*Corresponding author. Email: sjennewein@ufl.edu

Proc. Fla. State Hort. Soc. 136:11. 2023.

—Scientific Note—

Growth and Physiological Responses of Blueberry to Different Percentages of Reclaimed Water Application

Yasmeen Saleem[1], Shinsuke Agehara[2], and Davie Kadyampakeni[1]*

[1]*University of Florida, IFAS, Citrus Research and Education Center, Lake Alfred, FL*

[2]*University of Florida, IFAS, Gulf Coast Research and Education Center, Wimauma, FL*

Keywords. blueberry, Florida, NaCl, reclaimed water, salinity

Crop irrigation is considered a main cause of accelerating groundwater depletion. In the United States, a high percentage of agricultural production depends highly on groundwater for irrigation. Finding an alternative irrigation water source, not supplied from fresh surface water or groundwater, is crucial for water conservation and environmental sustainability. Reclaimed water is highly treated wastewater that can be used for agricultural irrigation. Although reclaimed water contains plant-available nutrients and has the potential to reduce fertilizer inputs, it may deteriorate soil conditions when not used properly.

In Florida, reclaimed water has been used in the agricultural sector mainly for irrigating citrus (peeled crops) and golf courses. There is insufficient information on the optimal and safe reuse amount of reclaimed water on the Southern Highbush blueberry commercial cultivars. This trial is the first use of reclaimed water to irrigate unpeeled crops such as blueberries in Florida.

Two greenhouse experiments were conducted to evaluate the potential of reclaimed water as an alternative water source for blueberry production. Two major blueberry cultivars in Florida, 'Arcadia' and 'Kestrel', were selected. The salinity threshold of the selected blueberry cultivars was also tested. In the first experiment, six treatments were used: well water as a positive control; deionized water as a negative control; and reclaimed water diluted at 25%, 50%, 75% and 100% with deionized water. The second experiment had the same positive and negative controls as the first experiment, as well as sodium chloride solutions prepared at four levels of electrical conductivity (EC) corresponding to the four reclaimed water treatments in the first experiment. Plants were grown for 3 months in 3-gallon pots, filled with 60% pine bark and 40% peatmoss. Plant biomass accumulation, nutrient leaching, substrate–plant nutrient dynamics, and plant physiological responses were examined. Preliminary results showed no effect of the treatments on leachate pH. Leachate pH declined over time in both experiments, probably because of the elemental sulfur applied to the substrate. Leachate EC generally increased proportionally to the concentration of reclaimed water or sodium chloride. Based on the rest of data, reclaimed water treatment optimized for blueberry production will be selected and subjected to further evaluation through additional greenhouse and long-term field experiments.

IFAS Early Career Seed Grant 00131246-212, Soil and Water Science lab members at the University of Florida/IFAS Citrus Research & Education Center and Plant Physiology lab members at the University of Florida/IFAS Gulf Coast Research & Education Center.

*Corresponding author. Email: dkadyampakeni@ufl.edu

Proc. Fla. State Hort. Soc. 136:12–14. 2023.

'A5-14-8': A Large Fruit Muscadine Breeding Selection for Table Grapes

Zhongbo Ren,* Violeta Tsolova, and Islam El-Shakarway

Center for Viticulture, Florida A&M University, Tallahassee, FL

Keywords. commercial productivity, fruit taste, flower sex, fruit skin

'A5-14-8' is a muscadine breeding line with good potential for fresh market. This female flower selection produces very large fruits (over 15g) with several preferred traits such as high yield, uniform ripening, and lower ripe rot rate. The fruits taste pleasant with remarkable crunchiness and thin skins. The vine is moderately vigorous and disease resistant. Its fruits ripen around September 1st in Tallahassee, FL.

Muscadine (*Vitis rotundifolia* Michx) grapes are widely grown in the southeastern United States for fresh market and processing. It has become a multimillion-dollar industry and is growing rapidly as more and more people have started to consume muscadine grapes and products in recent years.

Unlike the muscadine grapes for processing, muscadine table grapes have different requirements, especially for fruit size and taste. Fruit size is important for muscadine table grapes, larger fruits tend to command higher prices and are more attractive to consumers. This makes large fruits a primary requirement which has always been desired by the muscadine grape industry.

Over the past few decades, muscadine table grape size has been improved tremendously, fruits have become larger and larger. A few of the largest cultivars today such as 'Florida Onyx', 'Majesty',' Paulk', 'Rubycrisp', and 'Supreme' weigh 15g, while 5~7g fruits were considered as good table grape size few decades ago. It would not be surprising to see even larger fruits in the future.

Therefore, providing the industry with large fruited cultivars is crucial to the development of industry, and is one of the major objectives in muscadine grape breeding.

The grape breeding program at the Viticulture Center, Florida A&M University (FAMU) has been working intensively on muscadine table grape breeding for the past 2 decades with numerous efforts and various approaches. We obtained the breeding line 'A5-14-8' recently which has shown several characteristics preferred by industry which suggests it could have high potential value for the industry.

Origin

'A5-14-8' was developed by the grape breeding program at the Viticulture Center, FAMU, Tallahassee, FL. It is an F1 hybrid between 'Majesty' ('Supreme' × 'Triumph') and 'A19-3-10'. 'Majesty' is a very large fruit, patented female (pistillate) flower cultivar; 'A19-13-10' is an advanced selection from 'Florida Onyx' (Ren et al., 2020) open pollination. It is a self-fertile, highly productive large grape vine obtained from the grape breeding program. 'Florida Onyx' ('Supreme' × 'Ison') is a newly patented large fruited cultivar from the Viticulture Center, Florida A&M University (FAMU), Tallahassee, FL. It produces very large black-red colored fruits, weighing about 3–5 grams more than other large muscadine grape varieties.

'A5-14-8' was asexually propagated with soft cuttings during mid-summer. The vines were evaluated at the FAMU vineyard (30°28'42"N, 84°10'21"W). Vines were planted at a density of 10 feet in rows and 12 feet between rows. It was trained to a single-wire bilateral cordon system, and was annually pruned with 3~4 buds spur-pruning technics. Major horticultural traits were compared with leading muscadine table grape cultivars 'Fry' and 'Ison'.

Major Horticultural Characteristics

Female flowers. 'A5-14-8' bears female flowers. A flower cluster (inflorescence) has 40 individual flowers which typically grow at the 3rd and 4th nodes.

"Cap stick" or dry calyptra, where the petals do not fall off and pollination does not occur, was rare during evaluation, which could minimize its yield fluctuations.

Very large and pleasant taste fruits. 'A5-14-8' fruits are large, weighing 15.0g. Fruit are significantly larger than fruits of the current leading cultivar of 'Fry' (~11g) and 'Ison'(~9g) (Fig 1, Table 1), and are similar in size to a few of the largest fruit cultivars 'Florida Onyx,', 'Majesty', 'Paulk', 'Rubycrisp', and 'Supreme'.

Fig. 1. The fruit size, color, and cluster of 'A5-14-8' and leading muscadine table grape cultivars.

*Corresponding author. Email: zhongbo.ren@famu.edu

Table 1. Productivity, fruit characteristics of 'A5-14-8' and leading muscadine table grape cultivars in Tallahassee, FL

Cultivar	Fruit wt/spur (g)	Yield[z] (lb/vine)	Dry scar rate (%)	Ripe rot rate (%)	Commercial yield (lb/vine)	Fruit size (g)	SSC[y] (%)	TA[x] (%)	pH	No-rotten fruit[w] (%)
Fry	172	30.3	81.3	18.1	22.2	11.5	16.3	0.40	3.36	19.4
Ison	240	41.0	84.0	5.2	32.6	8.9	15.1	0.44	3.27	33.3
Majesty	202	35.0	81.4	6.4	26.7	15.6	15.5	0.38	3.58	35.8
A5-14-8	180	31.7	81.0	5.9	24.2	15.0	16.1	0.36	3.45	31.7

[z]10-foot canopy vines with single-wire bilateral cordon training system.
[w]2 weeks storage at 74°F .
[y]SSC = soluble solids content.
[x]TA = titratable acidity.

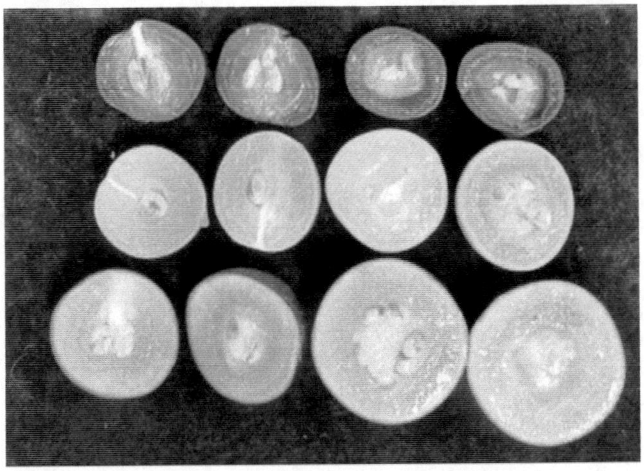

Fig. 2. Cut fruit of Ison (top row), Fry (middle row), and 'A5-14-8' (bottom row).

The fruits have excellent taste and are remarkably sweet and crunchy, thanks to their high sugar content and very thin skin (Fig 2, Table 1).

Satisfactory productivity. 'A5-14-8' is productive (Fig. 3), the average yield of a mature 'A5-14-8' vine is 31 lb which is comparable to the yields of 'Fry' and 'Majesty', but lower than 'Ison'. This was further confirmed by the spur productivity (total fruit weight/spur, Table 1). As a leading commercial cultivar, the yield of 'Fry' has been widely recognized by the industry, though it does not produce as much as the perfect flower cultivar 'Ison' generally produces, therefore, the similarity in yield of 'A5-14-8' to 'Fry' and 'Majesty' suggests that the yield of 'A5-14-8' is satisfactory as a female flowered vine.

Fig. 3. A5-14-8 fruit setting.

Higher commercial yield. muscadine grape vines vary in their rot fruit and wet/dry scar fruit rates, these can result remarkable differences in the amount of fruit sold commercially, as only fruit with healthy dry scars can be sold. 'A5-14-8' has higher commercial yield than 'Fry', apparently due to 'Fry' having a much higher rot fruit rate (Table 1).

Attractive fruit color. 'A5-14-8' fruits have a unique and beautiful red color, compared with the almost dark red skin color of other red muscadine grapes. This red color is not only attractive, but also makes the grapes look fresher (Fig. 1, Fig 3).

Semi-compact cluster and uniform ripening. 'A5-14-8' fruit clusters are semi-compact, which is an advantage by allowing for better air circulation and reducing fruit diseases. Fruits ripen uniformly around September 1 for a short period in Tallahassee, FL. This kind of ripening is a highly desirable trait.

Vine vigor. 'A5-14-8' has a moderately vigor vine. Its shoot internode length and circle are 3.8 cm and 1.8 cm respectively, which is similar with those of 'Fry' and 'Ison' (Table 2). The pruning weight of 'A5-14-8' from a 10-foot canopy vine with single-wire bilateral cordon straining system was 7.3 lb, which is close to that of 'Fry', while lower than the pruning weight of 'Ison' (Table 2).

Pierce's disease (PD). No PD symptoms were observed during evaluation.

Storage quality. After 2 week storge at 74°F, the commercial fruit rate of 'A5-14-8' was 31%, this was virtually the same as 'Ison' and 'Majesty' fruits, which were obviously higher than the 19% commercial fruit rates of 'Fry' (Table 1), which suggest a better storage life for'A5-14-8' than for 'Fry'.

Discussion and Summary

Muscadine grapes are produced by either female or self-fertile flower vines with the unique feature between flower sex and fruit size, in which female vines produce larger fruits with somewhat lower yields and yield fluctuations in some years. Self-fertile vines produce relatively smaller fruits with higher and more stable yields due to better fruit setting with less influence by environmental factors. Because self-fertile vines do not produce the fruits as large as those from female flower vines generally, female flower vines have been essential for muscadine table grape industry. That is why the most important fresh market cultivars

Table 2. Horticultural characteristics of 'A5-14-8' and leading muscadine table grape cultivars in Tallahassee, FL.

Cultivar	Flower	Fruit ripening	Uniform ripening	Vine vigor	Pruning wt[z] (lb/vine)	Internode length (cm)	Internode circle (cm)	PD[y] score (0–5)
Fry	f	begin. Sept.	even	moderate	7.9	3.9	1.8	0
Ison	p	end Aug.	even	moderate-high	10.9	4.1	1.9	0
Majesty	f	begin. Sept.	even	moderate	8.6	4.3	1.9	0
A5-14-8	f	around Sept. 1	even	moderate	7.3	3.8	1.9	0

[z]10-foot canopy vines with single-wire bilateral cordon training system.
[y]PD = Pierce's disease.

are all female flower vines today, through growers prefer self-fertile flower to female flower vines.

The major advantages of 'A5-14-8' are its excellent taste, very large fruit size, satisfactory yield, and attractive skin color, which are better the current leading cultivars 'Fry' and 'Ison' for commercial yield, fruit taste and size, respectively. These give this selection great potential for the industry.

Literature Cited

Ren Z., Lu J. & Tsolova V. 2020.'Florida Onyx' muscadine grape. US Plant Patent PP31,407 P2. Filed: Sept. 7, 2018. Issued: Jan. 28, 2020.

Proc. Fla. State Hort. Soc. 136:15. 2023.

—Scientific Note—

A Farm Level Econometric Demand Model for Florida Green-skin Avocados

Fredy Ballen*

Tropical Research and Education Center, University of Florida, IFAS, Homestead, FL

Keywords. Florida avocados, demand elasticity, Hass avocados

Despite its relatively low importance in terms of US domestic production (10%) and trade (2.5% of imports) in recent years, and the undeniable popularity of Hass avocados, the green-skin avocado industry represents a very important component of the Florida agricultural economy. There is a market for green-skin avocados, as they are favored over Hass by consumers with ties to the Caribbean, Central and South America, as well by consumers interested in a lower calorie-count. The Dominican Republic (DR) is emerging as the main overseas supplier of green-skin avocados to the United States, with imports more than tripling from 6,682 MT in 2015/16 to 25,912 MT in 2019/20 (USDA–FAS, 2022). In addition to the growing popularity of Hass avocados, and the increasing competition from the DR, the industry is now facing treats from the redbay ambrosia beetle (RAB) (*Xyleborus glabratus*) and its nutritional symbiont, a fungus identified as *Raffaelea lauricola* that causes the laurel wilt (LW) disease. LW is so aggressive that infected trees usually die within a matter of weeks. Under this new scenario, it is important to assess the competitive pressure of Hass and green-skin avocado imports and its potential impact over the industry's ability to cope with the LW disease. Therefore, is a need to conduct an econometric estimation for the demand of Florida avocados.

Findings

The US demand model for Florida green-skin avocados may be specified as:

$$QFLt = PFLt + PDRt + PHASSt + RDPIt + IMPt + Mex07$$

where: QFLt = quantity demanded of Florida green-skin avocados (lb/person); PFLt = real farm price of Florida green-skin avocados ($/lb); PDRt = real import price of DR green-skin avocados ($/lb; PHASSt = real price of imported and domestic Hass avocados ($/lb; RDPIt = real US disposable personal income ($/person); IMPt = green-skin avocado imports as a proportion of the total green-skin supply (%); and MEX = dummy variable that takes the value of 1 after year 2007, when all US states were open for year-round Mexican avocado imports.

The 2SLS IV estimator was used for the empirical estimation (Table 1). Instruments for the price of Florida avocados (PFLt) included the producer price index for fresh fruits (PPIF), real oil price (ROILP), and imports of avocado processed (AVOPQ). The empirical model uses annual data (USDA–ERS) for the crop seasons 1990–91 to 2019–20. Diagnostic tests indicate a good fit of the model, the estimated own price elasticity for Florida avocados is –0.998, meaning that avocados are very responsive to changes in prices. A 10% increase in the price of Florida avocados is likely to result in a 9.9% decrease in the quantity of Florida avocados demanded. Therefore, price increases for Florida avocados are not going to provide the extra revenue needed to offset the disease control costs.

Table 1. Empirical estimates for the Florida green-skin avocado demand model using IV/2SLS estimation.

	IV/2SLS	
Variables	Coeff.	Std. Err.
PFLt	–0.5198**	0.2146
PDRt	–0.0325	0.0902
PHASSt	0.0568**	0.0266
RDPIt	–5.05E-06	3.82E-06
IMPt	–0.2917	0.0441
MEX	–0.0273**	0.0131
Constant	0.4636	0.0642
Underid $\chi2$ *P*-value	0.0309	
Overid $\chi2$ *P*-value 0.1018		
Endogeneity $\chi2$ *P*-value	0.0837	
Cumby-Huiz. $\chi2$ *P*-value	0.3864	
Breusch-Pagan $\chi2$ *P*-value	0.1284	
IV Reset test $\chi2$ *P*-value	0.9895	
Adj R-sq	0.9887	
Centered R-sq	0.8881	

** Significant at 5%.

The expected substitution relationship between Hass and Florida avocados holds. A 10% increase in the price of Hass avocados might increase the demand for Florida avocados by only 3.1%. The openness of the US market to Mexican avocado imports have had a negative impact to the Florida industry (–0.028 lb/person). Results suggest that the Florida avocado industry may benefit from the growth in demand for avocados. However, the benefit seems limited as changes in the market conditions may marginally benefit the Florida industry.

Suggestions for Future Direction

It is important to keep updating the model to see how changes in the market may impact the viability of the Florida avocado industry.

Literature Cited

USDA–ERS. 2022. Fruit and Tree Nuts Yearbook, Noncitrus fruit. Accessed 15 Feb. 2023. https://www.ers.usda.gov/data-products/fruit-and-tree-nuts-data/fruit-and-tree-nuts-yearbook-tables/

USDA–FAS. 2022. Global Agricultural Trade System, Standard Query. Accessed 15 Feb. 2023. https://apps.fas.usda.gov/GATS/ExpressQuery1.aspx

*Corresponding author. Email: fredy.ballen@ufl.edu

Proc. Fla. State Hort. Soc. 136:16–18. 2023.

Partial Documentation of an Unknown Avocado Fruit Deformity

Jonathan Crane*, Romina Gazis, and Daniel Carrillo

[1]University of Florida, IFAS, Tropical Research and Education Center, Homestead, FL

Keywords. avocado fruit, fruit and leaf deformities

Abnormalities of avocado fruit have been described and documented previously. Their cause may be genetic, abiotic, biotic, and/or caused by cultural production practices. Mishappen fruit may result from seed abortion, incomplete closure of the carpel, fruit with two seeds, or fruit with one developed and one-aborted seed. In some cases, a longitudinal ridge of peel tissue may develop, or small individual scattered bumps may be observed. Other peel defects result from abiotic environmental conditions such as sunburn, dry winds, water stress, and mechanical injury. Biotic causes include damage from insect probing or mite feeding and disease symptoms caused by pathogens. None of these factors appeared to be responsible for the symptoms observed in the leaf and fruit samples in question. In most cases, only one or a few randomly distributed fruits throughout a grove may show peel deformities that cannot be attributed to biotic or abiotic causes. The deformities observed occurred on nearly all the fruit of an unnamed avocado scion selection on 40 top-worked trees. Fruit had superficial but pronounced protrusions one or more sections of the peel surface. There were no visible symptoms below the affected fruit epidermal layer. No cultural (e.g., chemical spray applications), fungi or bacteria, mites and/or insects cause was discernable. The terminal five to seven leaves on nearly every shoot showed mottling, chlorosis, some leaf distortion, and reduced leaf size. Symptomatic leaves were slightly deficient in calcium (Ca) and magnesium (Mg), whereas copper (Cu) levels were much higher in non-symptomatic leaves. Based on the available information, the cause of the observed leaf and fruit deformity may be genetic.

Abnormalities of avocado fruit have been described and documented previously (Fisher & Davenport, 1989; Hoshi & Arpaia, 2002; Schroeder, 1953–54). Seedless fruit (stenospermocarpy) may occur (usually cucumber-shaped) when fruit possess an aborted ovary. Mishappen fruit may also result from incomplete closure of the carpel (ovary, stigma, and style) in which an abnormal fruit is attached to a normal fruit, or from fruit with two seeds (called double-fruit), or from fruit with one developed and one-aborted seed (Hofshi & Arpaia, 2002). In these cases, a longitudinal ridge of peel tissue may develop. Symptoms associated with sunblotch viroid disease include depressed peel lesions with bright yellow or whitish or reddish streaks of coloration (Faber et al., 2015; Hodgson, 1935; Schroeder, 1954–54; Hofshi & Arpaia, 2002). Other peel defects are the result of environmental conditions such as intense sunlight/heat (sunburn), dry winds, water stress (e.g., crick-side, ringneck), mechanical injury (e.g., carapace spot, alligator-skin), and disease (e.g., scab).

Previous research in Florida has attributed the epidermal bumps seen on the surface of very young avocado fruit to the feeding of several species of flower thrips (*Frankliniella bispinosa* and *F. kelliae*) during the flowering and fruit setting period (Fisher & Davenport, 1989). These epidermal bumps are generally small in height and scattered over the fruit surface.

During November of 2022, avocado fruit from a commercial planting in Homestead, Florida with extremely bumpy peels were submitted to scientists at the University of Florida, Institute of Food and Agricultural Sciences (UF/IFAS) Tropical Research and Education Center. The deformed fruit came from 40 three- to four-year-old topworked avocado trees bearing their first crop since being top-worked. The scion used to top-work those 40 trees was an unnamed selection and the fruit morphology suggested that the scion material was a Guatemalan × West Indian hybrid, selected for late season fruit. Upon inspection, it was found that approximately 80% of the fruit on these trees showed deformity symptoms (Fig. 1) and that most of the terminal 5–10 leaves (last flush) were also deformed (Fig. 2). These 40 trees were surrounded by mature avocado trees of a different cultivar that had no fruit or leaves with deformity symptoms. The grove

Fig. 1. (**A–B**) fruit peel deformed outgrowths of the peel symptoms; (**C**) transversal cut into pulp; and (**D**) from left to right—two symptomatic fruits; one non-symptomatic fruit; and another symptomatic fruit—deformed, outgrowth of the skin, not deep into the pulp.

*Corresponding author. Email: jhcr@ufl.edu

Fig. 2. Terminal shoot and leaf symptoms. Note terminal leaves show virus-like and nutrient deficiency symptoms whereas subtending (older flush) leaves do not.

owner indicated that all the avocado trees in this block were under identical production practices and that the mother tree (source of the scion used to topworked the 40 trees) never shown deformity symptoms in leaves or fruit and produced abundant fruit of high quality. We were not able to inspect the mother tree because it had succumbed to laurel wilt disease (causal agent: *Harringtonia lauricola*). Symptomatic fruit and leaf samples were taken for disease diagnosis, symptomatic and non-symptomatic leaf samples were collected for nutrient analysis, and terminal shoots and symptomatic fruit were collected for arthropod inspection.

Results and Discussion

No insects or mites were found associated with the fruit malformation. Some flower buds were damaged by an unidentified borer (Fig. 3). Some lepidoptera and ambrosia beetles (e.g., *Xylosandrus compactus*) can bore into flower buds but none were detected. Some predatory thrips (*Leptothrips* spp.) and fungivorous tarsonemid mites were found inside the damaged flower buds but they are not known to cause leaf or fruit deformities.

Upon slicing through the peel, it was apparent that the bumpiness was superficial and there were no lesions or discoloration below the surface (Fig. 1). No pathogenic fungi or bacteria were recovered from the symptomatic tissue and only saprotrophic fungi were observed after samples were incubated in a moist chamber.

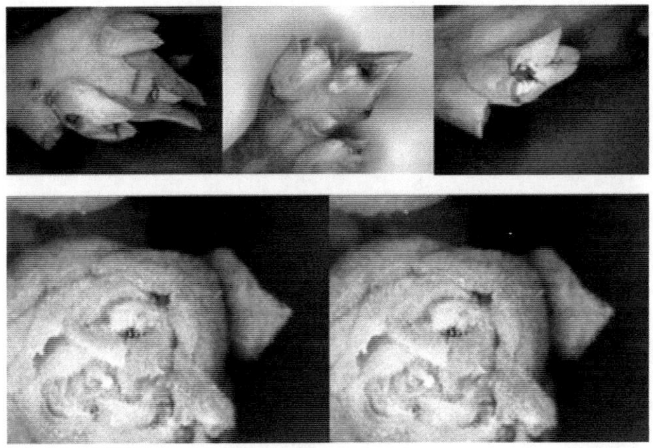

Fig. 3. Terminal/lateral bud symptoms (boring, piercing, necrosis) from trees with fruit and leaf distortion symptoms.

Consultation with the Florida Department of Agriculture and Consumer Services—Division of Plant Industry (FDACS-DPI) virologist (Dr. Kishore Dey) deemed that a viral/viroid cause was unlikely for the leaf and fruit symptoms observed.

The nitrogen (N), phosphorus (P), and potassium (K) leaf content of symptomatic and non-symptomatic leaves were within normal ranges reported for avocado trees growing in the calcareous, high pH soil of south Miami-Dade County (Table 1) (Ruehle, 1963; Young & Koo, 1976 and 1977; Koo & Young, 1977). The calcium (Ca) and magnesium (Mg) levels of symptomatic leaves were below the normal range. In contrast, boron (B) levels of non-symptomatic leaves were below the normal range whereas symptomatic leaves were within the normal range. Interestingly, the copper leaf content for symptomatic leaves was with the normal range but non-symptomatic leaf copper (Cu) was very high. Nutrient levels of iron (Fe), manganese (Mn), and zinc (Zn) for symptomatic and non-symptomatic leaves were within normal range.

Summary

No pathogenic fungi or bacteria were recovered from the symptomatic fruit. Fruit symptoms observed in the samples do not match those of trees affected with the sunblotch viroid (Bender et al., 2015; M.L. Arpaia, handout/personal communication). Fruit symptoms of sunblotch include reddish, yellow, or white depressed streaks on the peel. Leaf symptoms of the present samples and those caused by sunblotch viroid have some similarities, such as leaf deformity (puckering), interveinal chlorosis, and symptoms resembling Zn and/or Mg deficiency. However, sunblotch viroid associated leaf symptoms are not commonly or extensively expressed whereas the majority of the terminal shoots of the 40 trees in question were symptomatic. In addition, symptomatic leaves only occurred at the terminal-end of the shoots, which has not been reported as part of the sunblotch viroid symptomatology. Leaf nutrient content of symptomatic and non-symptomatic leaves was within normal range for N, P, K, sulfur (S), Fe, Mn, and Zn. Calcium and Mg levels of symptomatic leaves were below the normal range. In contrast, B levels of non-symptomatic leaves were below the normal range whereas symptomatic leaves were within the normal range. Interestingly, the Cu leaf content in symptomatic leaves was within the normal range but non-symptomatic leaf Cu was very high. The leaf stunting, distortion, and interveinal chlorosis

Table 1. Leaf nutrient results and comparison with published ranges from Florida.[z]

Element	Symbol	Unit	Leaf samples: Symptomatic	Non-symptomatic	Standard nutrient ranges (rockland soil)
Nitrogen	N	%	2.03	2.23	1.5–2.1
Phosphorus	P	%	0.15	0.13	0.08–0.15
Potassium	K	%	1.14	0.91	1.1–1.9
Calcium	Ca	%	1.12	2.30	1.79–3.62
Magnesium	Mg	%	0.22	0.35	0.31–0.57
Sulfur	S	%	0.20	0.29	–
Boron	B	ppm	30	23	40–50
Iron	Fe	ppm	55	53	28–199
Manganese	Mn	ppm	38	64	34–161
Zinc	Zn	ppm	36	49	14–160
Copper	Cu	ppm	39	182	19–75

[z]Koo & Young, 1977; Ruehle, 1963; Young & Koo, 1976.

observed in symptomatic leaves may be associated with Ca and Mg deficiency. The very high Cu levels in non-symptomatic leaves may be due to their exposure to more Cu applications than symptomatic leaves (which were terminal leaves i.e., younger). Differences in leaf nutrient content could also be the result of differences in leaf age i.e., non-symptomatic leaves were older than symptomatic leaves (Koo & Young, 1977).

Since there is a lack of strong evidence that the observed fruit and leaf symptoms were due to pest/arthropod damage, pathogen infection (fungi, bacteria, virus), or differences in plant nutrient content we speculate that the cause is the use of budwood that suffered from a genetic mutation (called a bud-sport, somatic mutation, or chimera) (Hodgson, 1935; Schroeder, 1953-54). Potentially the budwood that was inadvertently used to top-work (graft) these trees were from a branch that carried a genetic mutation (chimera). Avocado chimeras have been observed, suspected, and documented since the 1930s.

In most cases only one or a few fruits, throughout a grove or seen at a packinghouse, show peel deformities that cannot be attributed to environment, cultural practices (e.g., spray damage), and/or mechanical damage by biotic or abiotic factors. It is speculated that in these fruits a mutation occurred in a few cell layers immediately beneath the epidermis at the ovary level early-on in the fruit's development (Hofshi & Arpaia, 2002). As the fruit grows (cells expand) the deformity becomes more pronounced. These individual ovary-level mutations are not passed on through vegetative propagation (i.e., grafting and budding). However, if the mutation occurs in the woody tissue (i.e., stems and/or limbs), most stems and fruit would show the mutation (e.g., deformed leaves and fruit), which could be passed on through vegetative propagation.

As far as we know, the observed fruit symptoms have not been reported nor documented before and do not appear to match any previously documented deformities of avocado fruit. A definitive answer to the cause of these fruit and leaf deformities will not be possible as the grove owner destroyed the trees. However, it is important that the symptoms and our preliminary analysis be published for the record.

Literature Cited

Faber BA, Eskalen A, & Bender GS. 2015. Diseases (Chapter 3). p. 55–86. In: Bender G.S. (ed.). 2015. Avocado production in California: A cultural handbook for growers. Second Ed. Book two–Cultural care. Univ. of California Cooperative Extension, San Diego County and the California Avocado Society.

Fisher J.B. & Davenport T.L. 1989. Structure and development of surface deformations on avocado fruits. HortScience. 24:841–844.

Hodgson R.W. 1935. Avocado fruit abnormalities. California Avocado Association Yearbook 20:108–109.

Hofshi R. & Arpaia M.L. 2002. Avocado fruit abnormalities and defects revisited. California Avocado Association Yearbook 86:147–162.

Koo R.C.J. & Young T.W. 1977. Effects of age, position, and fruiting status on mineral composition of 'Tonnage' avocado leaves. J. Amer. Soc. Hort. Sci. 102:311–313.

Ruehle G.D. 1963. The Florida avocado industry. Bull. 602. Fla. Coop. Ext. Service, University of Florida, IFAS, Gainesville, FL.

Schroeder C.A. 1953-54. Abnormal fruit types in the avocado. California Avocado Association Yearbook 20:108–109.

Young T.W. & Koo R.C.J. 1976. Mineral composition of avocado leaves in Florida. Proc. Fla. State Hort. Soc. 89:238–241.

Young T.W. & Koo R.C.J. 1977. Influence of soil and cultivar on mineral composition of avocado leaves in Florida. J. Amer. Soc. Hort. Sci. 102:308-311.

Proc. Fla. State Hort. Soc. 136:19–25. 2023.

Influence of Rootstock Ecotype on Physiological Responses and Survival of Avocado Trees to Flooding

MELINDA H. YIN, ANA I. VARGAS, EDWIN A. GUTIERREZ-RODRIGUEZ, AND BRUCE SCHAFFER*

Tropical Research and Education Center,University of Florida, IFAS, Homestead, FL

Keywords, *Persea americana*, net CO_2 assimilation, transpiration, stomatal conductance, water use efficiency

Avocado (*Persea americana* Mill.) trees are classified into three ecotypes: Mexican (M), Guatemalan (G), and West Indian (WI) based on their geographical center of origin. In southern Florida, avocado trees are grown on WI seedling rootstocks, mainly 'Waldin', generally with WI or WI × G hybrid scions. Flood tolerance is influenced by the rootstock, and 'Waldin' seedling rootstock is sensitive to short-term flooding. However, effects of rootstock ecotype on flood tolerance have not been reported. This study compared responses of a WI avocado scion ('Simmonds') among clonal avocado rootstocks of each ecotype. 'Simmonds' was grafted onto clonally propagated 'Duke 7' (M), 'Reed' (G), or 'Waldin' (WI) rootstocks and subjected to either short-term (1.5 days) of flooding or not flooded (control). As an indicator of plant stress, leaf gas exchange [net CO_2 assimilation (A), stomatal conductance (g_s), transpiration (*Tr*), and water use efficiency (*WUE*)] was measured during the flooding period and after trees were unflooded until flooded plants lost too many leaves to measure. After new leaves emerged on defoliated trees, leaf gas exchange was measured, and trees were harvested for dry weight determination. Flooding reduced leaf gas exchange and leaf dry weight of trees on clonal M or G rootstocks, but not on the clonal WI rootstock. All trees on the clonal WI rootstock survived flooding, whereas 87% of trees on clonal M and 50% of trees on clonal G rootstock survived flooding. Thus, 'Simmonds' avocado trees on a clonal WI rootstock were more flood tolerant than trees on a clonal M or G rootstock.

Avocado (*Persea americana* Mill.) is grown commercially in tropical, subtropical, and mild Mediterranean climates around the world. As the global population continues to climb and due to global climate change, avocado production will be forced into increasingly marginal lands (Grüter et al., 2022). Root zone flooding due to poor soil drainage, imprecise irrigation and management, and increasingly variable precipitation events due to changes in climate patterns are issues that will negatively impact avocado orchards in many production areas worldwide (Grüter et al., 2022), including southern Florida (Her et al., 2017).

Avocado trees are vegetatively propagated and grafted, with the scion and rootstock selected for horticulturally important characteristics (Ernst et al., 2013). Fruit tree rootstocks impart pathogen resistance, abiotic stress tolerance, and scion size, and have notable impact on precocity, productivity, cold tolerance, shoot pathogen resistance, and fruit quality (Warschefsky et al., 2016). As the primary point of contact between the soil and plant, avocado rootstocks play a critical role in acclimation and tolerance to flooding (Schaffer et al., 1992).

Avocado trees emerged in three distinct habitats and are divided into three ecotypes (more often called "botanical races") based on their native habitats, the Mexican (M), Guatemalan (G), and West Indian (WI) ecotypes (Bost et al., 2013; Chanderbali et al., 2013). The WI ecotype is a misnomer as this ecotype originated in the lowlands of Central America and was later brought to the West Indies (Bost et al., 2013). The disparate ecogeographical centers of origin and distribution of the different avocado ecotypes have been associated with varying tolerances to temperature (Campbell et al., 1977; Krezdorn, 1970), salinity, (Kadman and Ben-Ya'acov, 1976) and nutrient deficiencies (Kadman and Ben-Ya'acov, 1982), but resilience to flooding stress has not been characterized in the context of ecotype.

In southern Florida, commercial avocado trees are grown on WI seedling rootstocks, mainly 'Waldin', and scion cultivars are primarily WI or WI × G hybrids, because 'Waldin' seedling rootstock and WI and WI hybrid scions are more adapted to Florida's high pH calcareous soil and subtropical climate than the other ecotypes, which are more suited to Mediterranean climates (Crane et al., 2013).

Southern Florida's avocado production area is prone to flooding due to frequent heavy rainfall, tropical storms and hurricanes (Schaffer, 1998). Thus, a considerable amount of research has been conducted in southern Florida to assess the response of avocado cultivars to flooding stress. These studies have found avocado trees to be very susceptible to short-term (1–7 days) flooding (Ploetz and Schaffer, 1987, 1989; Schaffer et al., 1992; Sanclemente et al., 2013; Sanclemente et al., 2014, Yin et al., 2023). However, flooding studies of avocado trees have been conducted primarily with trees on 'Waldin' (WI) seedling rootstock. The effect of avocado rootstock ecotype on tolerance to short-term flooding has not been reported. The objectives of this study were to compare responses to and recovery from short-term flooding among clonal rootstocks of each avocado ecotype, all with the same WI scion. This information should be useful for selecting rootstocks for avocado regions currently experiencing short-term flooding or waterlogging or which may experience these conditions in the future due to the impacts of climate change and sea level rise.

*Corresponding author. Email: bas56@ufl.edu

Materials and Methods

'Duke 7' (M), 'Reed' (G), and 'Waldin' (WI) avocado rootstocks were clonally propagated and grafted with 'Simmonds' (WI) scions at a commercial nursery (Brokaw Nursery LLC, Ventura, CA). Trees were transported as liners to the University of Florida, Tropical Research and Education Center (UF-TREC) in Homestead, FL, about one-year after grafting. At UF-TREC, trees were transplanted to 11.4-L plastic nursery pots filled with a commercial potting mix (PRO-MIX Premier Tech, Quebec, Canada) and grown in a shade house (20% light reduction). Trees were manually irrigated up to twice each day to field capacity and fertilized as needed to keep them vigorous.

When trees were about 3-years old from the time of grafting, they were moved to an air-conditioned greenhouse set at 28 °C and allowed to equilibrate to greenhouse conditions for three weeks prior to the beginning of the experiment. Trees in the greenhouse were divided into two treatments, flooded and non-flooded. For the flooded treatment, 18 72-L plastic utility tubs were filled with tap water and left for 3–4 d at ambient temperature to allow all oxygen to dissipate from the water; oxygen content was monitored using a YSI Pro 20 Galvanic Dissolved Oxygen probe (YSI Inc., Yellow Springs, OH). Water in the utility tubs was used for the flooded treatment. For the flooded treatment, each potted tree (replicate) was placed in a separate plastic utility tub by submerging the pot in water to 2.5 cm above the soil surface. Flooded trees were not irrigated. For the non-flooded (control) treatment, trees were not submerged, and irrigation was maintained as previously described.

For each treatment, there were six single-tree replicates of each rootstock, which was based on the variability observed in previous avocado flooding experiments (Sanclemente et al., 2013; Sanclemente et al., 2014). After treatments were initiated, leaf gas exchange variables were measured daily during the flooding period as described below.

The duration of the flooding treatment was determined from daily leaf gas exchange measurements. Trees were unflooded after there was a significant statistical difference ($P \leq 0.05$) for at least one cultivar between the flooded and control treatments for any gas exchange variable. Trees were unflooded by drilling holes in the plastic utility tubs and allowing for drainage to occur over the course of 2 hours.

Leaf gas exchange [net CO_2 assimilation (A), stomatal conductance of water vapor (g_s), transpiration (Tr)], a physiological indicator of flooding stress prior to the appearance of symptoms (Ploetz and Schaffer, 1989), was measured with a CIRAS-3 portable gas analyzer (PP Systems, Amesbury, MA, USA) prior to flooding (day 0), daily while trees were flooded and at 3–4 day intervals after trees were unflooded until several flooded plants lost too many leaves to continue measurements. For each plant, two fully expanded leaves with fully developed cuticles were measured and the means of the two leaves recorded for each replicate. During leaf gas exchange measurements, the reference CO_2 concentration was 400 $\mu m \cdot mol^{-1}$, the photosynthetic photon flux (PPF) was 1000 $\mu m \cdot m^{-2} \cdot s^{-1}$ in the leaf cuvette and the air flow rate into the cuvette was 200 $ml \cdot min^{-1}$. Water use efficiency (*WUE*) was calculated as A ÷ E. When new leaves emerged to enable leaf measurements of a sufficient number of plants of each rootstock, the number of surviving plants on each rootstock was determined and leaf gas exchange was measured on all surviving plants. Plants were then harvested, and leaf, stem and root dry weights were determined for all plants.

Statistical interactions between flooding treatments and ecotypes were determined by two-way analysis of variance (ANOVA; $P \leq 0.05$). For measurements over time, significant differences between flooding treatments or among ecotypes were determined by repeated measures ANOVA ($P \leq 0.05$). For measurements at the end of the study (dry weights), significant differences between of flooding treatments within each ecotype were determined by a non-paired T-test ($P \leq 0.05$) and significant differences ($P \leq 0.05$) among each ecotype within each flooding treatment were determined by one-way ANOVA and Waller-Tukey's K Ratio Test ($P \leq 0.05$). All statistical analyses were done with SAS Statistical Software (SAS Institute, Cary, NC, USA).

Results

For most leaf gas exchange variables, there was a significant ($P \leq 0.05$) statistical interaction between flooding treatment and rootstock (Table 1). Therefore, differences between flooding treatments were analyzed within each rootstock separately, and differences among rootstocks were analyzed separately within each flooding treatment.

After just 36 hours of flooding, A of trees on the G rootstock was significantly lower for trees in the flooded than non-flooded treatment (Fig. 1). Therefore, all trees were unflooded after 36 hours. For trees on the WI rootstock, by day 15 (13 days after trees were unflooded), A, g_s, and E of flooded trees returned to levels similar to those in the non-flooded treatment (Figs. 1 a–c, 3 a–c, and 4 a–c). Leaves wilted (Fig. 5) and eventually abscised on all flooded trees on M or G rootstocks, and all trees on these rootstocks were completely defoliated after day 15, with new growth appearing on these trees at day 35. Only two trees on the WI rootstock became totally defoliated, but also had new growth by day 35. There were significant differences in *WUE* between flooded and control trees on M and G rootstocks between day 4 and day 15, but surviving trees recovered by day 35. For trees on the WI rootstock, a significant difference in *WUE* was observed between flooded and control treatment trees, but this was not observed on any other days (Fig. 4 a–c).

For non-flooded trees, there were slight but significant differences in A, E, and g_s among rootstocks (Figs. 1 d, 2 d, and 3 d), probably reflecting differences in vigor among rootstocks. There were no significant differences in *WUE* among rootstocks of non-flooded trees (Fig. 4 d). For flooded trees, A, E, g_s and *WUE* were higher on trees with the WI rootstock than trees on the M or G rootstocks on most measurement days (Figs. 1 e, 2 e, 3 e, and 4 e).

Leaf dry weight was lower for flooded trees than non-flooded trees on all rootstocks, but these differences were only statistically significantly ($P \leq 0.05$; Table 3-2) for trees on the G rootstock. Leaf, stem, and root dry weights were not significantly different ($P > 0.05$) among rootstocks in any flooding treatment or between flooding treatments for any rootstock (Table 2). Plant dry weight was significantly different between flooded and non-flooded tress for the M and G rootstocks but not for the WI rootstock (Table 2).

By the end of the study, 87% of trees on the M rootstock and 50% of trees on the G rootstock survived, whereas all trees on the WI rootstock survived. Although some trees on M and G rootstocks survived, severe leaf epinasty and abscission were observed and recovery of leaf gas exchange did not occur for trees on these rootstocks, whereas all trees on the WI rootstock recovered by 47 days after plants were unflooded (Figs. 1–4).

Table 1. Factorial analysis of variance (ANOVA)results: physiological responses to flooding among rootstock cultivars of three different ecotypes.

Dependent variable	Source of variance	Degrees of freedom	Mean square	F-value	*P*-value
A	Days after flooding (DAF)	12	6047.06	16.58	< 0.0001
	Rootstock ecotype	1	205234.61	562.77	< 0.0001
	Flooding	2	36521.03	100.14	< 0.0001
	DAF*Rootstock ecotype	12	2433.65	6.67	< 0.0001
	DAF*Flooding	24	535.29	1.47	0.0730
	Rootstock ecotype*Flooding	2	14837.64	40.69	< 0.0001
g_s	DAF	12	111.59	23.54	< 0.0001
	Rootstock ecotype	1	2943.38	620.84	< 0.0001
	Flooding	2	513.03	108.21	< 0.0001
	DAF*Rootstock ecotype	12	37.16	7.84	< 0.0001
	DAF*Flooding	24	6.424	1.36	0.1238
	Rootstock ecotype*Flooding	2	397.73	83.89	< 0.0001
Tr	DAF	12	3.3166	10.78	< 0.0001
	Rootstock ecotype	1	202.20	657.38	< 0.0001
	Flooding	2	32.82	106.71	< 0.0001
	DAF*Rootstock ecotype	12	2.62	8.52	< 0.0001
	DAF*Flooding	24	0.59	1.94	0.0052
	Rootstock ecotype*Flooding	2	16.74	54.45	< 0.0001
WUE	DAF	12	37.22	17.01	< 0.0001
	Rootstock ecotype	1	252.29	115.28	< 0.0001
	Flooding	2	73.57	33.62	< 0.0001
	DAF*rootstock ecotype	12	8.58	3.92	< 0.0001
	DAF*Flooding	24	2.69	1.23	0.2090
	Rootstock ecotype*Flooding	2	100.78	46.05	< 0.0001

DAF = days after flooding, *A* = net CO_2 assimilation, g_s = stomatal conductance, *Tr* = transpiration, *WUE* = water use efficiency.

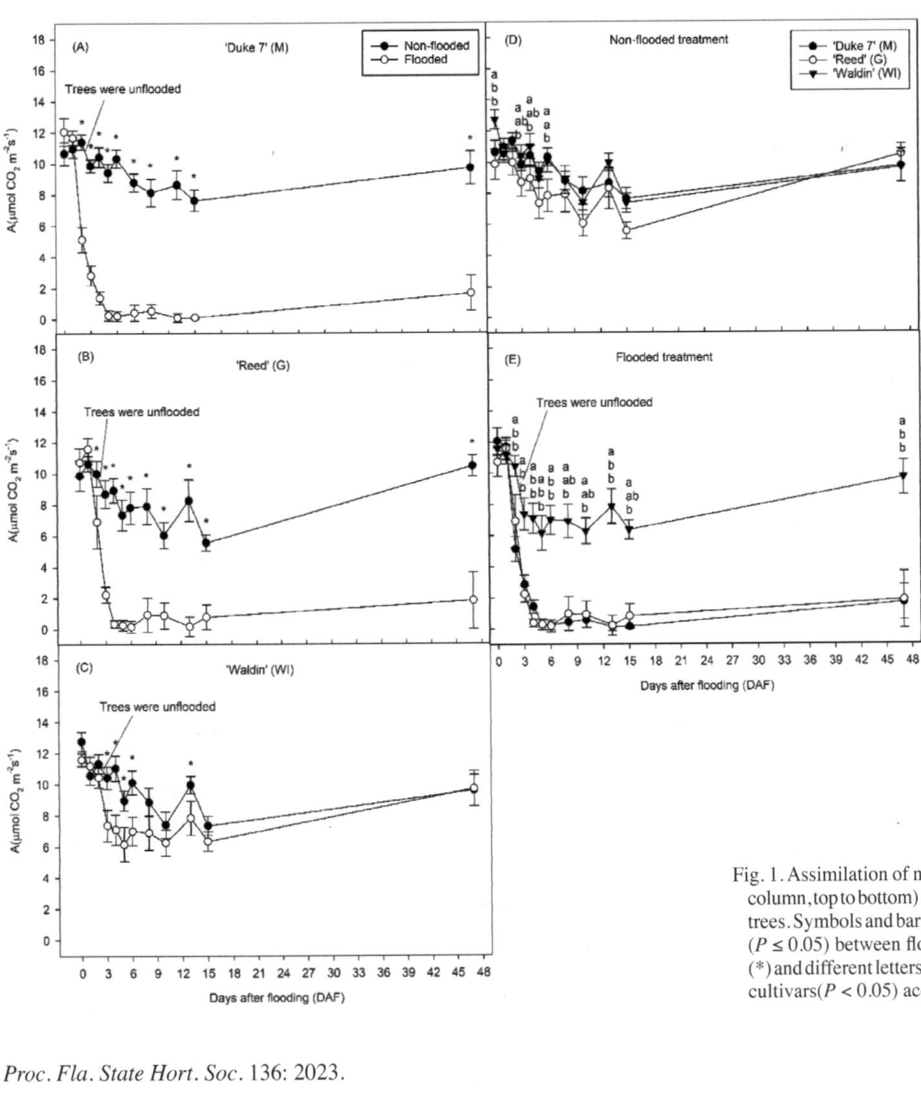

Fig. 1. Assimilation of non-flooded and flooded M, G, and WI trees (left column, top to bottom) (right column, top to bottom) effects of flooding on trees. Symbols and bars indicate means ± std. err. Significant differences ($P \leq 0.05$) between flooding treatments are indicated with an asterisk (*) and different letters indicate a significant difference among rootstock cultivars($P < 0.05$) according to repeated measures ANOVA.

Table 2. Leaf, stem, and root dry weight of 'Simmonds' avocado scion as affected by flooding and rootstock ecotype.

Rootstock	Leaf dry wt (g)		Root dry wt (g)		Plant dry wt (g)		Stem dry wt (g)	
cultivar	Non-flooded	Flooded	Non-flooded	Flooded	Non-flooded	Flooded	Non-flooded	Flooded
'Duke 7' (M)[z]	59.35 a[y]	25.95 b*	42.92 a	27.31 a	59.55 a	36.60 a	42.92 a	27.31 a*
'Reed' (G)	98.01 a	18.08 b*	48.60 a	23.67 a*	62.77 a	29.09 a*	48.60 a	23.67 a*
'Waldin' (WI)	77.20 a	51.52 a	50.25 a	36.56 a	58.31 a	43.54 a	50.25 a	36.56 a

[z]M = Mexican ecotype, G = Guatemalan ecotype, WI = West Indian ecotype.

[y]Asterisks indicate significant difference between flooding treatments according to a non-paired T-test ($P < 0.05$). Different letters within columns indicate significant differences among means according to a Duncan-Waller K-ratio test ($P \leq 0.05$).

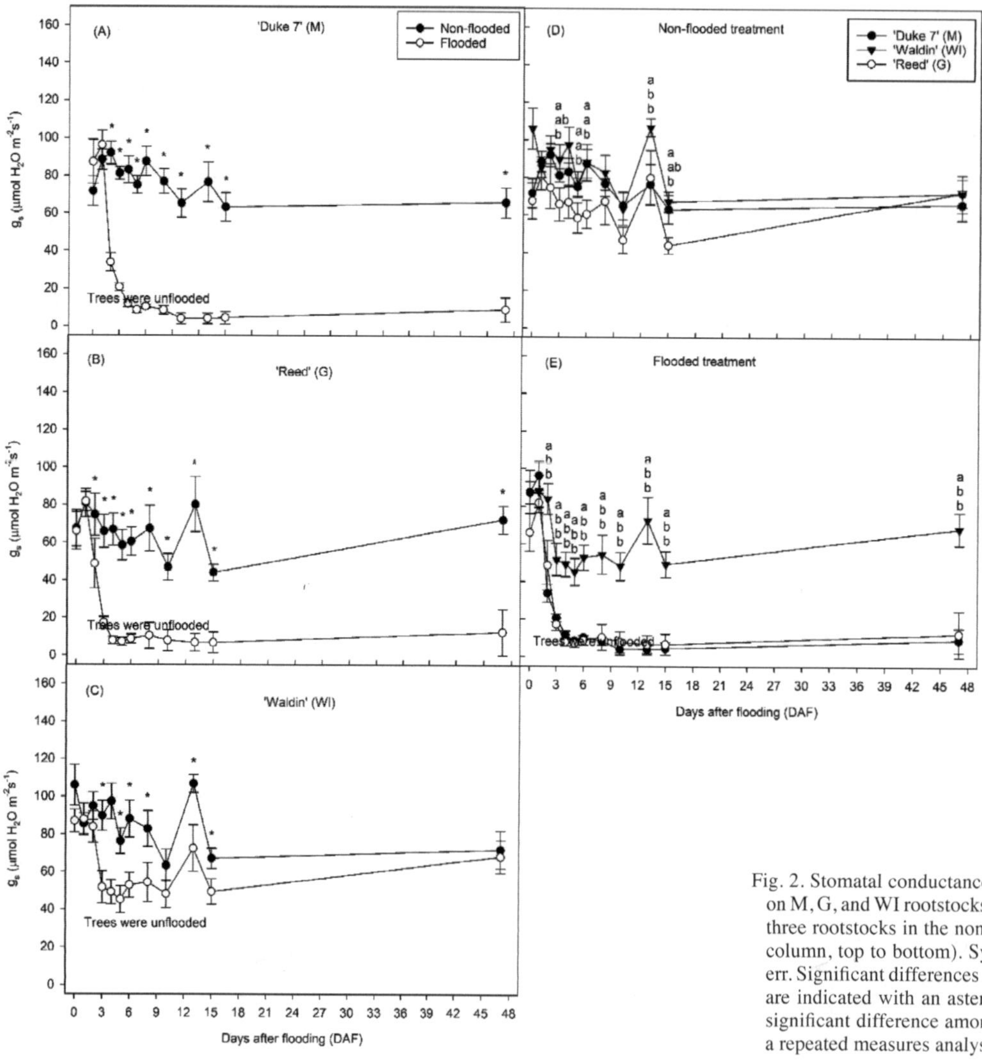

Fig. 2. Stomatal conductance (g_s) of non-flooded and flooded trees on M, G, and WI rootstocks (left column, top to bottom) and of the three rootstocks in the non-flooded and flooded treatments (right column, top to bottom). Symbols and bars indicate means ± std. err. Significant differences ($P \leq 0.05$) between flooding treatments are indicated with an asterisk (*) and different letters indicate a significant difference among rootstockss ($P < 0.05$) according to a repeated measures analysis of variance.

Discussion

In perennial horticultural crops, rootstocks and scions are often bred for different objectives, and rootstock-scion compatibility can be measured in several ways, including overall yield, dwarfing effect, and mitigation of soil and nutrient stresses (Irisarri et al., 2021). While our previous work with non-grafted clonal avocado trees (Yin et al., 2022), showed that that non-grafted, clonal propagated M ecotype avocado trees were the most resilient and non-grafted, clonally propagated G ecotype avocado trees were the sensitive to short-term flooding based on survival, appearance, and leaf gas exchange, the potential effects of ecotype should be viewed through a real-world applicable lens. In contrast, in the present study of grafted trees with a WI scion, those on the clonal WI ('Waldin') clonal rootstock were more flood tolerant than those on clonal M or G rootstocks. Interestingly, we found in a preliminary study with a 60-hour flooding period (M. Yin and B. Schaffer, unpublished data) that M rootstocks grafted with the same WI scion ('Simmonds') experienced 100% mortality, exceeding the purportedly susceptible G rootstock by 50%, while the WI rootstock had 100% survival. In the present study, with a 36-hour flooding period, though the rootstock-scion combina-

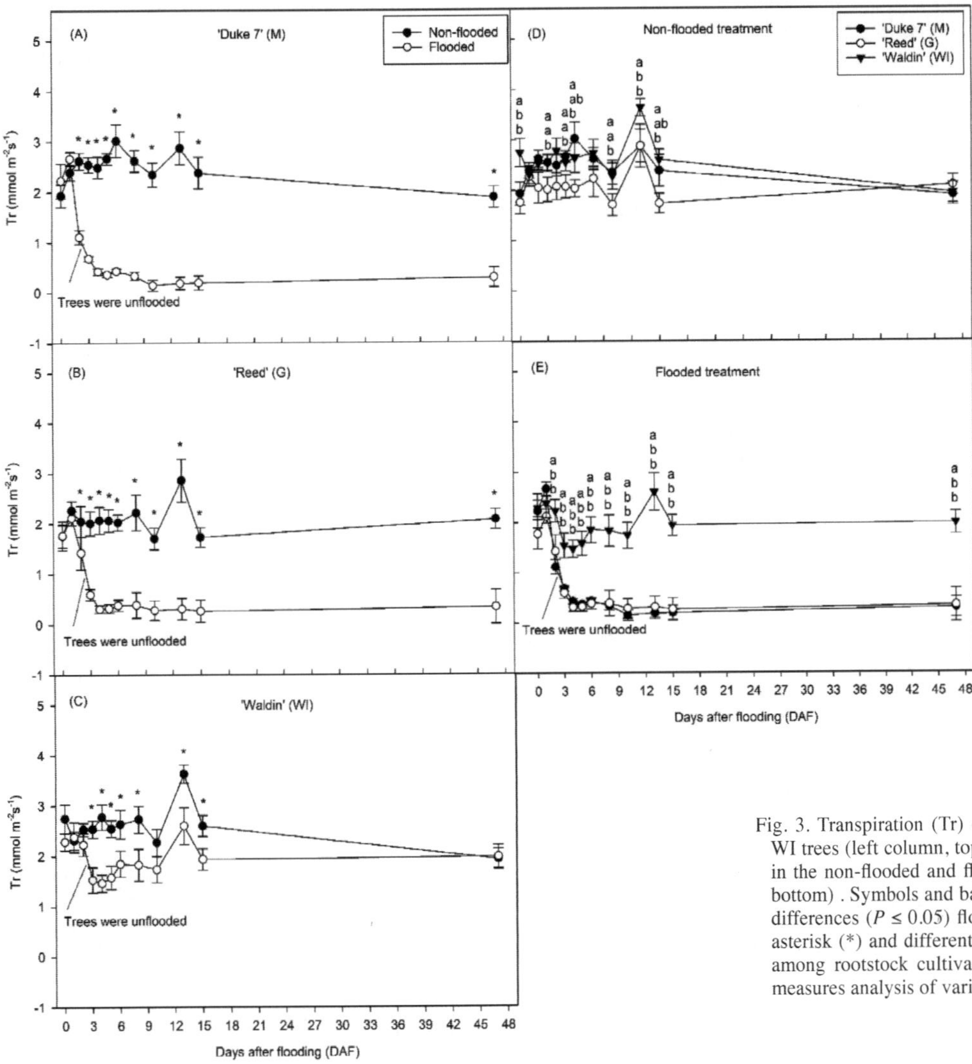

Fig. 3. Transpiration (Tr) of non-flooded and flooded M, G, and WI trees (left column, top to bottom) and of the three rootstocks in the non-flooded and flooded treatments (right column, top to bottom) . Symbols and bars indicate means ± std. err. Significant differences ($P \leq 0.05$) flooding treatments are indicated with an asterisk (*) and different letters indicate a significant difference among rootstock cultivars ($P < 0.05$) according to a repeated measures analysis of variance.

tions performed in a similar pattern, trees on clonal M, then G, then WI rootstock experienced the greatest to least percentage recovery and longest to shortest recovery periods, respectively.

Originating from higher altitudes and more northern latitudes, M avocados have historically performed well in warm Mediterranean climates such as California, and representative cultivars are able to withstand cooler temperatures than either G or WI ecotypes, whereas WI rootstocks have been found to be more tolerant of calcareous soils and salinity (Bergh and Ellstrand, 1986; Witney and Arpaia, 1991; Whiley, 1992; Ben-Ya'acov and Michelson, 1995). In our previous examination of clonally propagated, non-grafted avocado trees (Yin et al., 2023), we found that clonal WI trees had consistently higher leaf gas exchange compared to clonal M trees when not exposed to environmental stress, and that clonal G trees had slightly higher leaf gas exchange than either WI or M trees.

The WI ecotype can be traced back to lowland tropical rainforests in Central America, potentially accounting for its tolerance of poorer soils (Kadman and Ben-Ya'acov, 1976; 1982; Crane et al., 2013; Schaffer et al., 2013; Acosta-Rangel et al., 2019). The superior performance of WI-WI grafted trees in the current study suggests a potentially higher graft compatibility between the representative WI cultivars selected. Tropical trees are exposed to high temperatures, high humidity levels, and fluctuations in soil moisture in their native environments, and have evolved mechanisms for recovering hydraulic conductance in the context of these shifts (Tyree et al., 1999; Brodribb and Holbrook, 2004).

The maintenance of the soil-plant water continuum is inextricably linked to leaf gas exchange, carbon fixation, and ultimately energy production in plants. Though lower transpiration rates, like those observed in our representative non-grafted, clonally propagated M trees may be conducive to greater *WUE* under flooding stress, the addition of a subtropical-tropical WI scion with a higher transpirational demand could contribute to overall plant collapse. In the current study, the greater survival rate of WI–WI grafted trees compared to WI–M or WI–G trees could be partially attributed to ecologically based root and scion hydraulic-transpirational compatibilities. Xylem vessel diameter and number has been shown to vary by cultivar, and may ultimately have an effect on sap flow (Fassio et al., 2009, Castillo-Argaez et al., 2022) and consequently different graft combinations may have different vascular compatibilities.

This experiment only compared clonal rootstocks of one representative cultivar of each ecotype. Also, only one WI scion was used to represent a scion typically grown in southern Florida. Differences in leaf gas exchange as effected by biotic or abiotic

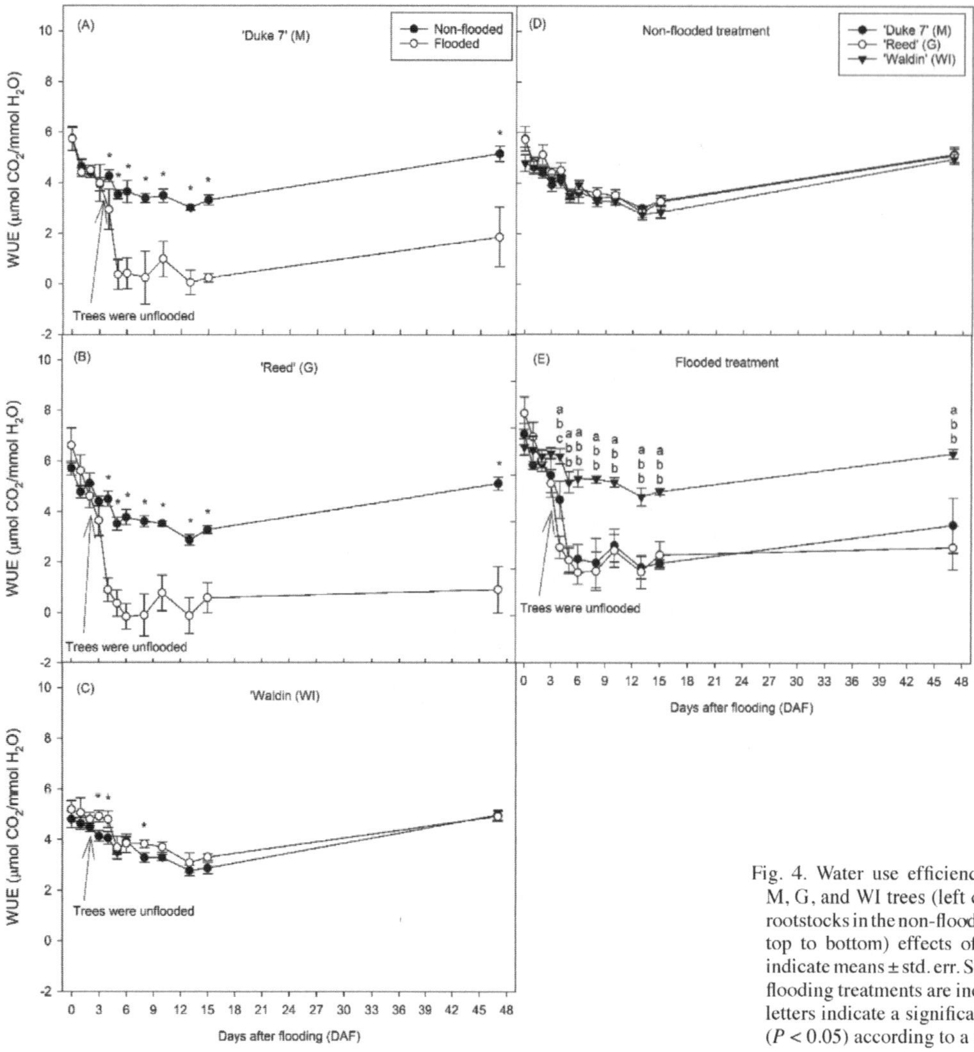

Fig. 4. Water use efficiency (WUE) of non-flooded and flooded M, G, and WI trees (left column, top to bottom) and of the three rootstocks in the non-flooded and flooded treatments (right column, top to bottom) effects of flooding on trees. Symbols and bars indicate means ± std. err. Significant differences ($P \leq 0.05$) between flooding treatments are indicated with an asterisk (*) and different letters indicate a significant difference among rootstock cultivars ($P < 0.05$) according to a repeated measures analysis of variance.

Fig. 5. Effects of flooding on 'Simmonds' (WI) avocado scion grafted onto clonal rootstocks of three different ecotypes. Non-flooded 'Simmonds' (West Indian) scion grafted onto: A) Mexican ecotype, 'Duke-7', B) Guatemalan ecotype, 'Reed', and C) West Indian ecotype 'Waldin'; flooded 'Simmonds' scion grafted onto: D) 'Duke-7', E) 'Reed', and F) 'Waldin', after 47 days of recovery. Photos courtesy of authors.

stress may be more related to cultivar than ecotype. For example, Castillo-Argaez et al. (2021), concluded that due to the variability among cultivars within each ecotype, leaf gas exchange and xylem sap flow responses to laurel wilt disease is probably more related to cultivar than ecotype (Castillo-Argaez et al., 2021). Therefore, to further explore the consequences of flooding of inter-ecotype grafted combinations, more representative cultivars of each ecotype should be evaluated and rootstocks and additional scions of different ecotypes should also be evaluated. Also, in the present study, trees on clonal 'Waldin' rootstock appeared to be much more flood tolerant compared to previous studies of trees on seedling 'Waldin' rootstock (Ploetz and Schaffer 1987, 1989; Sanclemente et al. 2013, 2014). Therefore, additional studies should be conducted to compare flood tolerance of clonal versus seedling 'Waldin' rootstocks with different scion cultivars.

Literature Cited

Acosta-Rangel A.M., Li R., Celis N., Suarez D.L., Santiago L., Arpaia M.L., & Mauk P.A. 2019. The physiological response of 'Hass' avocado to salinity as influenced by rootstock. *Scientia Hortic*. 256:108629.

Bost J., Smith N., & Crane J.H. 2013. History, distribution and uses. p. 10–30. In: Schaffer B., Wolstenholme B.N. & Whiley A.W. (eds.). The Avocado: Botany, production and uses, 2nd Ed. CAB International Press, Wallingford, UK.

Ben-Ya'acov A. & Michelson E. 1995. Avocado rootstocks. *Hort. Reviews*. 17:381–429.

Brodribb T. & Holbrook N.M. 2004. Diurnal depression of leaf hydraulic conductance in a tropical tree species. *Plant Cell Environ*. 27:820–827.

Campbell C., Knight R., & Zareski N. 1977. Freeze damage to tropical fruits in southern Florida in 1977 (varieties, injury ratings). *Proc. Fla. State Hort. Soc*. 90:254–257.

Chanderbali A., Soltis D., Soltis P., & Wolstenholme B. 2013. Taxonomy and botany. In: Schaffer B., Wolstenholme B.N. & Whiley A.W. (eds.). The avocado: Botany, production and uses, 2nd Ed. CAB International Press, Wallingford, UK. pp. 31-50.

Castillo-Argaez R., Konkol J.L., Vargas A.I., Ploetz R.C. & Schaffer B. 2021. Disease severity and ecophysiology of rootstock/scion combinations of different avocado (*Persea americana* Mill.) genotypes in response to laurel wilt. *Scientia Hortic*. 287:110250

Castillo-Argaez R., Vazquez A., Konkol J.L., Vargas A.I., Ploetz R.C., & Schaffer B. 2022. Laurel wilt susceptibility of three avocado (*Persea americana* Mill.) ecotypes in relation to xylem anatomy, sap flow and leaf gas exchange. *Trees*. 36:649–1663.

Crane J., Douhan G., Faber B., Arpaia M., Bender G., Balerdi C., & Barrientos-Priego A. 2013. Cultivars and rootstocks. p. 200-233. In: Schaffer B., Wolstenholme B.N. & Whiley A.W. (eds.),The avocado: Botany, Production and Uses, 2nd Ed. CAB International Press, Wallingford, U.K.

Ernst A.A, Whiley A.W., & Bender G.S. 2013. Propagation. p. 234-267 In: Schaffer B., Wolstenholme B.N. & Whiley A.W. (eds.), The avocado: Botany, Production and Uses, 2nd Ed. CAB International Press, Wallingford, U.K.

Fassio C., Heath R., Arpaia M L., & Castro M. 2009. Sap flow in 'Hass' avocado trees on two clonal rootstocks in relation to xylem anatomy. *Scientia Hortic*. 120:8–13.

Grüter R., Trachsel T., Laube P. & Jaisli I. 2022. Expected global suitability of coffee, cashew and avocado due to climate change. *PlosOne* 17:e0261976.

Her Y., Boote K., Migliaccio K.W., Fraisse C., Letson D., Mbuya O., Swamy A.A., Chi H., Ngatia L., & Asseng S. 2017. Climate change impacts and adaptation in Florida's agriculture. p. 235–267. In: Chassignet E.P., Jones J.W., Misra, V. & Obeyesekera J. (eds.). Florida's climate: Changes, variations, and impacts. Florida Climate Institute., CreateSpace Independent Publishing.

Irisarri P., Errea P., & Pina A. 2021. Physiological and molecular characterization of new apricot cultivars grafted on different *Prunus* rootstocks. *Agronomy* 11:1464.

Kadman A. & Ben-Ya'acov A. 1976. Selection of avocado rootstocks for salnie conditions. *Acta Hortic*. 57:189–198.

Kadman, A. & Ben-Ya'acov A. 1982. Selection of avocado rootstocks for calareous soils. *J. Plant Nutrition*. 5:639–643.

Krezdorn A. 1970. Evaluation of cold-hardy avocados in Florida. *Proc. Fla. State Hort. Soc*. 83:382–386.

Ploetz R.C. & Schaffer B. 1987. Effects of flooding and Phytophthora root ot on photosynthetic characteristics of avocado. *Proc. Fla. State Hort. Soc*. 100:290–294.

Ploetz R.C. & Schaffer B.1989. Effects of flooding and Phytophthora root ot on net gas exchange and growth of avocado. *Phytopathology*. 79:204–208.

Sanclemente M.A., Schaffer B., Gil P.M., Davies F.S., & Crane J.H. 2013. Leaf removal before flooding influences recovery of avocado (*Persea americana* Mill.) trees from flooding stress. *Scientia Hortic*. 150:154–163.

Sanclemente M.A., Schaffer B., Gil P. M., Vargas A. I., & Davies F. S. 2014. Pruning after flooding hastens recovery of flood-stressed avocado (*Persea americana* Mill.) trees. *Scientia Hortic*. 169:27–35.

Schaffer B. 1998. Flooding responses and water-use efficiency of subtropical and tropical fruit trees in an environmentally-sensitive wetland. *Annals Bot*. 81:475–481.

Schaffer B., Andersen P.C., & Ploetz, R.C. 1992. Responses of fruit crops t flooding. Hort. Rev. 13:257–313.

Schaffer B., Gil P., Mickelbart M., & Whiley A. 2013. Ecophysiology. p. 168–199. In: Ploetz, R.C. & Schaffer B. (eds.). The avocado: Botany, production and uses, 2nd dd. CAB International Press, Wallingford, UK.

Tyree, M., Sobrado M., Stratton L., & Becker P. 1999. Diversity of hydaulic conductance in leaves of temperate and tropical species: Possible causes and consequences. *J. Tropical Forest Sci*. 11:47–60.

Warschefsky E.J., Klein L.L., Frank M.H., Chitwood D.H., Londo J.P., von Wettberg E.J., & Miller A.J. 2016. Rootstocks: Diversity, domestication, and impacts on shoot phenotypes. *Trends Plant Sci*. 21:418–437.

Whiley A. 1992. Avocado varieties and rootstocks—A review. p. 15–20. In: Proceedings of the avocado research workshop. Queensland Department of Primary Industries, Brisbane.

Witney G., & Arpaia M. L. 1991. Tree recovery after the December 1990 freeze. *Calif. Avocado Soc. Yrbk*. 75:63–70.

Yin M.H., Vargas A.I., Fuentealba C., Shahid M.A., Bassil E. & Schaffer B. 2023. Differences in physiological and biochemical, responses to short-term flooding among the three avocado (*Persea americana* Mill.) races. *Plant Physiol. Biochem*. 196:925–939.

Proc. Fla. State Hort. Soc. 136:26. 2023.

The University of Florida's Grape Breeding Program: Historical Insights and New Directions

Manjul Dutt*

Citrus Research and Education Center, University of Florida, IFAS, Lake Alfred, FL

—abstract—

Bunch grapes have been grown in Florida since the days of the early pioneers. However, all initial attempts at growing the European *Vitis vinifera* wine and fresh market cultivars were unsuccessful as vines eventually succumbed to Pierce's disease (PD) infection. This xylem limited disease caused by *Xylella fastidiosa* is vectored by the glassy-winged sharpshooter and is endemic in Florida. Scientists identified several native grape species such as *V. aestivalis, V. munsoniana, V. rotundifolia, V. simpsonii,* and *V. shuttleworthii* and observed them to be thriving even under endemic PD conditions. This led to new breeding efforts at the University of Florida (UF), initially by Loren Stover, and subsequently by John Mortensen to introgress PD and other disease resistance genes from these wild species into *Vinifera* selections. The first bunch grape with adequate enological traits and disease resistance was the 'Lake Emerald' cultivar by Loren Stover in 1954. Since then, many other bunch grapes and muscadine hybrids have been released. The UF has not had an active grape breeding program since the retirement of Mortensen in 1991. To reinvigorate the conventional grape breeding program, promising bunch grape selections were identified from the old breeding program. These selections have several important traits such as abiotic stress tolerance, disease resistance, as well as loose, medium-sized clusters and berries. Replicated field trials are planned with several of these selections. We also identified several muscadine selections. The first crosses were made in 2022 and a large population of hybrid progeny is being evaluated. The key focus of the new breeding initiative will be to develop grape cultivars (both muscadine and bunch grape) that are productive with good enological properties and resilient even under changing environmental conditions.

*Corresponding author. Email: manjul@ufl.edu

Proc. Fla. State Hort. Soc. 136:27–28. 2023.

C13-15-1: A Promising Muscadine Grape Breeding Line for White Wine

Zhongbo Ren*, Violeta Tsolova, and Islam El-Sharkawy

Center for Viticulture, Florida A&M University, Tallahassee, FL

Keywords. debris precipitation, evaluation score, internode size, shoot growth

'C13-15-1' is a new muscadine grape breeding line selected for white wine at the Center for Viticulture, Florida A&M University. This selection is self-fertile and productive, it has produced fine white wine with a fruity aroma, pleasant flavor, and clear finish. Its moderately vigorous vines are disease resistant and have a remarkably lower fruit rot rate than other bronze muscadine grapes during ripening. 'C13-15-1' fruits ripen in early September at Tallahassee, FL, about a week later than the leading wine cultivars, 'Carlos' and 'Noble'.

A majority of muscadine grapes (*Vitis Rotundifolia*) are grown for making wines, which has become a multimillion-dollar industry in the southeastern United States today. 'Carlos' (bronze) and 'Noble' (red) are the predominant wine cultivars due to their wine profiles and vine horticultural characteristics that are preferred by the industry, although several muscadine cultivars (such as 'Doreen', 'Magnolia', 'Scuppernong', 'Welder', and 'Ison') in recent years have produced some good wines. 'Carlos' is by far the most widely planted muscadine grape for wines, accounting for 90 to 95% of the commercial muscadine production in North Carolina. This cultivar has consistent high productivity, and more importantly, it produces excellent premium wines. The self-fertile 'Carlos' can produce more than 50 lb of fruit per vine with a 10-foot single wire canopy training system. 'Carlos' grape vines are very vigorous and disease resistant, the uniform ripening of 'Carlos' also facilitates mechanical harvesting for the industry. 'Noble' is a self-fertile, very productive, and disease resistant wine cultivar. It produces fine muscadine wines with a distinctive and sweet taste. The purple pigments of 'Noble' are more stable than other red muscadines, which tend to brown over time.

'Carlos' and 'Noble' have been dominating the muscadine wine industry for decades with these advantages, while the drawback of the entire industry depending on only two cultivars has been recognized. As more and more people start to consume muscadine wines, diversity is necessary for the industry to develop further. Meanwhile, 'Carlos' has experienced high fruit rot incidents during ripening, which reduces juice quality, and could ruin the crop in some years. The excessive vigor and excessive lateral or secondary shoot growth of 'Carlos' can easily create overcrowded vine canopies. These result in poor air flow within canopies, and make vineyard management more difficult. In addition, the devastating Pierce's disease (PD) has been a newly emerged problem of 'Carlos' vines in recent years. Therefore, the industry has been requiring new cultivars, especially one which could solve the problems of 'Carlos' for muscadine white wines.

To respond the industry's requirements, the grape breeding programs at Viticulture Center, Florida A&M University (FAMU), has been dedicated to breeding new muscadine grapes for the industry with various approaches for two decades. A new breeding selection, 'C13-15-1', has great potential value for the industry with several preferred horticultural characteristics and the fine wine it produces.

*Corresponding author. Email: zhongbo.ren@famu.edu

Origin

'C13-15-1' resulted from a scientifically directed cross-hybridization of 'O21-2-4' × 'Carlos'. 'O21-2-4' is a large fruited advanced breeding selection from the opening pollination of 'Pam.' This bronze skin fruit selection has a remarkably sweet taste due to its high sugar content, which is a preferred trait for both fresh market and wine type grapes. Breeders hoped to integrate this high sugar level to F1 hybrids. 'Carlos', with its several horticultural and enological preferred traits, has been a primary choice as a white wine grape breeding parent.

After primary selection, 'C13-15-1' was asexually propagated with soft cuttings during mid-summer. The rooted vines were planted and evaluated at the vineyard in the Viticulture Center, FAMU, Tallahassee, FL (30°28'42"N; 84°10'21"W). These vines were planted at density of 10 feet in rows and 12 feet between rows, trained into single-wire bilateral cordon system, and were annually pruned with 3~4 buds spur-pruning technics. Standard vineyard management was applied to the vines. The grapes were processed following standard enology procedures of the FAMU breeding program. The major horticultural traits and wine quality of 'C13-15-1' were compared with the leading muscadine wine grape cultivars 'Carlos' and 'Noble'.

Major Characteristics

Flower. 'C13-15-1' is self-fertile, which means it may have constant yield without being affected by the environment. A flower cluster (inflorescences) on average consists of 101 individual flowers. These flower clusters typically grow at the 3rd and 4th nodes of shoots.

Satisfactory productivity. A mature 'C13-15-1' vine produced 36 LB fruit on average, which was lower than 'Carlos' and 'Noble'. Similar information was observed with spur productivity, where a 'C13-15-1' spur produced 204 g fruit, which was lower than the 274 g a 'Carlos' produced (Table 1). This indicates this selection has somewhat lower productivity than 'Carlos' and 'Ison', however, this is a satisfactory yield compared with the other commercial muscadine grape cultivars.

Table 1. Horticultural characteristics of C13-15-1 and leading wine muscadine cultivars in Tallahassee, FL

Cultivar	Vigor	Shoot internode (L × D, cm)	Pruning wt[z] (Lb/vine)	PD[y] (0–5)	Cluster/spur	Fruit #/spur	Fruit wt/spur[x]	Yield[z] (Lb/vine)	Fruit rot rate (%)
Carlos	v. vigor	4.2 × 0.58	19.5	occasional	4.6	59	274	48	14.9
Noble	vigor	4.2 × 0.59	7.8	0	4.3	71	233	41	3.5
C13-15-1	m. vigor	4.1 × 0.59	6.7	0	3.2	36	204	36	7.6

[z]10-foot canopy vines with single-wire bilateral cordon training system.
[y]PD = Pierces disease.
[x]Average of 20 random spurs labeled in the spring.

Fruit rot rate. The of fruit rot rate of 'C13-15-1' was 7.6%, which was remarkably lower than the 14.9% fruit rot rate of 'Carlos' (Table 1). High fruit rot incidents have been affecting muscadine grapes with bronze color all the time, caused mostly by ripe rot (*Glomerella cingulata*) and bitter rot (*Greeneria uvicola*) diseases. Grapes harvested with rotten fruits can negatively impact wine quality, and the lower incident of fruit rot in 'C13-15-1' could reduce this effect and improve wine quality, which is an advantage over 'Carlos'.

Vine vigor. The pruning weight of 'C13-15-1' was 6.7 Lb, it was close to that of the vigorous 'Noble. but was much lower than the 19.5 lb pruning weight of the very vigorous 'Carlos' (Table 1), which suggested the vine vigor of 'C13-15-1' could be similar to that of 'Noble', but lower than the very vigorous 'Carlos'. Because the shoot internode sizes of 'Carlos', 'Noble', and 'C13-15-1' are similar (Table 1), the heavier pruning weight of 'Carlos' means it has more internodes. This could be the result of longer shoots and greater secondary shoot growth. This could be responsible for the generally observed very crowded 'Carlos' canopies. The lower pruning weight of 'C13-15-1' could means relatively less secondary shoot growth, which would form less crowded canopies with better air flow and light exposure.

Fruit and cluster. The semi-compact clusters of 'C13-15-1' consist of bronze-colored fruits (Fig. 1), average fruit weight of 5.8 g, which is close to the 5.6-g fruit size of 'Carlos' (Table 2). Fresh juice acid and pH was not significantly different from 'Carlos' or 'Noble'. Its sugar level (15.9% SSC) is close to that of 'Noble' and higher than the 14.9% sugar content of 'Carlos' (Table 2). Higher sugar and adequate acid levels are advantages in producing fine muscadine wines.

Uniform ripening. 'C13-15-1' fruit ripen uniformly (Fig. 1) in early September at Tallahassee, FL, about a week later than the leading wine cultivars 'Carlos' and 'Noble'.

Enology and wine. 'C13-15-1' produces a fine white wine with the rich characteristics of muscadine grapes. The wine taste stays pleasant with a fruity aroma and flavor, and a clear finish. The color of 'C13-15-1' wine is a little lighter than 'Carlos' wine. Panel evaluation with American Wine Society (AWS) evaluation procedure information had wine scores of 13.2 for 'C13-15-1' and 13.6 for 'Carlos'—meaning these wines are similar (Table 2). This suggests that 'C13-15-1' produces a fine wine.

It has been noticed that 'C13-15-1' wine tends to precipitate fast and form sediment more compactly than 'Carlos' and other muscadine white wine grapes during racking, which is an advantage for wine racking and clarification procedures.

Table 2. Fruit and wine characteristics of C13-15-1 and premium wine muscadine cultivars in Tallahassee, FL

Cultivar	Fruit color	Size (g)	SSC[z] (%)	TA[y] (%)	pH	Wine score (0–20)
Carlos	bronze	5.6	14.9	0.41	3.2	13.6
Noble	red	3.3	16.0	0.44	3.3	–[x]
C13-15-1	bronze	5.8	15.9	0.45	3.4	13.2

[z]SSC = soluable solids content.
[y]TA = titratible acidity.
[x]Red wine is not comparable to white wine.

Fig. 1. 'C13-15-1' at harvest, showing uniformly ripe fruits and semi-compact clusters.

In Brief

The productivity of the FAMU breeding selection 'C13-15-1' was not as high as very productive 'Carlos', but it compared favorably with other muscadine cultivars. Combining the productivity, the advantages over 'Carlos' of lower fruit rot rate, moderately vigorous but not too vigorous vines, and the fine wine it produces, suggest this selection will be very valuable for muscadine white wine industry.

Proc. Fla. State Hort. Soc. 136:29. 2023.

—Scientific Note—

Using Information Technology Advances to Better Manage Economic Risks Along the Specialty Crop Supply Chain

KIMBERLY L. MORGAN*

Southwest Florida Research and Education Center, University of Florida, IFAS, Immokalee, FL

Keywords. competitiveness, market diversification, market risk, perishable and seasonal foods

Economic risk is the possibility of adversity or loss and refers to "uncertainty that matters." In agriculture, risks include production, financial, human resource, marketing, and legal and/or regulatory risk. Risk management involves choosing among available alternatives to reduce the effects of risk.

Challenges facing specialty crop producers exist between and betwixt, such as decreasing access to farmland, retail pressure and market power, decrease in produce consumption trends, labor supply concerns, rising interest rates, and continued input supply disruptions resulting in higher prices due to global conflicts. Yet opportunities abound and offer ways to differentiate fresh and value-added produce market offerings, changing consumer focus towards selection of fresh produce with proven health benefits, moving away from a side dish toward featuring veggies as the main course, or adoption of conservation-focused production, processing, and packaging practices.

Producers try to control risk for many reasons, such as reducing income variability, more accurate debt repayment planning, ensuring some level of income for family living expenses, and to drive business growth and survival. Controlling uncertainty requires evaluation of tradeoffs between choices, changes across and within areas of risk, knowledge of expected returns, degree of desired entrepreneurial freedom, and other measurable factors. Optimal decision-making occurs where the marginal gains meet or exceed marginal costs of each choice and drives an individual producer's ability to compete in the markets.

Southeastern US farms growing perishable and seasonal food continue to achieve gains due to economies of scale and scope, largely dependent on management decisions made based on market information. Many produce farms have a corporatestructure and grow vegetables nearly year-round on farms located throughout the US and abroad. Such operations have farms strategically located and follow the progression of seasons from south Florida to northern states to provide a year-round supply of produce as demanded by retail and foodservice buyers. Medium size farmers are joining forces, and technology-driven tools offer benefit in the form of savings in time and resources required to access market information. Given that market access and market share drive profitability, exciting new technologies are emerging that reduce the cost of KNOWING and empower the individuals making more informed decisions.

One such market-driven database is AgTools, (https://www.agtechtools.com) an online database that provides data for members of the produce supply chain and perishable industry. A subscription-based service specific to each fruit or vegetable and informed by experienced producers and retail buyers, AgTools find and organize relevant current and historical data into a user-freindly dashboard. Commodity data with over 76 variables and 29 years of records features market demands, labor trends, transportation costs, and more, may be used by growers to make daily production and harvest decisions to improve gains at the margin. Variables include market price data, import data and trends, fuel and labor costs over time, current and optimal weather, measures of sustainability (food miles), and up to the minute news specific to each commodity. A brand-new feature allows buyers to see real-time phenological growth stages of each crop and any reported disease issues unique to the production region. It is the intent of this article to introduce AgTools and to demonstrate why this technology offers time-saving ways to both improve marketing and production decisions and to level the playing field across the supply chain.

*Corresponding author email: kimorgan@ufl.edu

Proc. Fla. State Hort. Soc. 136:30–32. 2023.

Florida Bell Pepper Novel Treatments Profitability Risk Assessments

LONEGE OGISMA[2], TARA WADE[2]*, KELVIN K. AMON[1], AND PAMELA ROBERTS[2]

[1]*Food and Resource Economics Department, University of Florida-IFAS, Gainesville, FL*

[2]*Southwest Florida Research and Education Center, University of Florida-IFAS, Immokalee, FL*

Keywords. bell pepper, bacterial spot, risk analysis

Bacterial spot on pepper can severely impact yields and increase production costs thus threatening the long-term viability and profitability of the industry. Though many have proven effective, adoption of novel treatment techniques has been hindered due to the lack of information on the implementation costs. This study fills the gap by providing a risk analysis of adopting novel control methods for Florida. We conduct partial budgeting and stochastic modeling to assess the risks of adopting each treatment. Partial budgeting is used to reflect how each treatment affects the cost of production. The net returns for each practice are calculated and compared to determine the economic feasibility of each treatment.

Florida produced approximately 32% of the US fresh-market bell peppers, making the state the second largest production area in the US in 2022 (USDA–NASS, 2023). However, production has decreased over time: on average, growers harvested 17,000 acres/year between 2002 and 2011, which decreased to an average of 12,000 acres/year between 2012 and 2019. In 2022, Florida harvested 10,800 acres of bell peppers (Degner et al., 1997; Wade et al., 2020; USDA-NASS, 2023). Despite reduced acres, in 2021, the state's bell pepper production was estimated at $262 million due to a yield peak of 1287 bu/acre (USDA–NASS, 2023). Bell pepper production substantially contributes to the state's gross domestic product, however diseases like bacterial spot, caused by species of *Xanthomonas*, primarily *X. euvesicatoria*, can severely affect yields and the sustainability of the industry. The pepper industry, therefore, needs to develop strategies to minimize losses from bacterial spot.

The bacterial spot pathogen can infect stems, foliage, and fruit of pepper plants and can cause severe yield losses (Gupta & Prashad, 2011). Spray programs using conventional and alternative products have shown promise to manage bacterial spot on pepper (Šević et al., 2019). In the studies conducted in Southwest Florida by University of Florida, Institute of Food and Agricultural Sciences (UF/IFAS), the goal was to reduce economic losses in pepper production caused by bacterial spot while promoting sustainable and profitable production. Using partial budgeting technique and Monte Carlo analysis, we evaluate the risk of each spray program to assess which is economically viable given current costs of production and probabilistic prices. We find that five of the programs show positive net returns (NR) with varying levels of risk.

Materials and Methods

Experimental Design. Pepper seedlings were transplanted in Fall 2013, to field plots located at the UF Southwest Florida Research and Education Center, Immokalee, FL. Treatments were arranged in a randomized design with ten replications. Each plot consisted of 15 plants in a single row spaced 10 in. apart within a 12 ft row with 5 ft between each plot and 10 ft between each row. Guidelines established by UF/IFAS were followed for land preparation, fertility, irrigation, weed management, disease and insect control using the 2013 Pepper Production Guidelines, which is no longer available. Table 1 in the Guidelines describes the products and rates used in each spray program. The applications were done on weekly basis. Plants were inoculated with suspension the bacterial spot pathogen at approximately 6 and 8 weeks after planting. A modified Horsfall Barratt scale was used to estimate disease severity (Barratt and Horsfall, 1945). Fruits were harvested multiple times and determined to be marketable or non-marketable and weighed.

Risk Assessments. Our risk assessment began with evaluating the costs of production (COP) for conventional pepper farms. COP estimates came from Wade et al. (2020) and were updated to 2022–23 values using inflation factors for fixed and variable costs. The variable cost inflation factor is the average PPI for

Table 1. Spray program descriptions; weekly applications starting September through December.

Program no.	Product description, quantity/acre
Water-1	Non-treated Control or water control
2	Manzate Pro-Stik 75DG, 2.0 lb; Kocide 3000 DF, 2.0 lb
3	KNuCop 50HP, 1.0 lb; Manzate Pro Stik 75DG, 2.0 lb
4	NuCop 50HP, 1.5 lb; Manzate Pro-Stik 75 DG, 2.0 lb
5	NuCop 50HP, 2.0 lb; Manzate Pro-Stik, 2.0 lb
6	Serenade Max WP, 1.0 lb ; Kocide 3000 DF, 2.0 lb
7	Serenade Max WP, 1.0 lb; Kocide 3000 DF, 2.0 lb; Manzate Pro-Stik, 2.0
8	Biological a*, 5.0 lb
9	Biological b*, 1.25 lb
10	Kocide 3000 DF, 0.75 lb; Actinovate AG, 3 oz
11	Actigard 50WG, 0.33 oz; Actinovate AG, 3 oz

*Numbered compounds provided by the cooperator.

This work is supported by the USDA National Institute of Food and Agriculture, Specialty Crops Research Initiative project award 2019-51181-30010.

*Corresponding author. Email: tara.wade@ufl.edu

2009–23 (2.65%/year). The fixed cost inflation factor was calculated using the relative difference of the producer price index and was estimated at 2.61%/year. The variable cost inflation factor was applied to labor wages, fuel, chemicals, seed, materials, fertilizer, interest, and other costs that increase as planted land area usage increases. The fixed cost inflation factor was applied to rent, building ownership, depreciation, taxes, repairs, and machinery costs (Wade et al., 2020). The total per acre COP for conventional bell pepper production was estimated at $19,897 including $2,400.25 for conventional spray programs costs. Partial budgeting was used to assess the cost of each treatment relative to typical cost of production and the control (water).

The risk analysis incorporated changes in yield and pepper prices as both are expected to change over time due to changes in environment and market conditions. Total production costs were assumed to be deterministic. Monte Carlo simulations (1000 draws) were used to randomize prices and yields and deviations in NR are assessed to determine economic risks. Pepper prices come from USDA–NASS and are the basis for draws from the uniform distribution. Yields come from experiment station trials and are the basis for draws from the triangular random distribution (Arunraj, 2013). Revenue is the product of yield and prices. NR are the difference between revenue and cost. From the 1000 draws, we computed the expected average NR and the standard-deviation of NR for each spray program to determine profitability and risk. Breakeven prices were also computed.

Results

All treatments except treatment 8 significantly reduced disease severity calculated as Area Under the Disease Progress Curve (AUDPC) compared to the untreated control. Table 2 shows the spray costs, total COP, average marketable yield from the experiments, breakeven prices, and NR for each spray program. Conventional chemical applications account for 12.1% of Florida bell pepper production costs. The novel spray programs show potential savings for growers as spray costs are less than 2% of realted production costs ($17,702.18/acre). From $2,400.25/acre for conventional pesticides applications, spray program costs fell to an average of $266.65/acre. Growers could expect significant financial gains in modifying pesticide programs if yields remain the same as conventional practices. The table also shows that total COP varies from $17,496.75/acre to $18,186.75/acre with programs 2, 3, 4, 5, and 7 resulting in positive average NR. Relatively high yields in spray programs 2 and 5 resulted in the lowest breakeven prices: less than $23/bu.

Fig. 1 shows the deviations in NR for programs with positive NR. Here we see that spray program 2 has the highest average

Table 2. Spray program costs, yields, breakeven prices, and net returns.

Spray programs	Spray program costs ($/acre)	Total cost of production ($/acre)	Average marketable yield (bu/acre)	Average breakeven price ($/bu)	Average net returns ($/acre)
Water-1	0.00	$17,496.75	467.8	$37.40	($5,472.82)
2	$210.00	$17,706.75	784.1	$22.58	$4,021.99
3	$140.00	$17,636.75	730.1	$24.16	$774.57
4	$157.50	$17,654.25	748.8	$23.58	$574.45
5	$210.00	$17,706.75	792.4	$22.35	$2,614.13
6	$266.80	$17,623.55	504.1	$34.96	($5,178.91)
7	$336.80	$17,566.75	741.6	$23.69	$3,022.89
8	$690.00	$18,186.75	596.4	$30.49	($2,427.79)
9	$87.50	$17,584.25	579.8	$30.33	($2,040.20)
10	$244.20	$17,740.95	502.0	$35.34	($5,748.22)
11	$323.70	$17,820.45	621.2	$28.69	($2,889.88)
Conventional	$2,400.25	$19,897.00	-	-	-

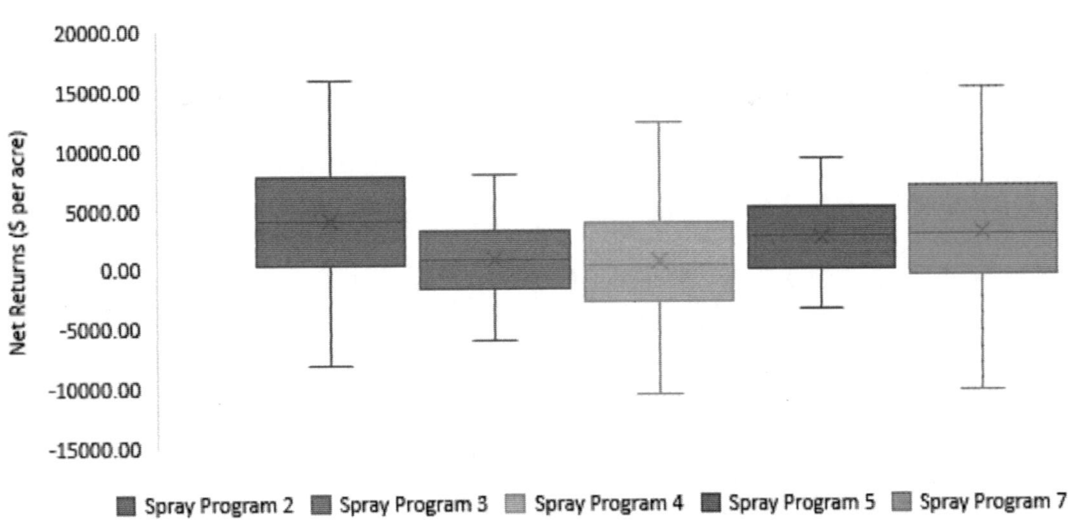

Fig. 1. Profitability risk analysis of Florida spray programs with positive net returns.

NR ($4,021.99/acre) but is also risky with a standard deviation of $4,565/acre and potential losses close to $8,000/acre. Spray program 7 is also very risky with an average NR of $3,022.89/acre and a standard deviation of $5,210/acre and potential losses of $9,210/acre. Spray program 5 presents as a less risky option with average NR of $2,614.13/acre (higher than programs 3 and 4 and similar to program 7) and standard deviation of $3,043/acre. Risk averse growers may prefer to adopt spray program 5 while risk takers may adopt program 2.

Summary

Bacterial spot on pepper can severely impact yields and increase production costs thus threatening the long-term viability and profitability of the industry. We examine the economic feasibility and risk of adopting 11 novel spray programs conducted at the UF/IFAS, Southwest Florida Research and Education Center, Immokalee FL. The spray programs examined do not have a large effect on costs of production but can have large effects on average marketable yields which may vary from 502 bu/acre to 792.4 bu/acre. Three programs were found to be economically viable when considering NR and potential risk: program 2 which used 2 lb of Kocide 3000 DF and 2 lb of Manzate Pro-Stik 75DG; program 5 which used 2 lb of NuCop 50HP, and 2 lb of Manzate Pro-Stik; and program 7 which used 1 lb of Serenade Max WP, 2 lb of Kocide 3000 DF, and 2 lb of Manzate Pro-Stik. With average NR of $4,021.99 per acre and a standard deviation of $4,565, spray program 2 was found to be the most profitable but a relatively high risk of the 11 programs examined. Spray program 5 had average NR of $2,614.13 with the lowest standard deviation of $3,078, then it is the least risky of the spray programs.

Literature Cited

Arunraj N.S., Mandal, S., & Maiti, J. 2013. Modeling uncertainty in risk assessment: An integrated approach with fuzzy set theory and Monte Carlo simulation. https://doi.org/10.1016/j.aap.2013.03.007

Barratt R. and Horsfall, J., 1945. An improved grading system for measuring plant disease. *Phytopathology*. 35:655.

Degner R.L., Moss, S.D., & Mulkey. 1997. Economic impact of agriculture and agribusiness in Dade County, Florida. Institute of Food and Agricultural Sciences, University of Florida, Gainesville.

Gupta S. K., & Prashad, D. 2011. Studies on bacterial spot of bell pepper caused by *Xanthomonas campestris* pv. *vesicatoria*. *Plant Disease Res*. 26(2):161.

Šević M., Gašić K., Ignjatov M., Mijatović M., Prokić A., & Obradović A. 2019. Integration of biological and conventional treatments in control of pepper bacterial spot, *Crop Protection,* 119:46-51. https://doi.org/10.1016/j.cropro.2019.01.006

Wade T., Hyman, B., & McAvoy, E. 2020. Constructing a Southwest Florida Bell Peppers Enterprise Budget. *EDIS*, *2020*. (6):6. https://doi.org/10.32473/edis-fe1088-2020

US Department of Agriculture National Agricultural Statistics Service (USDA-NASS) (2023). USDA/NASS QuickStats Ad-hoc Query Tool. Available at https://quickstats.nass.usda.gov/

Proc. Fla. State Hort. Soc. 136:33–34. 2023.

The Oriental Fruit Fly Host Database

JEFF WASIELEWSKI[1]*, CLIFF MARTIN[2], AND DANIEL CARRILLO[2]

[1]*Miami-Dade County Extension, University of Florida, IFAS, Homestead, FL*

[2]*Tropical Research and Education Center, University of Florida, Homestead, FL*

Keywords. agricultural industry, *Bactrocera dorsalis*, database, Florida, OFF

The Oriental fruit fly (OFF), *Bactrocera dorsalis*, is one of the most destructive pests in the world, with 437 species of known host plants. These include many of Florida's fruit crops, vegetables, and ornamental plants. Because of the potential of OFF to economically damage Florida's agricultural industry, an interactive database was created to help emergency programs focus their control efforts by using accurate fruit phenology for all potential hostplants in Florida at any time. For each hostplant, the database provides the counties where it is found, its common and scientific names, and months of fruit production. The database is easily searchable and hosted online, https://go.ufl.edu/offhosts. The database was created by a grant from the US Department of Agricultural, Animal, and Plant Health Inspection Service.

In late August 2015, several OFF were found in the agriculturally rich area of Miami-Dade County known as the Redland, an area unlike any other in the continental United States. It is home to an incredible array of tropical fruit and vegetables, with well-known crops such as avocados and squash, and unique ones like sugar apples and winged beans (USDA, 2017). Vegetable fields, fruit groves, packinghouses, produce stands, plant nurseries, and even homes are all intermixed and tended to by people as unique and diverse as the crops they grow. The oriental fruit fly is an insect with a massive host list of 437 species, which include most of Miami-Dade's fruit crops, many vegetables and even some ornamental plants; they are strong fliers, and one female fly can lay 1000 eggs in her lifetime (CDFA, 2018). Females lay their eggs in host fruits or vegetables; then, the young larvae hatch and feed on the fruit, effectively rendering it unpalatable and impossible to sell.

Previous detections of this destructive pest in Florida were limited to a small number of flies, which were quickly trapped and destroyed (Citrus Industry, 2022). In late 2015, however, an extraordinary number of male flies suddenly appeared (45 in one trap!), and live larvae were found in the fallen fruit. This triggered the Florida Department of Agriculture and Consumer Services–Division of Plant Industry (FDACS-DPI) and United States Department of Agriculture–Animal and Plant Health Inspection Service Plant Protection Quarantine (USDA–APHIS–PPQ) to initiate the "OFF Eradication Program" and a 99-mi^2 quarantine zone. Agricultural activities in Miami-Dade County produce $2.7 billion in annual revenue, and support 20,000 jobs, so the OFF problem, and needed quarantine, were taken very seriously (Miami-Dade County, 2022). If the OFF had been established, Miami-Dade County would have faced even stricter rules and regulations, and Florida's $7.5 billion agricultural industry may have had severe consequences (USDA, 2022).

The quarantine did not mean a total stoppage of sales or movement of products, but it did call for all growers, shippers, and producers within the quarantine zone, or doing business in the zone, to sign a compliance agreement with FDACS, and then follow the many steps needed to safely sell and move their products in and out of the zone.

Help for the Next Find

Recognizing the need to be better prepared for the next big OFF find, USDA–APHIS–PPQ awarded a grant to faculty and staff at the University of Florida (UF/IFAS) Miami-Dade Extension and the UF/IFAS Tropical Research and Education Center to create an interactive OFF database. It will help federal and state emergency programs focus their control efforts using accurate fruit phenologies of all potential host plants in Florida throughout the year. Host plant information will include the county where found, common and scientific names, and months of fruiting. The database will be hosted online and easily searchable. For instance, if an OFF outbreak appeared in Brevard County in October, that information would be entered into the database, and all host plants found in Brevard County that fruit in October would be listed.

How the Database was Produced

A list of 437 plant species which APHIS determined were hostplants of OFF was initially provided. Most of these plants had edible fleshy fruits and thus provided food for the larvae. Based on this list, an Excel database containing data for all plant species was established. The list was first reviewed for validity of each scientific name using World Florida Online (Miller & Ulate, 2022). This was the primary source for the updated names and authorship and was the "default" database for the majority of species because nearly all the world's plant taxa were listed; hence, it was a consistent source. Using (Miller & Ulate, 2022), 12 of the 437 species were found to be synonyms of other species listed in the database. *Citrus deliciosa* Ten., *C. depressa* Hayata, and *C. unshiu* (Tanaka ex Swingle) Marcow were all synonyms of the tangerine, *C. reticulata* Blanco.

Nine additional sources were reviewed for the presence and fruiting months of each hostplant species in each county for the database. The sources were chosen because of their known quality and reliability. For example, the USDA–APHIS supplied data for the locations where their traps were hung as well as the species

*Corresponding author. Email: jwasielewski@ufl.edu

Table 1. Plants that are recommended for addition to the database (from the comments for plant species in the database).

In our database[z]	Recommended closely related species not found in our database.[y]
(a) Antidesma ghaesembilla	*A. bunius* [x]
(b) Carissa carandas	*C. macrocarpa* [w]
(a) Cissus repens	*C. verticillata* [v]
(a) Ficus erecta	*F. aurea* [v], *citrifolia* [x], *altissima* [x], *drupacea* [x], and *lutea* [x]
(a) Garcinia subelliptica	*G. spicata* [w]
(b) Morinda citrifolia	*M. royoc* [v]
(b) Morus nigra and *(b) M. alba*	*M. rubra* [v]
(b) Opuntia ficus-indica	*O. humifusa* [v], *O. cochinellifera* [2], and *O. stricta* [u]
(b) Passiflora tripartita	*P. biflora* [v]
(b) Pereskia grandiflora	*P. grandifolia* [w]
(b) Sambucus javanica	*S. canadensis* [v], *S. nigra* [t], and *S. simpsonii* [t]

[z] *(a)* = not found in the nine source databases in Florida or *(b)* = found in at least one of these sources but may be uncommon.
[y] Established or successfully cultivated and fruiting in Florida.
[x] Occasionally found and may produce abundant fruit in Miami-Dade County.
[w] Common in cultivation.
[v] Common in wild.
[u] Occasional in the wild.
[t] Other species cultivated in Florida.

of trees. Sources also included the Plant Atlas website and input from University of Florida/IFAS Extension agents throughout the state. Not all of Florida's 67 counties are fully represented in the database. The twenty-four counties are minimally represented are mostly in northern Florida and have small populations. However, this subset notably includes Leon County, which has a large population compared with the others. The other 43 counties are fully represented and are included in searches for the presence or absence of each plant species and their fruiting months,

The database was in Excel and very large and unwieldy. Data was then formatted and simplified to permit entry into the PowerBI platform. This platform allows the user to easily search by county and month to find out which host plants are in the county and when they set fruit.

Room for Improvement

Although the database is useful, several problems have arisen.

1. The original list of 437 plant species included many species found in Asia and nearby islands (native OFF range), but not in Florida. Only 297 (68%) of these OFF hosts were found in any database in any part of Florida. The remaining 140 species (32%) were absent from all databases and counties considered.

2. Some species that fruit profusely in the wild, but are not present in the database are likely hosts of OFF, especially if they share the same genera as the listed hosts. OFF is very polyphagous, and fruits of plants both within and outside the database are often very similar. A glaring example is *Cissus verticillata* (possum grape) in south Florida. This plant is abundant on fences, where it fruits profusely, yet is absent from our list of 437 plant species. However, its relative *Cissus repens*, which is on the list but was not found in any of the nine databases. There are probably many other common OFF hosts that are not in the database (Table 1). They should be investigated to determine if they are indeed OFF hosts or warrant additions to the database.

Literature Cited

California Department of Food and Agriculture (CDFA). 2018. Oriental fruit fly factsheet. Accessed September 8, 2023. https://www.cdfa.ca.gov/plant/factsheets/OFF_FactSheet.pdf

Citrus Industry. 2022. Oriental fruit flies found in Florida (June 28, 2022). Accessed Sept 8, 2023. https://citrusindustry.net/2022/06/28/oriental-fruit-flies-found-in-florida/

Miami-Dade County. 2022. Agriculture in Miami-Dade County. Accessed September 10, 2023. https://www.miamidade.gov/global/economy/agriculture.page

USDA. 2017. Census of Agriculture, County Profile, Florida, Miami-Dade. Accessed Sept 10, 2023. https://www.nass.usda.gov/Publications/AgCensus/2017/Online_Resources/County_Profiles/Florida/cp12086.pdf

USDA. 2022. Economic Research Service. Accessed September 20, 2023. https://www.ers.usda.gov/data-products/chart-gallery/gallery/chart-detail/?chartId=104917#:~:text=At%20%247.5%20billion%2C%20Florida%20accounted,received%20from%20agricultural%20commodity%20sales.

Miller C. & W. Ulate W. 2022. World flora online project: An online flora of all known plants. Biodiversity Information Science and Standards. Accessed November 18, 2022. http://www.worldfloraonline.org/

Proc. Fla. State Hort. Soc. 136: 35–41. 2023.

David Fairchild's Plant Introduction Garden in Miami Celebrates 100 Years of Horticultural Research

Raymond G. McGuire*

Miami, FL

Keywords. Florida history, subtropical horticulture research

One hundred years after its creation by David Fairchild as a research garden for the introduction of tropical plants, the Subtropical Horticulture Research Station in Miami has survived the very real threat of closure and continues to provide cutting edge research to aid South Florida growers and homeowners. Over the past 25 years, one focus of research has continued to be the genetic improvement and distribution of new horticultural plants such as varieties of *Portlandia*, *Lagerstroemia speciosa*, *Handroanthus impetiginosus*, and *Hippeastrum brasilianum*. Plant explorations have been conducted in search of new species of cycads, palms, and amaryllids that could be tested for ornamental merit in South Florida. Another research focus has been the genetic characterization and improvement of tree crops such as cacao, mango, lychee, longan, jaboticaba, and avocado. Breeding productive cultivars of cacao resistant to fungal diseases such as frosty pod, witch's broom, black pod, and *Ceratocystis* wilt has been a priority, as well as breeding avocado for resistance or tolerance to the laurel wilt pathogen now invading South Florida. Entomological research has focused on the development and commercialization of lures for the entrapment of invasive pests of fruits preharvest. Attractant chemicals have been identified and tested for control in the field of insects such as the Caribbean and other tephritid fruit flies in several tropical fruits and the redbay ambrosia beetle in avocados. A fourth area of research was agro-hydrology. Scientists sought to understand the effects resulting from the re-engineering of the flow of water in the Everglades. They sought to model water balance, crop production, and pesticide and nutrient movement within the water management system to evaluate the effect of alternative designs and operational changes on the sustainability of agricultural production and environmental quality in South Florida.

I was a boy of only nine living in California in 1960 when I first heard of the Plant Introduction Garden in Miami, Florida. My mother and I were planting calla lilies (*Zantedeschia aethiopica*), the giant white cup-shaped flowers made famous in paintings by Frida Kahlo, as well as the newly available pink and yellow species (*Z. rehmannii* and *Z. elliottiana*, respectively), all originally from South Africa. She told me how men would travel around the world collecting exotic plants like these and bring them back to the U.S. to multiply them and then distribute them to people around the country. She told me that one such plant explorer was David Fairchild, who collected tropical plants and brought them to a garden in Miami. It seemed to me then that this must be a wonderful occupation and this garden must be a fantastic place to work. I could not conceive that 30 years later I would be working there myself.

The year 2023 is the quasquicentennial of the start of U.S. Department of Agriculture (USDA) research in Miami, FL, and the 100th anniversary of the Plant Introduction Garden conceived by Fairchild at the site of a decommissioned World War I army airbase along Biscayne Bay. One-hundred-twenty-five years ago, in 1898, Fairchild started with just 7 acres south of the Miami River, but by 1923 he had been able to convince the War Department to lease 160 acres of the Army's Chapman Field Military Reservation for the propagation of cold-sensitive tropical plants. In 2023 this USDA property now encompasses 197 acres and is known as the Subtropical Horticulture Research Station (SHRS) of the USDA's Agricultural Research Service. Every 15 to 25 years since its inception, a paper has been written detailing the work of the USDA in Miami (Fairchild, 1923; Fairchild, 1938; Shrum, 1959; Burditt et al., 1973; McGuire et al., 1999); this is the sixth installment in that series.

As a USDA employee in the late 19th and early 20th centuries, Fairchild hoped to improve the diets of Americans and increase the variety of crops they farmed. He explored the world for new crops, and many tropical species, such as avocado and mango, were brought to Miami (Fairchild, 1934). Over 100 years, these introductions have come to include a huge variety of tropical fruit species, as well as horticultural species destined for the gardens of South Florida residents. During the latter half of the 20th century, large fruit and nursery industries developed locally as a result, in part, of this research. These industries were hit very hard in 1992 as a result of Hurricane Andrew, and the SHRS was threatened with closure in 1995 as more and more damaged agricultural land was converted to housing. With the help of the local scientific community and local homeowners the station was saved. An organization of concerned citizens was created, the Friends of Chapman Field, to lobby for the station with USDA officials and advise them of local agricultural concerns.

During the decade before Hurricane Andrew, the major focus of the SHRS had been threefold: the improvement of tropical fruit species such as mango, carambola, and passionfruit (Knight et al., 1995; Schnell and Knight, 1998); germplasm research to preserve the diversity within agriculturally important plants (Ronning & Schnell, 1995); and postharvest research to improve the market quality of fruits for local consumption and for export that focused

The author was a Research Plant Pathologist at SHRS from 1989 through 2000. He would like to thank Dr. Ricardo Goenaga, the Acting Location Coordinator at SHRS, who provided invaluable assistance at a time when COVID made access to resources at this station otherwise virtually impossible to obtain.

*Corresponding author. Email: miarmc@comcast.net

Fig. 1. The hurricane-resistant laboratory building constructed in 2010 at SHRS is of reinforced concrete, with an area approximately 32,500 square feet, comprising 32 administrative offices, one large conference room, one library archive/reading room, 18 laboratory spaces, and a cold room.

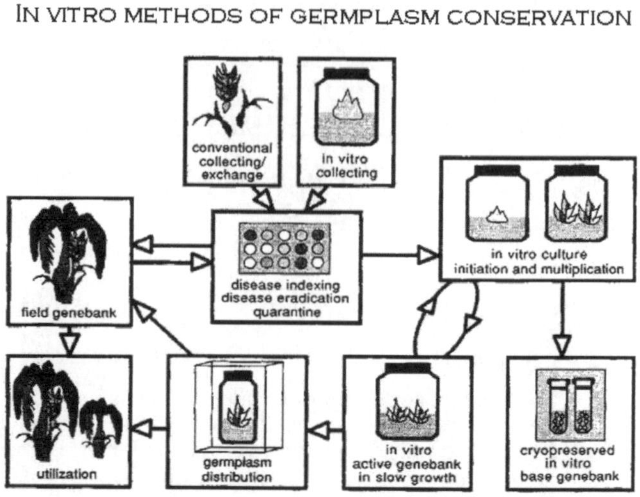

Fig. 2. In vitro methods of germplasm conservation.

on elimination of insect pests within the fruits through hot and cold treatments, fumigation, and irradiation (Sharp & Hallman, 1994).

Damage to local agriculture after Hurricane Andrew forced a change in the objectives of the research at the SHRS. Germplasm and horticultural research continued, and new research into improving the hydrology of the Everglades was initiated, but postharvest market quality research of tropical fruits was largely replaced by new studies into chemical attractants to eliminate fruit pests before they could damage fruits. A great deal of postharvest insect eradication research had already been accomplished, and after the hurricane a great deal less fruit were being exported from South Florida. But the station's budget was secured, and a new research lab was completed in 2010 that would bring all the scientists together into one building (Fig. 1) where previously many had been individually scattered in quaint coral rock buildings bult by the Civilian Conservation Corps in the 1940s; these coral rock buildings were subsequently leased to cooperators.

National Germplasm Repository

The SHRS in Miami is one of 18 national germplasm repositories and shares responsibility with Mayaguez, Puerto Rico, for maintaining the U.S. clonal collections of mango, avocado, banana and plantain, tropical citrus, annonas, sugarcane and related grasses, palms, *Tripsacum*, and a few other relatively minor tropical crops. It maintains approximately 6000 accessions: 3500 are in the major fruit and grass collections, and the remaining 2500 are ornamental, chemurgic, and spice introductions from tropical and subtropical areas of the world. These plants are a unique collection; those that are found to be infected with viral diseases can be purified through tissue culture (Fig. 2). Requests for material come from many scientific disciplines, and small quantities of germplasm are distributed to bona-fide scientists for research purposes.

Horticulture Research

Plant introduction and improvement had been a major goal of the USDA's Chapman Field site since its inception in 1923, and has continued throughout the past 100 years. From 1999 through 2020, Alan Meerow oversaw research on the conservation, genetics, systematics, characterization, and breeding of species of palms, and subtropical and tropical ornamentals. He bred and released new cultivars of *Portlandia*, *Lagerstroemia speciosa*, *Handroanthus impetiginosus,* and *Hippeastrum brasilianum* (Meerow & Ayala-Silva, 2008; Meerow et al., 2010; Meerow, 2014). He also sought to identify the biochemical compounds involved with floral fragrance and corolla color, and their genetic control, in *Hippeastrum brasilianum* and species of *Greffinia*. Like David Fairchild, Meerow also traveled in search of new varieties of subtropical and tropical species that could be tested for ornamental merit in South Florida. He specifically sought out new species of cycads, palms, and amaryllids (Fig. 3). Besides determining optimum fertilization and light conditions for species of palms and cycads, he was also tasked with finding an integrated systematic approach to the phylogeography and conservation genetics of the Caribbean *Zamia* clade (Calonje et al., 2019). This work included the only native (and endangered) Florida cycad, *Zamia integrifolia* (also referred to as *Z. pumila*).

Fig. 3. Dr. Alan Meerow with *Griffinia intermedia* (Amaryllidaceae) in Brazil. Photo by Judy Dutilh.

Genetic Research

Over the past 25 years, genetic analysis of subtropical and tropical fruit and grass species has become a major focus for the improvement of these crops at SHRS. Ray Schnell directed several programs that focused on the genetic improvement of cacao (*Theobroma cacao*) with the goal to develop and disseminate new productive disease-resistant cultivars. His group sought to identify DNA-based markers for resistance to the fungal pathogens causing frosty pod (*Monileophora roreri*), witch's broom (*M. perniciora*), black pod (*Phytophthora* sp.), and Ceratocystis wilt (*Ceratocystis fimbriata*) and to use these markers in a selection program for resistance (Schnell et al., 2005; Schnell et al., 2007). He was able to determine the genes involved with disease resistance and develop a genome map of cacao. In 2005 he was able to secure a 5-year commitment from Mars, Inc. of $500,000 per year to support field testing of selected seedlings in Costa Rica, Ecuador, Ghana, and Papua New Guinea.

Also working on genetic research for improvement of cacao was Osman Gutierrez, who took over from Schnell in 2011 when Schnell left the USDA to become the Director of Cacao Plant Science for Mars, Inc., stationed at SHRS. As such, Gutierrez continued the gene identification and mapping work begun previously (Stack et al, 2015). By this time some potentially disease-resistant plants had been produced, and these were to be evaluated for characteristics such as yield, flavor and fat content, and tree architecture, as well as disease resistance in field trials (Fig. 4) (Gutiérrez, 2021). Cooperative agreements were secured with those countries where trials would be held. This work continued through 2018 with additional emphasis on breeding for biotic stress and tolerance to environmental extremes with new phenotypic trials in Puerto Rico and Hawaii. Genetic assays were developed for use in facilities where no genetic lab was available. Ultimately, eight clones were distributed to breeding programs in West Africa, East Asia, and Central and South America.

David Kuhn also participated in the cacao research (Kuhn et al., 2011) but worked on several other programs as well. By 2010 he had completed the major sequencing of the cacao genome and helped to create an integrated online database (cacaogenomedb.org). Since the SHRS is a germplasm repository for avocado (*Persea americanum*), Kuhn was able to screen over 100 accessions and apply genomic tools to begin breeding for resistance or tolerance to the laurel wilt pathogen (*Raffaelea lauricola*), which was introduced from overseas and has the potential to obliterate the avocado industry in Florida, as well as decimate a native Florida tree species in the same genus (Kendra et al. 2013; Kuhn et al., 2019). Other research focused on analyses of jaboticaba, mango, lychee, and longan to learn the genetic diversity within these crops in South Florida and try to improve that through breeding with plants obtained through exchange with other repositories worldwide (Ayala-Silva et al., 2014). By improving the genetic variation within these plants, it was hoped growers could avoid a repetition of the devastation predicted to result from the fungal blight of avocado. Another project overseen by Kuhn involved the characterization, evaluation, regeneration and genetic analysis of sugarcane (*Saccharum* sp.) *Miscanthus* sp., and *Tripsacum* sp., for which SHRS is also a germplasm repository.

Entomology Research

Bob Heath was hired in 1998 to become the new Location Coordinator and Lead Scientist for a new project that focused on the use of chemical attractants to lure invasive insect pests to their deaths before they could attack fruit preharvest. Over the next 25 years, this project has also come to include entomologists Nancy Epsky and Paul Kendra. Mr. Heath had been working for several years in this field at another ARS facility and immediately started seeking chemicals to attract the Caribbean fruit fly (*Anastrepha suspensa*) that infested many local fruits. Lures were developed, and one was commercialized and is in use now locally.

Kendra was tasked with developing an attractant lure for the redbay ambrosia beetle (*Xyleborus glabratus*) that attacked avocado (Fig. 5) and the native redbay (*Persea borbonia*) (Kendra et al., 2018). He was able to identify the volatile chemicals from the host plants that attracted these beetles and then create a synthetic attractant. Several volatiles were purified, and the olfactory and behavioral responses of the beetles were measured. Electroantennography, a technique for measuring the average electrical

Fig. 4. Harvesting cacao.

Fig. 5. Avocado tree affected by the laurel wilt pathogen (*Raffaelea lauricola*).

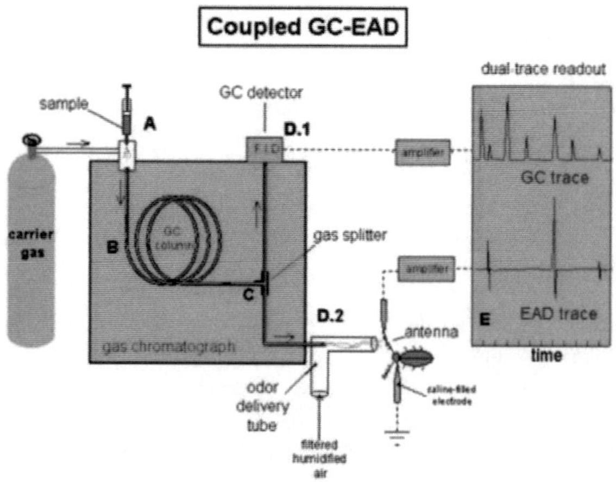

Fig. 6. A coupled gas chromatograph–electroantennographic detector (GC-EAD). (A) A mixture of several compounds is injected into a heated chamber (the injector) where the sample vapor is mixed with a current of inert gas and is swept into the entrance of the column. (B) As the vapor is carried along in the flow of inert gas, compounds with greater affinity for the column lining will travel more slowly and separate from compounds with less affinity for the lining. (C) The column effluent is split in two, with one half transmitted to a flame ionization detector (FID) and the other half carried to the insect antennal preparation. (D.1) The FID produces a voltage proportional to the amount of organic compound in the column effluent (D.2) Column effluent is mixed with purified air and blown over an insect's antenna attached to an amplifier. (E) The amplified electrical outputs of both the FID and antenna are plotted simultaneously.

output of an insect antenna to its brain for a given odor, will be used to evaluate the chemical attractants (Fig. 6) (Kendra et al., 2012). Field tests are being performed to confirm the efficacy of the attractants, and an attractant system was commercialized.

Epsky also worked with Kendra to develop improved trapping systems for invasive insect pests (Epsky et al., 2014). In addition to developing baits for the redbay ambrosia beetle (Epsky et al., 2011a), she also developed baits for the Mediterranean fruit fly (*Ceratitis capitata*) (Fig. 7) (Epsky el al., 2011b) and the cocoa pod borer (*Conopomorpha cramerella*). Besides chemicals that could lure insect pests to their deaths in traps, she sought chemicals that could disrupt pest behavior so fruits would not become infested preharvest. Epsky also continued to explore postharvest methods for controlling insect pests within the fruit prior to export. This work initially began at SHRS in the 1980s under Jennifer Sharp and others with the Caribbean fruit fly. Epsky tested the effect of time and temperature of hydrocooling mango on the mortality of *A. ludens*, the Mexican fruit fly. With many acres of mango trees in South Florida now gone due to Hurricane Andrew and urban encroachment, more of these fruit were being imported, potentially containing invasive pests. U.S. growers would not want to deal with one more invasive fruit fly that attacked many plant species.

Agro-hydrology Research

Beginning in 1997 and continuing through 2009, Reza Savabi directed a project that dealt with agricultural hydrology of the Everglades. After decades in which the Everglades had become criss-crossed with roads that acted like dikes, an extensive network of canals dug to divert water to provide suitable agricultural land, and more and more water being pumped for an expanding population, the River of Grass was drying up. Federal and State agencies sought to reverse this process by replumbing the Everglades while maintaining agricultural production and improving environmental quality that could sustain water supplies for the neighboring cities. This research included field experiments on soil water balance, infiltration, and non-point source water pollution to assess how these interrelationships affected regional hydrology and water quality (Fig. 8) (Savabi, 2001). A farm-scale computer model, called the USDA Everglades Agro-Hydrology Model (EAHM), was developed to simulate water balance, crop production, and pesticide and nutrient movement within the unsaturated zone. This model was linked with regional computer models (SFWMD-RSM and MODBRANCH) to evaluate the effect of alternative designs and operational changes in the

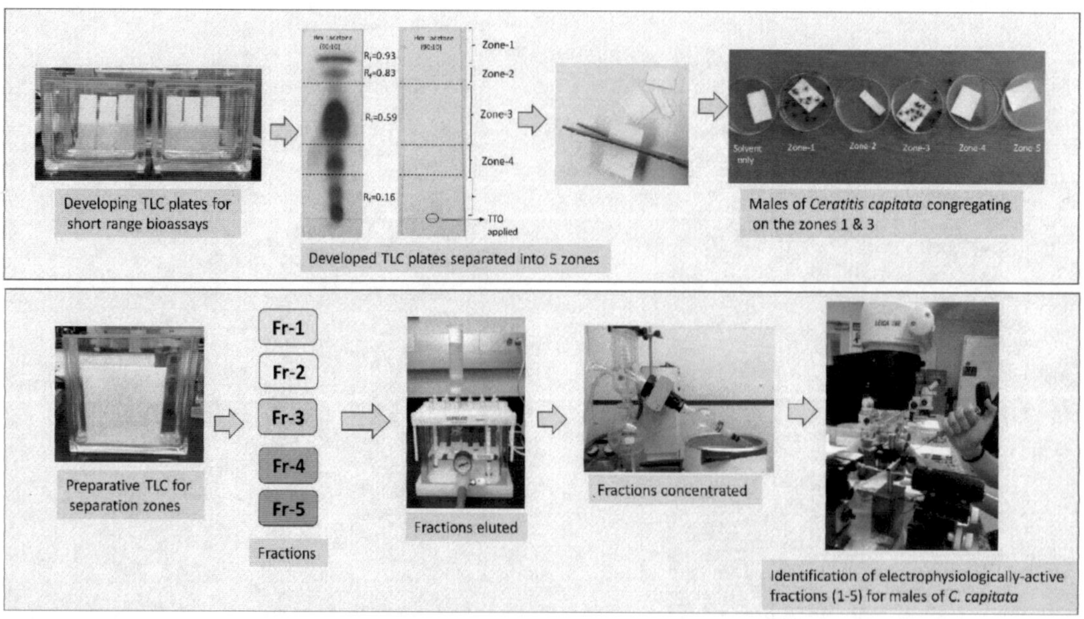

Fig. 7. Thin-layer chromatography (TLC) with a bioassay to isolate fractions of tea essential oil that are physiologically attractive to male Mediterranean fruit flies, *Ceratitis capitata*. These fractions could be used to lure insects to traps. From Tabanca et al., 2020.

Fig. 8. With the re-plumbing of the Everglades, many agricultural fields could flood at different times of the year. It was necessary to know how agricultural chemicals and contaminants would move through the soils.

water management system on the sustainability of agricultural production and environmental quality in south Florida (Savabi, 2007). It was anticipated that a more freely flowing Everglades would likely elevate the water table and possibly saturate many agricultural fields during certain times of the year. This effect on farm production was also studied at SHRS. Working initially with Savabi in hydrology was Stuart Reed. The two scientists also monitored the height of water in wells and the effect that flooding would have on levels of phosphorus (from field fertilization) in the well water (Reed et al., 2006). Well water is not only used for agriculture; many home owners also get their water from wells. Phosphorus in wells was not affected by periodic field flooding. The scientists also evaluated the effect of field flooding on arbuscular mycorrhizal fungi that form symbiotic relationships with plants and aid nutrient uptake (Sah, 2005). The aerobic mycorrhizae thrived in soil intermittently flooded up to three months a year. It was determined that flooding does not destroy established plant root/mycorrhizal associations. By 2006 the agro-hydrology program had split into a hydrology group that subsequently was moved to the ARS station in Tifton, Georgia in 2009. An agronomy group that remained at SHRS then merged with horticulture. Reed continued his work with the effects of flooding on crops and the bioaccumulation of nutrients and contaminants such as arsenic on agricultural and horticultural crops.

Cooperative Research and Extension

During this period, as before, the SHRS also provided facilities for research with several cooperators. As part of the Mars Cocoa Genetics & Breeding Program, Mars, Inc., has worked with SHRS geneticists to develop new cultivars of cacao that would be resistant to fungal pathogens and produce high quality cocoa. Their commitment has lasted over 15 years and contributed hundreds of thousands of dollars to the SHRS budget for equipment and personnel.

The Customs and Border Patrol (CBP), Agriculture Air Passenger Operations (the Beagle Brigade), as part of an intergovernmental program, maintains a permanent kennel on SHRS grounds. This building currently houses 10 USDA-trained detector dogs, which are kept safe while they are not working at the Miami airport. On SHRS grounds the dogs are groomed and exercised. Canine teams assist CBP agriculture specialists deployed to U.S. ports of entry to detect, and intercept prohibited agriculture items.

Several bureaus of the Florida Department of Agriculture and Consumer Services' Division of Plant Industry (FDACS-DPI) operate out of the SHRS. The Bureau of Plant and Apiary Inspection works to protect Florida's agricultural industry and the public from the introduction and spread of serious plant pests. Plant Inspection's Region 3 offices are located on the SHRS property due to its convenient location. This region has the most plant nurseries per square mile of any area in North America. Dog handlers from the Detector Dog Inspection Program maintain their office and training area on the SHRS. These highly trained canines and their handlers inspect parcel shipping facilities for regulated plant and agricultural materials inside packages as well as assist in pest eradication programs throughout the state. The Florida Cooperative Agricultural Pest Survey (CAPS), a cooperative team of USDA–APHIS and FDACS–DPI staff, also operates a South Florida office and laboratory out of the SHRS. With Miami being a high-risk introduction area, it is important for CAPS to maintain a staff in this part of the state to detect and respond quickly to pest introductions. Over the years, FDACS–DPI has conducted many emergency eradication programs in which the incident command system was utilized and headquartered out of SHRS. These eradication programs have included the giant African land snail (*Lissachatina fulica*), horntail snail (*Macrochlamys indica*), and many exotic fruit flies, such as the Oriental and Mediterranean. The SHRS has the resources to be able to accommodate these types of operations without interfering with the daily operations of other SHRS activities.

Conclusion

Just over 100 years ago, during World War I, a swampy tract of mangroves along Biscayne Bay south of Miami was developed into an army air base and named for the first US pilot killed in the war, Victor Emmanuel Chapman (McGuire, 2001). When decommissioned after the war, David Fairchild persisted in claiming much of Chapman Field for the USDA to be used as a plant introduction garden (Fairchild, 1934). Since 1923, tens of thousands of plant accessions have passed through the gates of what is now called the Subtropical Horticulture Research Station of the USDA's Agricultural Research Service.

Prior to Hurricane Andrew in 1992, the SHRS boasted a large staff including one horticulturalist, one geneticist, six entomologists, a plant pathologist, and a chemist, plus post-doctoral researchers, and an administrative and technical staff that numbered 24. Over the next three years, with uncertainty that the facility would remain open, all but three of those scientists either retired or transferred to new locations. By 1999, however, with new funding assured, and the new hydrology and entomology programs instituted, along with a reinvigorated horticulture program and a commitment to build a new $11 million laboratory building where all scientists could be housed, things were looking up again. Nevertheless, new scientific positions were often hard to fill due to budget constraints. Now, 30 years after the great hurricane that devastated South Florida agriculture, there is once again a sizable staff of five geneticists, three entomologists, one chemist, and one physical scientist, with nineteen working as technicians or in administration, and field support positions (Fig. 9). With recent retirements, eight of the ten scientists have been hired in only the past two years and have just recently taken their posts after covid work restrictions were removed. These

Fig. 9. SHRS Staff Picture, 2023, left to right: First row: Mike Winterstein, Kathleen Martinez, Monica Blanco, Brandon Rodriguez, Nurhayat Tabanca, Karla Agurto, Douglas De Stefano. Second row: Barbara Freeman, Aime Vazquez. Third row: Barukh Rohde, Steven Wagner, Madhugiri Nageswara-Rao, Kevin Cloonan, Micah Gill, Elena Schnell, Osman Gutierrez. Fourth row: Wayne Montgomery, Frank Hadfeg, Xiangbing Yang, Wil Quintanilla, Ricardo Goenaga, Xiomara Marte. Fifth row: Jonathan Miller, Gul Shad Ali, Paul Kendra, Ricardo Gonzalez, Chis Dunn, Hector Caceres, Richard Miller.

include Gul Ali, Kevin Cloonan, Kathleen Martinez, Madhugiri Nageswara Rao, Kyoko Nakamura, Barukh Rohde, Nurhayat Tabanca, and Xiangbing Yang. Ricardo Goenaga was named Acting Location Coordinator in 2017 in addition to his duties as Research Leader at the Tropical Agriculture Research Station of the ARS in Mayaguez, Puerto Rico.

COVID-19 had a huge impact on the nation and the world from 2019 through mid-2022. Like many entities, the USDA-ARS ordered those employees that were not considered critically necessary at the jobsite to work from home. Growing plants must be maintained, and a few field-support positions at the SHRS continued to see that germplasm and field trials survived. Scientists and technicians could hardly continue labwork from home. Instead, scientists planned future research, wrote research papers, and prepared reports. Active research on site resumed in Spring 2023.

If things continue as they have over the past 125 years, another installment in this history of the USDA in Miami should appear in the year 2048. What will South Florida be like at that time? Surely there will be less land in agriculture and more homes, but people will still want mangoes and avocadoes, and other tropical fruits. More invasive insect pests will invade the U.S. from those parts of the world that will increasingly export their commodities to the U.S. These will need to be controlled in some way. The need for research in this area likely will not fade over the next 25 years. Over the past 25 years, understanding the genomes of plants as a way to develop new cultivars resistant to pathogens, pests, and environmental stresses and to improve quality characteristics, has become essential, and this will only become more critical. As the world warms, South Florida can provide a suitable place to develop heat-tolerant and salt-tolerant species; the oceans will rise, and the SHRS sits just 10 feet above sea level. Indeed, all of South Florida is projected to lie below sea level by the turn of the next century. But for the next quarter century at least, USDA research at the Subtropical Horticulture Research Station in Miami continues to hold promise.

Literature Cited

Ayala-Silva T., Gubbuk H., Kuhn D., Winterstein M., Salas A., & Gozlekci S. 2014. Characterization and Evaluation of Five Jaboticaba Accessions at the Subtropical Horticulture Research Station in Miami, Florida. *HortScience*. 49(9S):171. Abstract.

Burditt Jr. A.K., Soderholm P.K., Spalding D.H., & Knight, Jr R.J. 1973. Seventy-five years of USDA research at Miami. *Proc. Fla. State Hort. Soc*. 86:303–308.

Calonje M., Meerow A., Gritfith M., Salas-Leiva D., Vovides A., Coiro M., & Francisco-Ortega J. 2019. A time-calibrated species. Tree phylogeny of the new world cycad genus *Zamia* L. (Zamiaceae, Cycadales). Int. *J. of Plant Sci*. 180:286–314.

Epsky N.D., Kendra P.E. & Schnell E.Q. 2014. History and development of food-based attractants. p. 75–118. In: Shelly T.E., Epsky, N., Jang, E.B., Reyes-Flores, J., & Vargas, R.I. (eds.). Trapping and detection, control, and regulation of Tephritid fruit flies: Lures, area-side programs, and prade implications. 2014 edition. Springer: New York, NY.

Epsky N.D., Niogret J., Kendra P.E., & Heath R.R. 2011a. Comparative analysis of terpenoid emissions from Florida host trees of the Redbay Ambrosia beetle, *Xyleborus glabratus* (Coleoptera: Curculionidae: Scolytinae). *Fl. Entomol*. 94:1010–1017.

Epsky N.D., Niogret J., Montgomery W.S., Kendra P.E., & Heath R.R. 2011b. Attraction and electroantennogram responses of male Mediterranean fruit fly to volatile chemicals from *Persea*, *Litchi* and *Ficus* Wood. *J. Chem. Ecol*. 37:483–491.

Fairchild D. 1923. What Chapman Field Plant Introduction Garden can do for tropical horticulture. *Proc. Fla. State. Hort. Soc*. 36:40–46.

Fairchild D. 1934. Reasons for a large general plant introduction garden in southern Florida. *Proc. Fla. State Hort. Soc*. 47:117–120.

Fairchild D. 1938. Reminiscences of early plant introduction work in South Florida. *Proc. Fla. State Hort. Soc*. 51:11–33.

Gutiérrez O.A., Puig A.S., Phillips-Mora W., Bailey B.A., Ali S.S., Mockaitis K., Schnell R.J., Livingstone D., Mustiga G., Royaert S., & Motamayor J.C. 2021. SNP markers associated with resistance to frosty pod and blackpod rot diseases in an F_1 population of *Theobroma cacao* L. *Tree Genetics & Genomes*. 17:28.

Kendra P.E. Montgomery W. S., Niogret J., Deyrup M.A., Guillén L.& Epsky N.D. 2012. *Xyleborus glabratus*, *X. affinis*, and *X. ferrugineus* (Coleoptera: Curculionidae: Scolytinae): Electroantennogram responses to host-based attractants and temporal patterns in host-seeking flight. Environmental Entomology. 41:1597–1605.

Kendra, P.E., Montgomery W.S., Niogret J., & Epsky N.D. 2013. An uncertain future for American Lauraceae: A lethal threat from redbay ambrosia beetle and laurel wilt disease (A Review). *Amer. J. of Plant Sci*. 4:727–738.

Kendra P.E., Montgomery W.S., Niogret J., Tabanca N., Owens D., & Epsky N. 2018. Utility of essential oils for development of host-based lures for *Xyleborus glabratus* (Coleoptera: Curculionidae: Scolytinae), vector of laurel wilt. *Open Chemistry*. 16:393–400.

Knight R.J., Payne J.A., Schnell R.J. & Amis A.A. 1995. 'Byron Beauty', An ornamental passion vine for the temperate zone. *HortScience*. 30:1112. https://doi.org/10.21273/HORTSCI.30.5.1112

Kuhn D.N., Livingstone III D., Main D., Zheng P., Saski C., Feltus F.A., Mockaitis K., Farmer A.D., May G.D., Schnell R.J., & Motamayor J.C. 2011. Identification and mapping of conserved ortholog set (COS) II sequences of cacao and their conversion to SNP markers for marker-assisted selection in *Theobroma cacao* and comparative genomics studies. *Tree Genetics & Genomes*. 8:97–111.

Kuhn D.N., Livingstone D., Richards J.H., Manosalva, P. van den Berg N. & Chambers, A.H. 2019. Application of genomic tools to avocado (*Persea americana*) breeding: SNP discovery for genotyping and germplasm characterization. *Scientia Hortic*. 246:1–11. https://doi.org/10.1016/j.scienta.2018.10.011.

McGuire R.G., Schnell R.J., & Gould W.P. 1999. A century of research with USDA in Miami. *Proc. Fla. State Hort. Soc*. 112:224–232.

McGuire R.G. 2001. Chapman Field: The evolution of a South Dade army airdrome. *Tequesta*. 61:58–85.

Meerow A.W. 2014. The Florida series of hybrid amaryllis: Five new *Hippeastrum* cultivars. *HortScience*. 49:1102–1107.

Meerow A.W. & Ayala-Silva T. 2008 'Miami Sunrise', 'Miami Sunset', 'Tangelo': Three cultivars of *Tecoma guarume*. *HortScience*. 43:546–548

Meerow A.W., Ayala-Silva T. & Irish B.M. 2010. *Turnera diffusa* 'Luisa', a drought-tolerant small shrub for warm climates. *HortScience*. 45:1895–1896.

Reed S., Shinde D., Konomi K., Jayachandran K., Nkedi-Kizza P., & Savabi M 2006. Phosphorus leaching potential from compost amendments in a carbonatic soil. *Soil Sci*. 171:865–873.

Ronning C.M. & Schnell R.J. 1995. Inheritance of random amplified polymorphic DNA (RAPD) markers in *Theobroma cacao* L. *J. Amer. Soc. Hortic. Sci*. 120:681-686.

Sah S., Reed S., Jayachandran K., Dunn C., & Fisher J. 2005. The effect of repeated short-term flooding on mycorrhizal survival in snapbean roots. *HortScience*. 41:598–602.

Savabi M.R. 2001. Determining soil water characteristics for application of WEPP model in South Florida. Trans. of the ASAE. 44:59–70.

Savabi M.R., Cochrane T., German E., Ikiz C., & Cockshutt N. 2007. Modeling evapotranspiration in a sub-tropical climate. *J. of Environ. Hydrol*. 15:1–15.

Schnell R.J. & Knight R.J. 1998. Phenology of flowering among different mango cultivars. *Proc. Fla. State Hort. Soc*. 111:320–321.

Schnell R.J., Kuhn D.N., Brown J.S. Olano C.T., Phillips-Mora W., Amores F.M., & Motamayor J.C. 2007. Development of a marker assisted selection program for cacao. *Phytopathology*. 97:1664–1669.

Schnell R.J., Olano C.T., Brown J.S., Meerow A.W., Cervantes-Martinez C., & Motamayor J.C. 2005. Retrospective determination of the parental population of superior cacao (*Theobroma cacao* L.) seedlings and association of microsatellite alleles with productivity. *J. Amer. Soc. Hortic. Sci*. 130:181–190.

Sharp J.L. & Hallman G.J., Eds. 1994. Quarantine Treatments for Pest of Food Plants. Westview Press: San Francisco. 290 p.

Shrum J.J.E. 1959. U.S. plant introduction station at Coconut Grove, Florida. *The Garden Journal of the New York Botanical Garden* .9:203–205.

Stack C., Royaert S., Gutierrez O., Nagai C., Holanda I., Schnell R., & Motamayor J.C. 2015. Assessing microsatellite linkage disequilibrium in wild, cultivated, and mapping populations of Theobroma cacao L. and its impact on association mapping. *Tree Genetics & Genomes*. 11:19.

Tabanca N., Niogret J., Kendra P.E., & Epsky N.D. 2020. TLC-based bioassay to isolate kairomones from tea tree essential oil that attract male Mediterranean fruit flies, *Ceratitis capitata* (Wiedemann). Biomolecules 10:683.

Proc. Fla. State Hort. Soc. 136:42. 2023.

Microgreens: Potential for Innovative Farming in Florida

Amir Rezazadeh*

University of Florida, IFAS, St. Lucie County Extension

—abstract—

Microgreens are the young growths of leafy greens, vegetables, and herbs that are picked when only the first pair of genuine leaves has formed. There is a definite need for more diversification alternatives in agriculture as there is increasing demand to shift away from intensive conventional food production systems. Microgreens have recently been in high demand due to their high value and reduced carbon footprint. Consumers are increasingly interested in foods (often referred to as super foods) with great nutritional and medicinal benefits. Microgreens have significant potential as a new crop for Florida's future because of characteristics like: being easy to grow with short cultivation period (7–14 days), low startup cost, fast turnaround time, year-round growing, higher nutrition, and high-value crop. Increasing demand for microgreens information and workshops is another factor contributing to the growing popularity of microgreens. Consumers are increasingly growing microgreens in their homes in Florida. The market for microgreens is expected to reach $2.5 billion by the year 2030. As people's interest in organic farming and healthy food options continues to rise, Florida has a great opportunity to become a market leader and generate large demand for a wide range of microgreens. The rising interest in organic foods and the growing health awareness are likely contributing factors. In order to encourage potential consumers to produce this crop, the marketing features of microgreens require extensive investigation.

*Corresponding author. Email: amir2558@ufl.edu

Proc. Fla. State Hort. Soc. 136:43–45. 2023.

Florida's Changing Mango Industry

Jonathan Crane[1]*, Jeff Wasielewski[2], Richard Campbell[3], Noris Ledesma[4], and Gary Zill[5]

[1]*University of Florida, IFAS, Tropical Research and Education Center, Homestead, FL*

[2]*Miami-Dade County Extension Service, University of Florida, IFAS, Homestead, FL*

[3]*Ciruli Brothers, LLC, Tubac, AZ*

[4]*For The Love of Mangos, Miami FL*

[5]*Zill High Performance Plants, Lake Worth, FL*

Keywords. *Mangifera indica,* cultivars

Mangos were introduced into Florida in the mid-19th century. During the early and mid-20th century acreage and production grew dramatically. Numerous institutional (USDA–ARS, UF/IFAS, Fairchild Tropical Botanic Garden) and private citizen mango introductions and seedling plantings resulted in numerous new and some superior mango cultivars (e.g., Haden, Palmer, Tommy Atkins, Keitt, Osteen, and Kent), which became the international trade standards. These mangos were primarily marketed through traditional packinghouses and handled in modern cold chain transport systems. The Florida mango industry had ~1600 acres by 1965; acreage peaked in 1989 at ~2900. However, destruction from Hurricane Andrew in 1992 left only ~1600 acres. Up until the implementation of the North American Trade Agreement (NAFTA) in 1994, Florida was the dominate domestic mango supplier for the small U.S. market. After the implementation of NAFTA and other trade agreements, however, Florida producers could not compete with the volume of offshore production (Mexico, Peru, and Brazil) to meet demand by the U.S. chain store market. This resulted in further acreage decline to ~1100 acres by 2003. Fortunately, Florida has historically been a source for innovation and entrepreneurship in mango culture. This included organized (the Mango Forum, Zill High Performance Plants) and informal mango cultivar selection and research into cultural practices. Today, the Florida mango industry has transformed into a highly diversified industry that includes production for the mature green (raw/vegetable) market and the fresh, direct marketing of select high quality specialty mango cultivars. The marketing options for producers has increased from traditional marketing outlets (packinghouses, traditional and boutique supermarkets) to direct and on-farm sales and electronic-mail-order sales of high-quality novel cultivars, mature green, and cooking mangos marketed to the diverse United States market through internet sales.

The history of mango introduction, acreages, and cultivars in Florida has been extensively documented (Campbell & Zill, 2009; Knight et al., 1984; Knight & Schnell, 1993; Malo, 1970; Wasielewski et al., 2022; and Wolfe, 1962). During the past ~130 years, numerous cultivars have been introduced, selected and evaluated in Florida (Campbell & Campbell, 1993; Campbell & Campbell, 2017; Campbell & Ledesma, 2015; Campbell & Zill, 2009; Knight, 2006; Knight & Schnell, 1993; Ledesma, 2020; Ledin, 1954; Ruehle, 1947; Sturrock, 1969; Young & Ledin, 1954). This has resulted in the potential opportunity for established and new mango producers and nurseries in Florida.

Commercial mango acreage in Florida grew slowly from ~1600 acres in 1965 to ~1900 acres in 1981, then increased dramatically to ~2900 acres in 1989, then declined slightly until 1992 when category-5 hurricane Andrew reduced the acreage to ~1600 acres (Fig. 1). The devastating effect of hurricane Andrew on mango groves along with the increasing volume of imported mango during the 1980s and especially after the implementation of the North American Trade Agreement (NAFTA) in 1994 resulted in a steady decline in mango acreage (Knight et al., 1984). Currently, there is some uncertainty about the extent of Florida's mango acreage because a state agricultural statistics survey has not been conducted since 2002. The USDA-National Agricultural Statistics Service (USDA-NASS) estimates there were 1212 acres, 2575 acres, and 2672 acres of mango in Florida for the 2007, 2012, and 2017 national agricultural census, respectively. However, an increase of over 1300 acres between 2007 and 2012 and 2017 has not been seen by state and private experts. A best estimate on the acreage was that it fluctuated around 1100 to 1200 acres from ~2003 to 2017, then increased to ~1500 acres by 2023.

Currently, the majority of mangos consumed in the U.S. are imported from Mexico and Central and South America and the Caribbean. 'Tommy Atkins', 'Keitt', 'Kent', and 'Ataulfo' dominate the import trade (NMB, 2023). Some others imported include 'Haden', 'Madame Francis', 'Palmer' and 'Nam Doc Mai','Kheio Savoy', and 'Mallika' These fresh fruit cultivars are mostly sold through large supermarket chains. While these imports have greatly increased mango consumption in the U.S., they do not cover the huge diversity in flavors, aromas, textures and colors of mangos.

As Florida's mango acreage increased during the 1970s and 1980s, the major commercial cultivars in Florida were Tommy Atkins, Keitt, Kent, Van Dyke, Palmer, and Irwin (Campbell &

*Corresponding author. Email: jhcr@ufl.edu

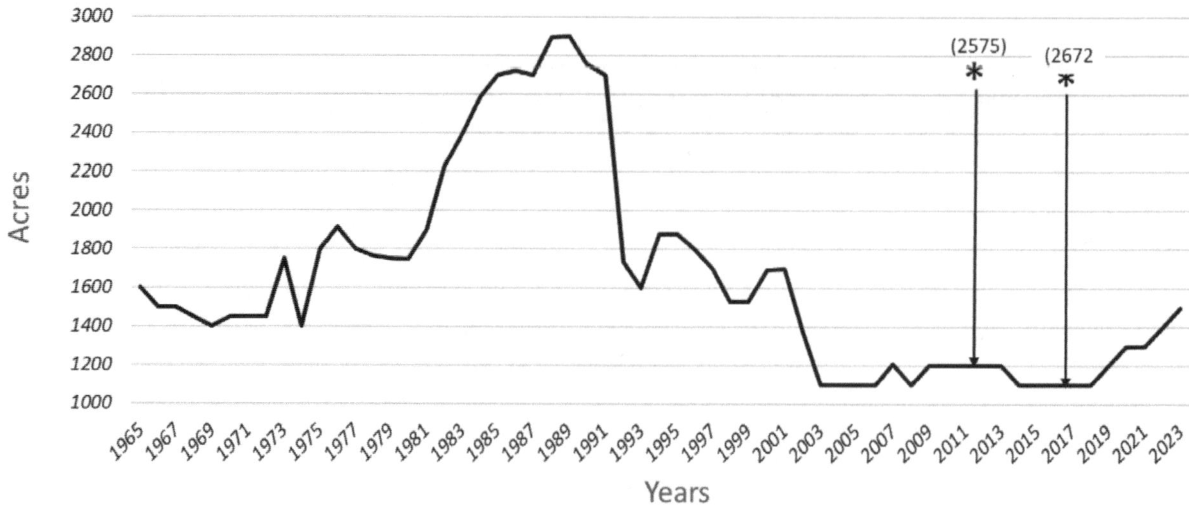

Fig. 1. Historic commercial mango acreage in Florida from 1965 to 2023. Data sources: USDA-National Agricultural Statistics Service (NASS) Census of Agriculture and USDA-NASS Florida Agricultural Statistics Service.

Campbell, 1993). As the cost of production in Florida is high relative to producers in dryer climates and with limited land available in Florida for mango production the acreage continued to decline during the 1990s to early 2000s. Other factors that depressed mango production in Florida was the poor eating quality of 'Tommy Atkins', the poor color of 'Keitt', internal breakdown issues with 'Van Dyke' and 'Kent', issues with determination of when to harvest 'Palmer' and the short postharvest life of 'Irwin'.

Fortunately, several significant events occurred from the mid-1990s to the present time. This included institution of a dedicated cultivar introduction and evaluation program by Drs. Richard Campbell and Noris Ledesma at the William's Farm of Fairchild Tropical Botanical Garden (FTBG), the breeding program of Gary Zill, Zill High Performance Plants, the increase in Asian-American consumers and producers in the U.S., and the phenomenal increase in popularity and consumption of mangos by the U.S. population (NMB, 2023).

The mango introduction and evaluation program at FTGB along with their promotion and educational programs on cultivar diversification and consumption, their annual Fairchild International Mango Festival (begun in 1993), the curator's Choice Program, and multi-media outreach facilitated the desire for and expansion and consumption of mangos with diverse flavors, aromas, textures, and appearance in the U.S. (Campbell, 2004; Campbell & Ledesma, 2015; Campbell & Zill, 2009). Simultaneously, Gary Zill's 30+ year mango selection and evaluation program has dramatically increased the number of unique and high-quality mango cultivars available to grow and consume. These two events along with the increase in digital communication (internet and social media) increased the interest of people familiar or new to mangos. The enhancement of direct sales (farm-stands, farmers markets), internet mail-ordering, and farm direct to store sales have facilitated established growers to be more profitable and launched new producers to grow these specialty—unique, mangos with diverse flavors and aromas.

Coincident with the FTGB and Zill cultivar program, mango consumption in the U.S. has more than doubled since 2000, from 1.75 lb per capita consumption to 3.66 lb per capita consumption (Statista, 2023). Part of this has been the result of an increase in the ethnic diversity (Latin Americans, Caribbean Americans) in the U.S. and the population growth of Asian-Americans, up about 17% from 2016 (Census, 2023). Anecdotally, the number of producers of specialty mangos appears to have increased as established and new producers' plant or top work to alternative cultivars.

Historically minor cultivars now being grown for the specialty markets include Bombay, Carrie, Duncan, Edward, Glenn, Graham, Irwin, Nam Doc Mai, and Valencia Pride. 'Haden'—once the dominant Florida commercial cultivar (1940s–1950s)—is also being grown on a small scale. 'Keitt' is still a major Florida cultivar and a major fresh fruit export mango that continues to be grown, but primarily for the green market. An example of some cultivars now being grown for specialty markets include Angie, Ah Ping, Carabao, Chokanan, Cogshall, Fairchild, Graham, Jumbo Kesar, Kesar, Mahachanok, Mallika, Phimsen Mun, Po Pyu Kalay, Rapoza, Rosigold, Southern Blush, and Tong Dam. Examples of cultivars being grown and sold developed by Gary Zill and the Campbell family include Coconut Cream, Harvest Moon, Lemon Meringue, Little Gem, Julliette, Orange Essence, Pineapple Pleasure, Sweet Tart, Venus, and Diamond (Table 1). In the last few years, several interspecific hybrids of mango with wild species (called wangos) have been bred, selected, and marketed by the Campbell family using the same specialty channels.

The marketing and sale options for specialty producers have increased during the past 30-odd years. Whereas fruit stands and farmers markets have always been common outlets and many producers traditionally marketed mangos through local packinghouses, many specialty mango producers today market their specialty mangos through direct sales to small to medium sized grocery chains, high-end produce outlets (stores), and more recently through electronic mail-order sales. The rise in social media has been a boon to specialty crop producers, with consumers having access to instant information and buying opportunities. Specialty mango producers have used the media to educate consumers, tell the story of their farms and increase interest in new mango cultivars. Whereas Florida may not become a producer of large volumes of mango fruit, it has become a major consumer source for a diverse collection of unique and interesting and high-quality mangos consumers can enjoy.

Table 1. New mango cultivars and selections from Florida.

Origin	Cultivar or selection name
Gary Zill	Butter Cream
	CeciLove
	Coconut Cream
	Cotton Candy
	Diamond
	Edgar
	Fruit Punch
	Fruit Cocktail
	Giselle
	Harvest Moon
	Julliette
	Kathy
	Lady's Choice
	Lemon Zest
	Little Gem
	M-4
	O-15
	Orange Essence
	Orange Sherbet
	P-22
	Phoenix
	Pina Colada
	Pineapple Pleasure
	Sugar Loaf
	Sunrise
	Super Alphonso
	Super Julie (Fairchild Ruby)
	Sweet Tart
	Tara Lay
	Ugly Betty
	Venus
Richard Campbell	Bolt
	Carioca
	Deimos
	Emerald
	Ruby
	King Lion
	Montego
	Nebula
	Nectar of Neptune
	Orion

Literature Cited

Campbell R.J. 2004. A new generation of mangos for Florida. *Proc. Fla. State Hort. Soc.* 117:205–205.

Campbell R.J. & Campbell C.W. 1993. Commercial Florida mango cultivars. *Acta Hortic.* 341:55–59.

Campbell T. & Campbell R.J. 2017. 'Rosigold', and early maturing cultivar for the Florida estate farm. *Proc. Fla. State Hort. Soc.* 130:4–5.

Campbell R.J. & Ledesma N. 2015. Mango cultivars with potential for commercial development. *Acta Hortic.* 1075:33–40.

Campbell R.J. & Zill G. 2009. Mango selection and breeding for alternative markets and uses. *Acta Hortic.* 820:189–193.

Census. 2023. Asian alone or in combination with one or more other races, 2021. U.S. Census Bureau. Accessed May 19, 2023. https://data.census.gov/

Ledesma N. 2020. The genetic diversity of mangoes. In: Galán Saúco, V. and Ping, L. (eds.), Achieving sustainable cultivation of mangoes. Burleigh Dodds Science Publishing, Cambridge, UK, 2018, (ISBN: 978 1 78676 132 3; www.bdspublishing.com).

Ledin R.B. 1954. Mango varieties. Proc. Fla. State Hort. Soc. 67:284–290.

Knight, R.J., Jr. 2006. Tropical fruit—Mango. In: Register of New Fruit and Nut Cultivars, List 43. Clark, J.R. & Finn C.E. (eds.). *HortScience*. 41(5):1101–1133

Knight R.J., Jr., Lamberts M., & Bunch J.S. 1984. World and local importance of some tropical fruit crops grown in Florida. *Proc. Fla. State Hort. Soc.* 97:351–354.

Knight R.J., Jr. & Schnell R.J. 1993. Mango introduction in Florida and the 'Haden' cultivar's significance to the modern industry. *Economic Botany*. 48:139–145.

Malo S.E. 1970. Mango and avocado cultivars present status and future devlopments. Proc. Fla. State Hort. Soc. 83:357–362.

NMB. 2023. Mango varieties and availability. National Mango Board. Accessed May 19, 2023. https://www.mango.org/professionals/industry/varieties-availability/

Ruehle G.D. 1947. Report of subtropical fruit committee. *Proc. Fla. State Hort. Soc.* 60:188–194.

Statista. 2023. Per capita consumption of fresh mangoes in the United States from 2000 to 2021. Economic Research Service, & U.S. Department of Agriculture. (November 7, 2022). Per capita consumption of fresh mangoes in the United States from 2000 to 2021 (in pounds) [Graph]. In: *Statista*. Accessed May 19, 2023. https://www.statista.com/statistics/257231/per-capita-consumption-of-fresh-mangoes-in-the-us/

Sturrock D. 1969. Final report on some mango hybrids—1969. *Proc. Fla. State Hort. Soc.* 82:318–321.

Wasielewski J., Crane J.H., & Campbell R.J. 2022. A brief summary of the current mango industry in Florida. Proc. Fla. State Hort. Soc. 135:36.

Wolfe H.S. 1962. The mango in Florida—1887–1962. *Proc. Fla. State Hort. Soc.* 75:387–391.

Young T.W. & Ledin R.B. 1954. Mango breeding. *Proc. Fla. State Hort. Soc.* 67:241–244.

Proc. Fla. State Hort. Soc. 136:46. 2023.

—Scientific Note—

Nondestructive Techniques to Determine Optimal Harvesting Time in Exporting Mango Varieties in Mexico

JORGE ALBERTO OSUNA-GARCÍA[1]*, MARÍA JOSEFINA GRACIANO-CRISTÓBAL[1], AND RICARDO GOENAGA[2]

[1]*National Institute for Forestry, Agriculture and Livestock Research. Santiago Ixcuintla Experimental Station. Km. 6. Carret. Internacional México–Nogales, Entronque a Santiago Ixcuintla, Santiago Ixcuintla, Nayarit, México. C.P.*

[2]*USDA-ARS, Tropical Agriculture Research Station. Mayaguez, PR*

Keywords. dry matter, mango varieties, spectrometer F-751®

In Mexico, the optimal harvesting time for mango is determined by pulp color and total soluble solids content (°Brix), both destructive procedures. Some countries use dry matter (DM) content, which is a nondestructive technique. During 2019, a unique model to determine mango maturity was developed for cultivars Ataulfo, Tommy Atkins, Kent, and Keitt. A software application was developed containing two artificial neural network models that estimate DM and °Brix using the second derivative of calculated absorbance from diffuse reflectance in the NIR spectra from 684 to 990 nm (Fig. 1). The resulting model was validated during 2020 and 2021 in two commercial orchards of each variety, located in Nayarit and Sinaloa states, Mexico. The DM content was measured nondestructively using an F-751 spectrometer preharvest and at harvest. This model provided a good prediction of DM content but there were varietal differences. However, this technique may substitute the current method.

Our objectives were to build a unique model to nondestructively predict DM content of the main mango exporting varieties in Mexico, as well as, to evaluate and validate the model under commercial conditions.

Findings

The DM model had an $R^2 = 0.8487$; an RMSE = 0.9059 and a PRESS of 1.2039 whereas the °Brix model had an $R^2 = 0.9097$; an RMSE = 0.8389 and a PRESS of 0.7827.

For the validation process, results showed that when comparing the F-751 DM in all varieties at both ripening levels, the DM for the device was 15.6% while it was 15.2% for the oven and $R^2 = 0.4227$ (Fig. 2). Conversely, differences among varieties were detected. The best F-751 model prediction was for 'Ataulfo' with a DM content of 15.9% vs 15.6% for the oven method with an $R^2 = 0.6466$ (Fig. 3).

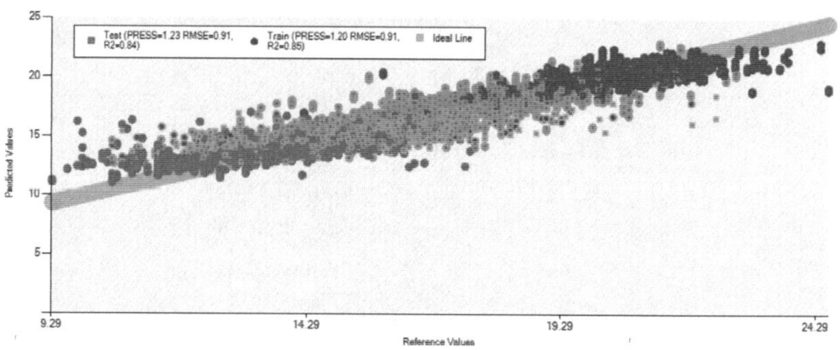

Fig. 1. Full model prediction for the four varieties at both harvesting stages.

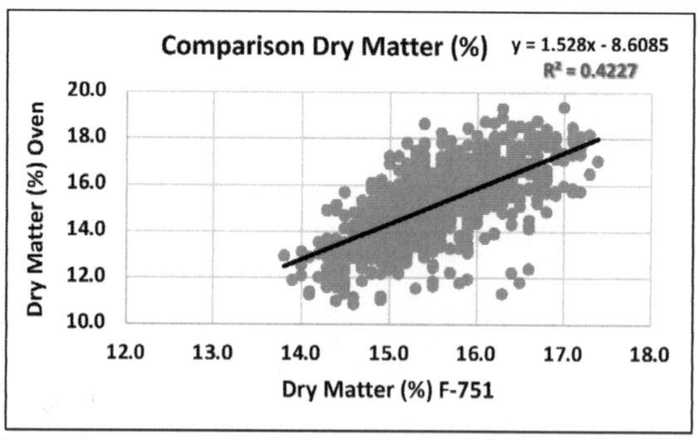

	Dry Matter (%)	
	F-751	Oven
Max	17.4	19.3
Min	13.8	10.8
Average	15.6	15.2
EstDev	0.7	1.6

Fig. 2. Dry matter model performance ($R^2 = 0.84$).

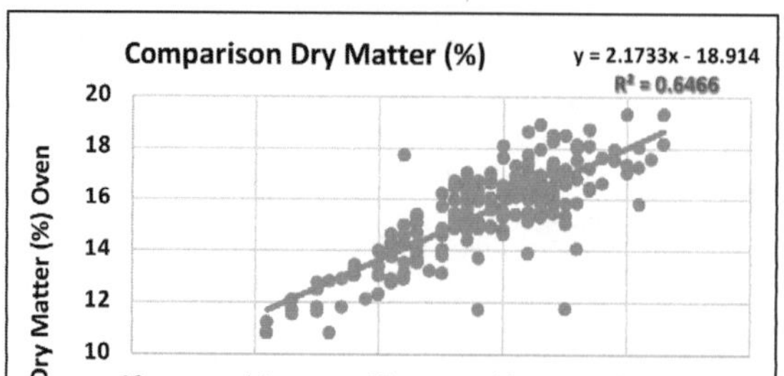

	Dry Matter (%)	
	F-751	Oven
Max	17.3	19.3
Min	14.1	10.8
Average	15.9	15.6
EstDev	0.6	1.8

Fig. 3. Model prediction for 'Ataulfo' at both harvesting stages.

*Corresponding authors. Email: osuna.jorgealberto@inifap.gob.mx; josunaga2@hotmail.com

Proc. Fla. State Hort. Soc. 136:47–49. 2023.

Botanical Characterization and Herbarium Specimens of *Mangifera* Species

Noris Ledesma[1]*, Lyndon Carvajal[2], and Carlos Andres Paz Lopez[2]

[1]*For the Love of Mangos, Homestead, FL*

[2]*Distrital University Francisco Jose de Caldas, Bogota, Colombia*

Keywords. conservation, ecology, collection, *Mangifera*, mounted plants

Mangifera **spp. belongs to the Anacardiaceae family. Twenty-two members of this genus have been collected at different living collections in Florida, including USDA Chapman Field–Subtropical Horticulture Research Station, the University of Florida Tropical Research Center, Fruit and Spice Park, and private collections. Individual plants, or parts of plants have been preserved and are stored at the Forestry Herbarium of the Distrital University Francisco Jose de Caldas, Bogota, Colombia UDBC and the Fairchild Tropical Botanical Garden Herbarium in Miami, FL. These materials may include pressed and mounted plants, seeds, pollen, frozen DNA extracts, and fluid-preserved flowers. This research is aims to preserve samples so that current and future generations can identify *Mangifera* spp., study its biodiversity, and use the collection to support conservation, ecology, and sustainable development.**

The mango, *Mangifera indica* L. (Anacardiaceae), is the best known and most widely cultivated species in the genus *Mangifera* (Iyer and Schnell, 2009). There are over 69 *Mangifera* species currently recognized in Southeast Asia, with many species locally rare and/or included on the IUCN Red List of Threatened Species. Bompard (1993) outlined the potential of *Mangifera* species as edible fruit crops, rootstocks, and as sources of genetic diversity for the future breeding of disease resistance, and desirable horticultural traits in the modern mango.

The establishment of these *Mangifera* species through ex-situ conservation in different living collections around the word has been mainly through grafting on *Mangifera indica* rootstock using local polyembryonic cultivars (Ledesma, 2015). Several *Mangifera* species with edible fruit have been collected and introduced to South Florida. Some of these trees have been reported in private collections, botanical gardens, institutional and private experimental farms. *Mangifera casturi*, *M. odorata*, *M. lalajiwa*, and other species of the genus have been propagated and distributed at the International Mango Festival at Fairchild Tropical Botanic Garden over several years. Today they are growing in backyards and private collections in south Florida. Some *Mangifera* species have been reported in private collections and botanical gardens in Guatemala, Peru, Brazil, Hawaii, Puerto Rico, and Florida (Ledesma, Campbell et al., 2019). New superior clonal selections of *Mangifera* species have been established in living genetic collections in Florida since 1991. Living accessions have been assessed for survival, growth and development, and genetic characterization and for fruit characteristics when fruiting started. This living collection, with different species growing side-by-side under controlled field conditions, will allow for the study and characterization of important phenological aspects of each species. This is essential for developing base line information of species' phenology, given that these *Mangifera* species, in their native habitat(s), may flower and/or fruit once every 15 years or more (Ledesma, Lopez Vasquez et al., 2019).

*Corresponding author. Email: norisledesma.mango@gmail.com

Our objectives were to collect, preserve, and curate botanical specimens of *Mangifera* species available in South Florida as herbarium samples. The archive samples are a critical resource for biodiversity, ecological, and evolutionary research studies. They are a primary data source of dried and labeled plant specimens that are arranged to allow for easy retrieval and archival storage.

This library is and will continue to be stored in biological form—as pressed, dried, and annotated plant specimens of *Mangifera* species chosen for their economic importance in their native or other regions, quality fruit characteristics and tolerance of constant warmth and humidity.

Materials and Methods

Project location and collecting samples. Twenty-two members of the *Mangifera* genus were collected from different living collections in Florida, including the USDA Chapman Field–Subtropical Horticulture Research Station, University of Florida, IFAS Tropical Research Center, Miami-Dade County Fruit and Spice Park, and private collections.

Individual plants, or plant parts have become part of the Forestry Herbarium of the Distrital University Francisco Jose de Caldas, Bogota, Colombia UDBC and Fairchild Tropical Botanical Garden Herbarium in Miami, Florida. The UDBC is an academic unit of the Faculty of Environmental and Natural Resources that conducts research, teaching, and extension activities related to Colombian flora including 47,400 specimens, 270 families, 2052 genera, and 8162 species.

Mango is an important fruit for consumption in Colombia. International and domestic market demand is also on the rise, and economically, mangos are part of a growing sector that is helping to alleviate poverty and hunger by providing jobs and new opportunities. The Distrital University Francisco Jose de Caldas is interested in providing baseline information on ultra-tropical *Mangifera* species suitable for the unique environmental conditions of Colombia. These include various climatic zones such as warm–dry, warm–semi–dry, warm–humid, warm influenced by cool fronts, and other factors such as different elevations, temperature, relative humidity, and precipitation that are ideal for mango propagation and production. Colombia is considered one of only a few countries where mango can be produced year round. Both flowering and harvesting can occur from one to three times per year depending on the region and cultivar. Great opportunities and big challenges exist for diversification of *Mangifera* species in the tropics.

Gathering specimens. Selected specimens in good condition, free of insect and/or disease damage, include mature parts such as well-developed leaves, stems, roots, flowers, or other reproductive structures (Figs. 1 and 2). Basic equipment includes pruners and plastic bags to store branches or carry individually bagged samples. Large branches are divided for pressing and collected extra flowers for later dissection. As each specimen is collected, it is assigned a unique collection number.

Preservation of specimens. Most plants deteriorate after two or three days if they are not dried or preserved in some fashion. Preservation follows Fosberg techniques (Fosberg., et al., 1965, and BCMF, 1996). After plants are collected, the terminal branches with leaves and inflorescences are pressed using newspaper and carboard between samples and numbered. Samples are fumigated in a microwave oven before drying. Flowers are preserved in liquid using a 70% alcohol; 28% distilled water; and 2% glycerol solution

Shipping. The specimens are shipped to the Herbarium of Distrital University in Bogota, Colombia for mounting, and accession.

Digitizing. The specimens are digitized in the collection to make the data freely available online to anyone, anywhere. This includes images and data, while helping scientists around the world conserve plants and find solutions to some of the most critical challenges facing humanity, for example the next generation of climate-resilient crops.

Fig. 1. *Mangifera casturi*.

Future Research and Conclusions

The importance of conserving these species and their genetic potential has been clearly recognized by the scientific community, but there is limited information available (Ledesma, et al. 2015 and Campbell, 2007). Herbarium samples can demonstrate how species are related to one another.

Identification and classification of species must be reviewed. Herbarium specimens and taxonomic review are proceeding. The DNA analysis using several laboratory teams and research groups is also underway to aid in the proper identification of these accessions.

There is a possibility of wild hybridization between species, and this may be detected with genetic analysis to potentially identify new species.

A collection of plant samples with associated data preserved for long-term study. This study of specimens in the Herbarium may contribute to scientists to determine how climate change is affecting the *Mangiferas* throughout different periods in our history, scientists can determine if plants are flowering earlier in the season than they were long ago comparing samples in different herbariums and collections. Studying Herbarium specimens enables us to document the distribution of plants at different times, we learn where plants grew in the past compared to where they are abundant today. Study the evolution of a wide range of taxa.

These materials may include pressed and mounted plants, seeds, dry fruits, wood sections, pollen, microscope slides, silica-stored

Fig. 2. *Mangifera indica* 'Cambodiana' in flower.

materials, frozen DNA extractions, and fluid-preserved flowers or fruits. It also may store and manage data sets, botanical illustrations, photographic slides, images, maps, libraries of relevant literature needed for consultation by researchers working with the specimens.

Literature Cited

British Columbia Ministry of Forests (BCMF). 1996. Techniques and procedures for collecting, preserving, processing, and storing botanical specimens. Res. Br., B.C. Min. For., Victoria, B.C. Work. Pap., Province of British Columbia Ministry of Forests Research Program.

Bompard J.M. 1993. The genus *Mangifera* re-discovered: The potential contribution of wild species to mango cultivation. *Acta Hortic*. 341:69–77.

Campbell R.J. 2007. The potential of new *Mangifera* species in Florida. *Proc. Fla. State Hort. Soc*. 120:11–12.

Fosberg F.R. & Sachet M. 1965. Manual for tropical herbaria. I.B.P.T. and N. Regnum Veget. 39:1–132.

Iyer C.P.A. & Schnell R.J. 2009. Breeding and genetics. p. 68–89. In: Litz, R.E. (ed.). The mango: Botany, production and uses. CABI Publishing, Wallingford, UK. https://doi.org/10.1079/9781845934897.0067

Ledesma N. 2015. Economic feasibility of small-scale specialty mango production in South Florida, *Proc. Fla. State Hort. Soc*. 128:25–27.

Ledesma N & Campbell R.J. 2014. Conservation and commercial development of *Mangifera* species [wild mangos] in Florida, USA. *Proc. Fla. State Hort. Soc*. 127:122–125.

Ledesma N., Campbell R.J., Hass M., & Campbell T.B. 2017. Interspecific hybrids between *Mangifera indica* and related species. *Acta Hortic*. 1183:83–88. https://doi.org/10.17660/ActaHortic.2017.1183.12

Ledesma N, Campbell R.J., Campbell T., Suárez S.J., & Aron Y. 2019. Preliminary field adaptation and fruit characterization of *Mangifera* species in Florida. *Acta Hortic*. 1244:33–38. https://doi.org/10.17660/ActaHortic.2019.1244.5

Ledesma N, Lopez Vasquez R, del Llano Rivas P, & Roberto Marquez S. 2019. Mango pollen collection and preservation method for Mangifera *M. odorata, M. lalijiwa y M. laurina. Proc. Fla. State Hort. Soc*. 132:12–14.

Proc. Fla. State Hort. Soc. 136:50–53. 2023.

Mango Production in Mexico

Sergio Roberto Márquez-Berber[1]*, Gustavo Almaguer-Vargas[1], Alma Velia Ayala-Garay[2], Diana América Reyna-Izaguirre[3], and Abdul Khalil Gardezi[4]

[1]*Crop Science Department, Chapingo Autonomous University, Texcoco, Estado de México, Mexico*

[2]*Head of OCIMA. National Institute for Agricultural, Livestock and Forrest Research, Texcoco, Estado de México, Mexico*

[3]*Agricultural Industries Department, Chapingo Autonomous University, Texcoco, Estado de México, Mexico*

[4]*Hidroscience Center, Postgraduate College, Texcoco, Estado de México, Mexico*

Keywords. economic value, exports, *Mangifera indica,* production, planted area, yield

Mango is the sixth most important fruit crop in Mexico in terms of production value. The country is the world's third-largest producer of mangoes. The fruit is primarily cultivated along the coasts of the Pacific Ocean and the Gulf of Mexico. The southernmost state of Chiapas has the highest planted area, while Guerrero leads in mango production and value. Other significant mango-producing states include Sinaloa, Nayarit, Oaxaca, and Michoacán, all located along the Pacific Ocean coast. In 2022, a total of 201,000 hectares were cultivated across 23 out of 32 states, resulting in an annual production of over 2.18 million tons of mangoes. The wide cultivation area allows for year-round availability of the fruit, with the peak harvest season occurring from April to August. The average yield is 10.85 tons per hectare, with the highest yield achieved in the northwestern state of Sonora at 17.25 tons per hectare. The main mango cultivars in Mexico are Ataulfo, Manila, Tommy Atkins, Kent, and Haden. In 2022, domestic mango production was valued at over 674 million U.S. dollars. Approximately 485,395 tons were exported, accounting for 22.5% of the total production and generating 509 million U.S. dollars in revenue. Of these exports, more than 84% were destined for the U.S. market, owing to Mexico's competitive advantages such as proximity to the largest import market, robust infrastructure, favorable costs, and availability of labor. As the world's largest exporter, Mexico faces significant risks from pests like the Mexican fruit fly and diseases such as anthracnose, which pose threats to exports. Additionally, better organization for small farmers and the consolidation of their agribusinesses are required. The introduction of new varieties is also essential to maintain competitiveness in the market.

The cultivation of mango in Mexico can be traced back to the late 18th century or early 19th century when the fruit was first introduced to the country from the Philippines by the Spanish colonizers aboard the "Nao of China" ship. Mexico's climate and soils provided an excellent environment for the mango trees to thrive. However, the mangoes initially introduced had a thin epicarp and soft pulp, making them unsuitable for long-distance transportation. This limited their marketing to local markets and hindered their expansion to distant places, thus restricting the growth of the crop. It was not until the mid-20th century that mango cultivation became a significant industry in Mexico, with the development of new varieties, improved production techniques, and expanded markets.

In the 1950s, the Tommy Atkins mango cultivar was introduced to Mexico, which proved to be a game-changer for the industry. Tommy Atkins mangoes had a long shelf life, which allowed for extended transportation and opened export markets. This led to a boom in mango production in the 1970s and 1980s, particularly in the states of Guerrero, Nayarit, Sinaloa, and Michoacan, which are still major producers of mangoes today.

In the following decades, other mango cultivars were introduced to Mexico, including Haden, Kent, Keitt, and the discovery of Ataulfo, in the southwestern tip of Mexico, which added variety to the industry and expanded the range of flavors and textures available to consumers.

Mango is a significant part of both Mexican cuisine and culture. The fruit is commonly eaten fresh as a snack or dessert, used to make juices or flavored water, and included in several dishes. In fact, Mexicans have an annual consumption of 13 kg of mango, according to SIAP (2022).

The mango industry provides employment and economic benefits to thousands of farmers and workers in Mexico, while Mexican mangoes are enjoyed both domestically and in export markets worldwide.

The objective of this article is to analyze the development, current situation, and perspectives of the mango crop in Mexico.

The authors would like to express their gratitude to Thelma Guadalupe Valdez Reyes, María Guadalupe Meras Rodríguez, Marvella Orduña Trujillo, and Margarita Duarte González for their valuable assistance in conducting information searches and creating graphics. Additionally, the authors extend their thanks to Dr. Sergio Ayvar Serna for his insightful contributions regarding the challenges faced in mango production.

*Corresponding author. Email: smarquezb@chapingo.mx

Materials and Methods

Official statistics from the Agri-Alimentary and Fisheries Information Service (SIAP), part of the Mexican Ministry of Agriculture and Rural Development (Secretaría de Agricultura y Desarrollo Rural), and the Food and Agriculture Organization (FAO) of the United Nations were utilized to analyze the development of the mango crop from 1980 to 2022.

In the statistics of FAO, the item used was mangoes, guavas, and mangosteens, item code 1316.

For the conversion of production and export values, an exchange rate of 17.9114 Mexican pesos per U.S. dollar was applied. This exchange rate represents the average for the months of April and May 2023, which are the latest available rates according to BANXICO (2023).

Results and Discussion

Mango is an important crop for Mexico, both economically and culturally. In 2022, Mexico was the third-largest producer of mangoes in the world, with over 2.18 million tons harvested (SIAP, 2022 and 2023), but the fifth in 2021 according to FAO (2023), after India, China, Indonesia, and Pakistan (Fig. 1). Furthermore, mango is the sixth most important fruit crop in Mexico in terms of production value, 674 million U.S. dollars in 2022, as reported by SIAP (2023).

The planted area of mango in Mexico has shown an overall trend of growth, although there have been some years with slight declines, such as 1999 and 2000 (Fig. 2). In 1980, the planted area was 71,767 hectares, which increased by 188.6% to 207,086 hectares in 2021 (SIAP, 2023). The average annual growth rate of 2.7% has not been consistent over the years, with varying levels of growth in different periods. The initial period from 1980 to 2001 experienced the highest growth rate (137.7%) with an annual growth rate of 4.4%. In contrast, the subsequent years from 2002 to 2021 had lower growth (21.4%), with an annual growth rate of only 1%. This rate was one-fourth of the growth rate during the initial period.

Considering this information, it is evident that the North American Free Trade Agreement has not had a noticeable promoting effect on mango cultivation in Mexico. However, it is important to note that the overall trend has been an increasing planted area, which highlights the profitability and significance of mango as a crop in Mexico.

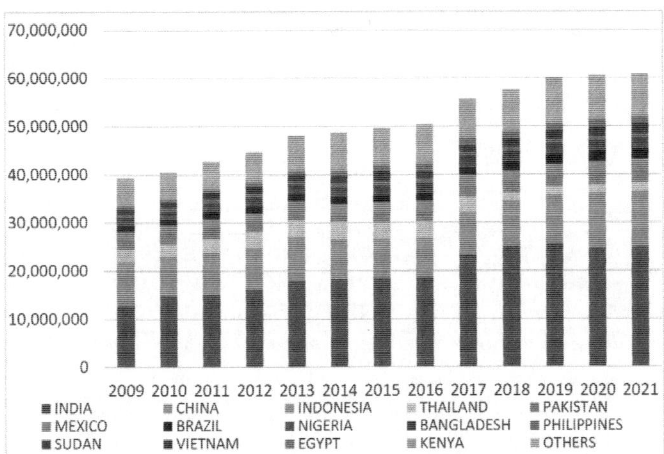

Fig. 1. Mango world production in tons. Source: FAO, 2023.

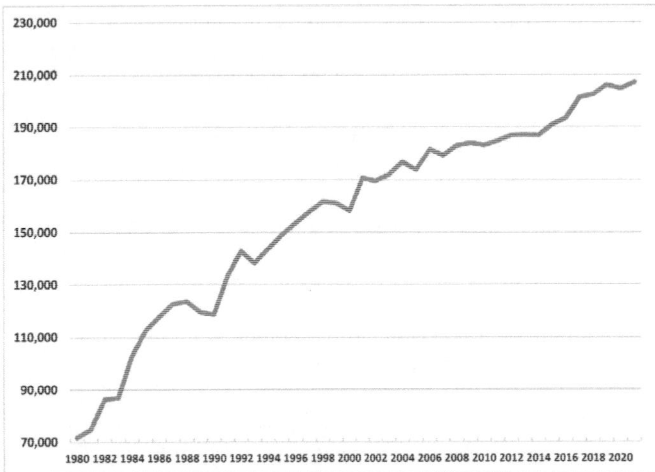

Fig. 2. Mango planted area in Mexico in hectares. Source: SIAP, 2023.

This growth can be attributed to several factors, including increasing demand for the fruit both locally and internationally, favorable weather conditions, and the resourcefulness of producers. It is worth noting that besides the National System of Plant Protection (SENASICA), almost no government policies currently support this crop.

There is also an upward trend in production. In the same period (1980–2022), mango production increased at a faster pace compared to the planted area, with a growth rate of 245.6%. The production rose from 629,567 tons in 1980 to 2,176,046 tons in 2022 (Fig. 3, SIAP, 2023), with an average annual growth rate of 3.15%, primarily due to higher yields. However, there is significant variability in yearly production due to the alternating yield of the planted varieties, particularly Keitt, but also in Kent, Tommy Atkins, and to a lesser extent, Ataulfo.

Organic mango production accounts for 1.6% of the total production (33,943 tons in 2021). The main variety cultivated within this system is Haden, representing 42% of the organic production, followed by Tommy Atkins at 22%, and Kent at 21% (SIAP, 2023).

Mango is cultivated in 23 out of the 32 states in Mexico, with the largest plantations located along the coasts of the Pacific Ocean and the Gulf of Mexico. The state of Chiapas, situated in the southernmost region, has the highest planted area. However, it is Guerrero that leads in mango production and holds the greatest

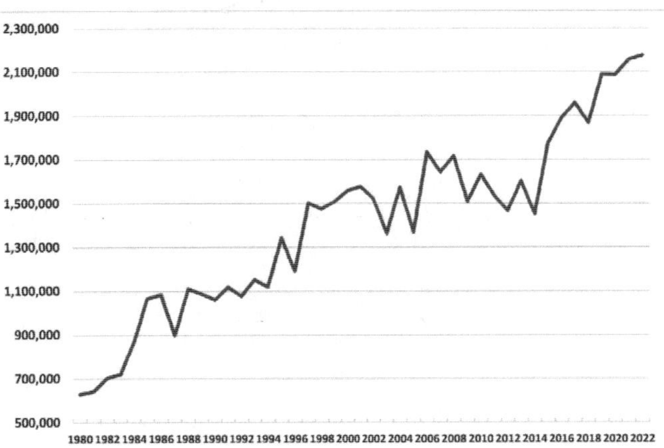

Fig. 3. Mango production in Mexico. Source: SIAP, 2023.

value of production. Other significant mango-producing states include Sinaloa, Nayarit, and Oaxaca, all located along the Pacific Ocean coast. In 2022, these five states accounted for 72% of the harvested area, 76% of the production, and 74% of the total value of production (Fig. 4, SIAP, 2023).

This wide distribution of mango plantations ensures a year-round supply of the fruit, with peak harvest season from April to August, which accounts for 80% of the harvest (Fig. 5, SIAP, 2022). The seasonality of mango production gives rise to significant challenges regarding the price of the fruit. As production increases, prices tend to decrease. Early productions in November and December typically command the highest prices. However, as the hot season commences in April, prices experience a decline of 50 to 60%, and by June, prices may plummet by 90%.

This has required producers to seek ways to force production to occur either earlier or later in the season. Unfortunately, research on practices for forced production is very scarce in Mexico, and technical assistance in this regard is non-existent.

Among the different mango cultivars, the Ataulfo variety holds the highest production share at 30%, followed by Kent at 15%, Manila at 14%, Tommy Atkins at 14%, Haden at 10%, and Keitt at 7% (SIAP, 2023).

The Ataulfo mango cultivar, discovered in a small orchard in the southernmost city of Tapachula, Chiapas (Mendoza-Hernández et al., 2020), has gained popularity as the most sought-after variety for commercial production. This is primarily due to its small size, rich flavor, and creamy texture. Over the course of two decades (2001–21), its production has experienced remarkable growth, increasing by 387% with an extraordinary average annual rate of 8.7% (Fig. 6, SIAP, 2023).

Over the past decade, mango exports have experienced rapid growth, with a remarkable increase of 147%. The export value has risen from 205.6 million U.S. dollars in 2011 to 509 million U.S. dollars in 2021 (FAO, 2023; SIAP, 2022), exhibiting an impressive average annual growth rate of 10.6%.

Mexico is currently the largest exporter of mangoes in the world (SIAP, 2022), with an estimated volume of 485,395 tons (22.5% of total production) exported annually to 55 nations as of 2021. These exports were valued at 509 million U.S. dollars, as reported by SIAP (2022).

The United States is the primary destination for Mexican mango exports, with 409,812 tons (84.4% of total exports) imported in 2021 due to Mexico's competitive advantages, such as proximity to the largest import market, solid infrastructure, adequate costs, and labor availability, generating significant income for the country at 406.4 million U.S. dollars (79.8% of total exports). China, the Netherlands, Germany, and the United Arab Emirates also make large external purchases of Mexican mangoes (SIAP, 2022).

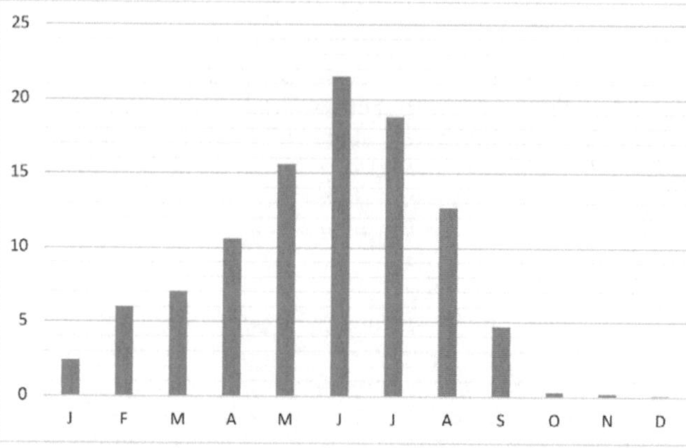

Fig. 5. Production of mangos per month in Mexico as a percentage. Source: SIAP, 2022.

The mango crop in Mexico faces several challenges, including:

1. Pests and diseases. Mango trees are susceptible to a variety of pests and diseases, which can cause significant yield losses and reduce fruit quality. Common mango pests in Mexico include Mexican fruit flies, and thrips, while common diseases include anthracnose and powdery mildew.
2. Climate change. Changes in temperature and rainfall patterns can affect mango production in Mexico, leading to reduced yields and quality. Extreme weather events such as hurricanes and droughts can also cause damage to orchards and disrupt harvests.
3. Market competition. Mexico faces stiff competition from other mango-producing countries such as Brazil, Dominican Republic, and Peru, which can affect prices and market access for Mexican mango growers.
4. Labor shortages. Mango production requires significant labor input, and there is a shortage of skilled workers in some areas of Mexico. This can lead to higher labor costs and difficulties in finding workers during peak harvest periods.

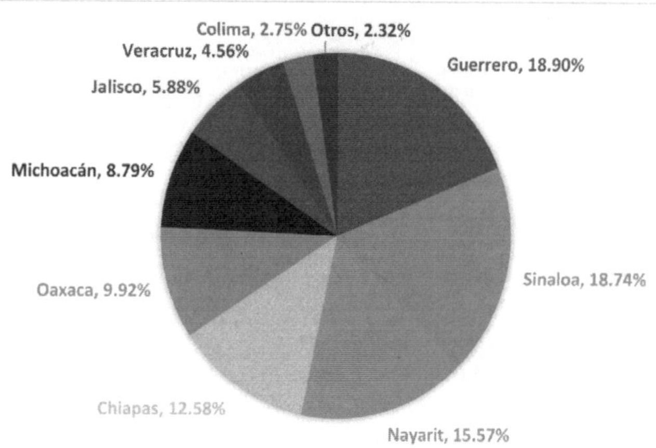

Fig. 4. Mango production per state in Mexico. Source: SIAP, 2023.

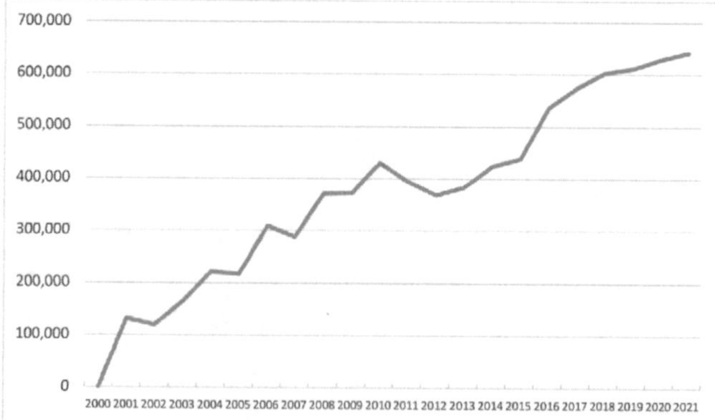

Fig. 6. Production of the mango cultivar Ataulfo in Mexico (tons). Source: SIAP, 2023.

5. Access to technology and financing. Many small-scale mango farmers in Mexico face challenges in accessing the technology and financing needed to improve their production and compete in the market. This can limit their ability to invest in irrigation, pest control, and other necessary inputs.
6. The introduction of new varieties is also essential to maintain competitiveness in the market.

Efforts need to be done to address these challenges, through research and development, extension services, and government support programs.

Literature Cited

BANXICO. 2023. Average exchange rate of the period—(CF86). Accessed 16 Jun. 2023. https://www.banxico.org.mx/SieInternet/consultarDirectorioInternetAction.do?sector=6&accion=consultarCuadro&idCuadro=CF86&locale=es

FAO. 2023. FAOSTAT. Accessed June 17, 2023. https://www.fao.org/faostat/en/#data/QCL

Mendoza-Hernández C., Rosas-Quijano R., Alfredo Vázquez-Ovando A., & Gálvez-López D. 2020. Retos y controversias del mango Ataulfo. *Revista Mexicana Ciencias Agrícolas* 11(7):1633–1645.

SIAP, 2022. Panorama Agroalimentario 2022. Accessed January 8, 2023. https://nube.siap.gob.mx/gobmx_publicaciones_siap/pag/2022/Panorama-Agroalimentario-2022

SIAP, 2023. Sistema de Información Agroalimentaria de Consulta. Accessed March 29, 2023. https://www.gob.mx/siap/documentos/siacon-ng-161430

Proc. Fla. State Hort. Soc. 136:54–60. 2023.

Feasibility of Modified Atmosphere Packaging (MAP) Plus Ethylene Scrubbing for Extended International Shipping of Mangos

Jeffrey K. Brecht[1]*, Steven A. Sargent[1], Faisal Shahzad[1], Moshe Doron[1], and Sergio Tonetto de Freitas[2]

[1]*Horticultural Sciences Department, University of Florida, IFAS, Gainesville, FL*

[2]*Brazilian Agricultural Research Corporation, Embrapa, Petrolina, PE, Brazil*

Keywords. *Mangifera indica*, chilling injury, optimum atmosphere, quality, shelf life, tree ripe

The feasibility of incorporating modified atmosphere packaging (MAP) technology with or without ethylene scrubbing into existing mango handling systems was evaluated as a way to allow shipment of fully mature (tree-ripe) fruit from South America to the United States in order to improve the quality of mangos available to U.S. consumers. In preliminary research, the optimum atmospheres (low O_2; high CO_2) for slowing 'Tommy Atkins', 'Kent' and 'Keitt' mango ripening were determined using controlled atmosphere (CA) treatments. We then replicated the CA lab tests using Breatheway™ (BW) MAP with and without It's Fresh™ (IF) ethylene scrubbers that had been custom designed for us by the companies based on the results of the CA experiments. In November 2022, we conducted two simultaneous shipping trials from Brazil using 'Tommy Atkins' and 'Kent' mangos. The mangos were shipped by ocean from Brazil to New Jersey, then by refrigerated truck to the University of Florida Postharvest Lab in Gainesville, all at 9 °C/48 °F for a total of 28 days, followed by a 4-day shelf life period at 20 °C/68 °F, to evaluate a 4.5-kg carton-scale BW MAP +/– IF in commercial conditions. A third shipping trial with 'Kent' mangos from Peru to California was conducted in February 2023, with evaluations conducted in California by Hazel Technologies (Breatheway's parent company). The preliminary research established that tree-ripe mangos (Stage 3.5–4) can be stored as low as 7 °C/45 °F for 4 weeks without showing chilling injury symptoms or deterioration of fruit internal quality when held in minimum 4% O_2 + 5 to 10% CO_2 ('Kent' and Keitt') or minimum 6% O_2 + 5 to 10% CO_2 ('Tommy Atkins'). The shipping test results indicate that MAP with IF slowed ripening-related softening, color changes, and losses of sugars and acids, while sensory panelists rated the treated fruit as sweeter, less acidic, and with an overall higher eating quality than the untreated controls, which were overripe. Overall, our results lead us to recommend MAP plus ethylene scrubbing for long distance (4-week) shipping if the goal is to export/import tree-ripe mangos. A pallet-scale MAP that is available would facilitate better temperature management by allowing the packed fruit to be pre-cooled before application of the MAP plus ethylene scrubbers, promising even better results.

The long shipping distances from mango growing regions in South America to the export markets in the United States (U.S.) result in shipping durations that can exceed 3 weeks. This forces exporters in Ecuador, Peru, and Brazil to ship fruit of minimal maturity and at lower than recommended temperatures so as to manage the onset and progression of fruit ripening and thereby minimize arrival problems due to soft, ripe, or decayed fruit. However, these practices increase problems with hot water injury and chilling injury (CI), to which less mature mangos are more susceptible. Fruit harvested early are also more susceptible to damage from water loss and mechanical damage (abrasion). Even if symptoms of hot water injury or CI do not become visible in imported mangos, the mangos may still fail to ripen or ripen with poor sensory quality. We documented this situation in our previous National Mango Board-funded research project, "Monitoring and Evaluation of the Mango Supply Chain to Improve Mango Quality" as reported in our *Mango Postharvest Best Management Practices Manual* (Brecht et al., 2020). In that study, we identified shipment of immature mangos at low temperatures as the most important problem that reduces mango consumer quality in the U.S. market.

Mangos that are harvested too early (immature physiological state) are more sensitive to hot water injury and CI, causing failure to ripen or development of poor sensory quality when ripe, while those fruit harvested at too late a stage of maturity have reduced shelf life from over-ripening and increased susceptibility to mechanical damage (i.e., bruising) and postharvest diseases (Brecht et al., 2020; Sivakumar et al., 2011). Chilling injury susceptibility decreases as mango fruit develop, mature, and ripen. Thus, immature fruit are more susceptible to CI than

We thank the National Mango Board for their support of this research by providing funding under agreement AGR00020789. We also thank our industry cooperators, including Michael Warren of CAPCO Farms for procuring the mangos used for the lab research. Thanks to Dr. Shehbaz Singh of Curation Foods-Breatheway® for designing and providing custom Breatheway™ MAP for the lab tests; Adam Preslar and Teresa Scattini of Hazel-Breatheway® for supplying the Breatheway™ MAP for the shipping tests, and Angel Reyes for constructing the Breatheway™ MAP bags; also, Andrea Rivera of Hazel Technology for conducting the California shelf-life evaluations. Thanks to Justin Creasy of It's Fresh™ for modifying and providing ethylene scrubbing filters. Thanks to Clark Golden of Amazon Produce, Helena and Emilia Medeiros of AgroBras, and Rocco Cifelli of Sunshine Export for collaborating on the shipping tests from Brazil and Peru to the United States. Thanks to Moshe Doron, Sonya Stahl, and Zafar Iqbal for assisting Faisal Shahzad in conducting the lab CA and MAP research.

*Corresponding author. Email: jkbrecht@ufl.edu

mature fruit, and fruit that are mature but have not yet begun to ripen are more susceptible to CI than fruit that are undergoing ripening (Medlicott et al., 1990; Mohammed & Brecht, 2002). Commonly, mango fruit that are exposed to chilling temperatures before they have begun ripening are never able to ripen normally. Often this involves ripening without the development of normal flavor and aroma—in other words, chilled mangos may ripen, but they are tasteless. Hydrocooling mangos after the quarantine hot water treatment was developed by mango handlers as a procedure to reduce incidence of hot water injury in low maturity fruit. The application of hydrocooling after hot water treatment is now proscribed by the U.S. Department of Agriculture Animal and Plant Health Inspection Service (USDA–APHIS) protocol that specifies the allowed timing and water temperatures for both the hot water treatment and the hydrocooling (T102-a http://www.aphis.usda.gov/import_export/plants/manuals/ports/downloads/treatment.pdf). Another procedure that is used by many mango handlers to avoid damage due to the hot water treatment is to impose a period of "rest" at ambient temperatures following the hot water treatment and prior to grading and packing. The rest period is intended to allow symptoms of hot water injury to develop before the fruit are graded on the packing line so that damaged fruit can be removed. If tree-ripe mangos could be successfully shipped the necessary distances required for export from South America to the U.S., these issues with CI and hot water injury in low maturity mangos could be largely eliminated.

Many years of research have shown that controlled and modified atmosphere (CA and MA) technology can significantly extend the postharvest life of mangos and have led to well-recognized and accepted recommendations for the optimum O_2 and CO_2 concentrations for maintaining mango quality (Brecht and Yahia, 2009; Thompson, 2010). Commercial CA and MA systems have been developed and are in widespread use for storage, transport and marketing of many crops (Brecht et al., 2009; Yahia, 2009). Controlled and modified atmospheres reduce respiration (Kelany et al., 2010) and ethylene production, which is how they slow the ripening process and extend the storage life of the fruit. Research has been conducted demonstrating that elevated concentrations of CO_2 (5 to 10%) in the atmosphere also alleviate CI symptoms in mangos, but reduced O_2 concentration (5%) has no significant effect (O'Hare and Prasad, 1993). Our work at the University of Florida, showed that mangos could be shipped for 3 weeks in modified atmospheres at 8 °C (46 °F) for tree-ripe (Stage 4) fruit or 12 °C (54 °F) for mature-green (Stage 2) fruit without developing CI (Bender, et al., 2000 a, b). Selection of the atmosphere must be done with care, for extended exposure to levels of O_2 or CO_2 that are too extreme can induce physiological damage in the fruit (Brecht, 2018). However, while CA and MA systems are available for marine containers, the technology is quite expensive and the availability of CA-equipped containers is insufficient to handle mango export volumes.

Modified atmosphere packaging (MAP) is a low-tech application of CA that works by balancing the O_2 and CO_2 permeability of a sealed package with the respiration rate of the enclosed product, which consumes O_2 and releases CO_2 (Tasneem, 2004). The concentration of gases within the package depends on several factors, including the film's gas permeability properties, the relative area and volume of the packaging, and the quantity and temperature of the product (Oliveira et al., 2005). Thus, at equilibrium, the MAP headspace has an increased level of CO_2 and a reduced level of O_2 that is appropriate for the product, which must be designed into the package to match the tolerance of the fruit to those gases. Using MAP for mangos has until recently been hampered by the lack of availability of proper package materials to achieve optimum atmospheres for mangos. However, we identified the MAP provider Hazel Breatheway® as having developed a MAP system for mangos, and who had, along with TransFresh (Tectrol™), jointly developed a pallet-scale MAP system being used commercially for blueberries exported from Argentina to the United States. They provided a similar MAP designed specifically for mangos for our use in this research. Additionally, we have seen that many MAP systems (including Breatheway™) are relatively impermeable to ethylene gas, which can therefore accumulate to high levels inside the package. Therefore, since ripening mango fruit produce ethylene, we realized that scrubbing the ethylene may be an essential system component for maximum effectiveness of MAP for mangos.

The objective of this research was therefore to determine the feasibility of incorporating BW MAP technology with or without IF ethylene scrubbing into existing mango handling systems as a way to allow shipment of fully mature (tree-ripe) fruit from South America to the U.S. in order to improve the quality of mangos available to U.S. consumers. To accomplish this, we conducted laboratory CA studies to determine the optimum atmospheres (low O_2; high CO_2) for slowing 'Tommy Atkins', 'Kent', and 'Keitt' mango ripening. We then replicated the CA lab tests using Breatheway™ MAP with and without It's Fresh™ ethylene scrubbers that had been custom designed for us by the companies based on the results of the CA experiments. To evaluate BW MAP plus/minus IF ethylene scrubber filters in commercial conditions, we conducted shipping trials from Brazil to New Jersey, with 'Tommy Atkins' and 'Kent' mangos, with the mangos then shipped by refrigerated truck from New Jersey to Gainesville, FL for shelf life evaluation; and from Peru to California with 'Kent' mangos, with the shelf life evaluation performed at the Hazel Technologies lab in Fresno, CA.

Materials and Methods

Laboratory determination of optimum atmospheres and evaluation of Breatheway™ MAP and It's Fresh™ ethylene filters. For experiments to evaluate CA +/– ethylene scrubbing, imported mango fruit at maturity stage 3.5 on a 1 to 5 flesh color development scale from 1 = no yellow color to 5 = 100% yellow or orange color were obtained from Central American Produce (CAPCO) in Pompano Beach, FL, on the day of arrival from the exporting country (Mexico, Brazil, Ecuador, or Peru for 'Tommy Atkins', 'Kent', and 'Keitt', respectively). For experiments to evaluate MAP +/– ethylene scrubbing, imported 'Tommy Atkins' mango fruit at maturity stage 3.5–4 were obtained from CAPCO on the day of arrival from Mexico or Ecuador.

The fruit were transported in an air-conditioned vehicle to the Postharvest Horticulture Laboratory at UF, Gainesville (approx. 4 hours) and stored at 7 °C and 85% relative humidity (RH) for 28 days, then transferred to 20 °C and 95% RH for up to 6 days. The CA treatments were established using a system employing needle valve flowmeters and manifolds to mix nitrogen, air and CO_2 delivered at equal pressure and the resulting gas mixtures were humidified by bubbling through water. The Breatheway™ MAP bags were large enough to hold 8 or 9 mangos weighing 4.5 kg and were designed based on results from previous respiration measurements to establish an atmosphere of 4 to 6% O_2 plus 10 to 15% CO_2 at 7 °C. Similarly, the It's Fresh™ filters were designed with sufficient capacity to adsorb the ethylene produced by the mangos for the entire duration of the storage plus shelf-life periods (types A5 and A6 had different areas for ethylene adsorption).

There were three replicates for each treatment on each timepoint, and each replicate consisted of a carton of eight or nine fruit. Fruit evaluations were done on Days 0 and 28 at 7 °C; also on Days 2, 4, and 6 at 20 °C for shelf-life evaluation. The MAP bags were kept sealed during the first 4 days of the shelf-life period since the Breatheway™ MAP adjusts its permeability to accommodate temperature changes. A completely randomized experimental design was used for all studies and data were analyzed using one-way analysis of variance (ANOVA) on each time point.

The following controlled atmosphere (CA) conditions were applied:

CA1 = Air control

CA2 = 4 kPa O_2 + 5 kPa CO_2 CA5 = 6 kPa O_2 + 5 kPa CO_2

CA3 = 4 kPa O_2 + 10 kPa CO_2 CA6 = 6 kPa O_2 + 10 kPa CO_2

CA4 = 4 kPa O_2 + 15 kPa CO_2 CA7 = 6 kPa O_2 + 15 kPa CO_2

The following Breatheway™ MAP and It's Fresh™ (ethylene scrubbing) treatments were applied:

T1 = Air control + No It's Fresh™ T6 = Breatheway™ MAP + No It's Fresh™

T2 = Air control + A5 It's Fresh™ T11 = Breatheway™ MAP + A5 It's Fresh™

T3 = Air control + A6 It's Fresh™ T12 = Breatheway™ MAP + A6 It's Fresh™

Evaluations of treatment conditions and fruit physicochemical and sensory quality. Gas concentrations (O_2, CO_2, and ethylene) were measured on a weekly basis during CA and MAP storage at 7 °C and on alternate days in MAP at 20 °C using gas chromatography. The gas concentrations were also used to determine the respiration and ethylene production rates by placing fruit in sealed jars for 1 hour. Fruit susceptibility to CI, decay incidence, lenticel discoloration, and CO_2 injury were also scored based on the skin area affected by each disorder. Rating scales of 1 to 5 were used: 1 = 0%; 2 = < 5%; 3 = 5 to 10%; 4 = 10 to 25%; and 5 = > 25% of the affected fruit area. Fruit firmness as resistance to compression and puncture forces were measured on both sides of fruit using a Texture Analyzer (Texture Technologies) with a flat plate and 8-mm diameter convex tip probe, respectively, with the average value from both readings for each fruit used for statistical analysis. Color attributes were determined (two readings per fruit on opposite sides): skin ground color; subepidermal flesh color; at the stone color; the measured L*, a*, and b* values were converted to hue (shade of color); and chroma (color purity). The total soluble solids (TSS), titratable acidity (TA), TSS/TA, and pH of combined fruit juice were measured using refractometry and titration. Sensory evaluation was done during shelf life after 4 days at 20 °C once the fruit were ripe/ready to eat. Attributes evaluated were mango aroma, texture, juiciness, sweetness, acidity, overall eating quality and ripeness appearance. Photographs were taken on both sides of the fruit skin, the subepidermal flesh, and at surface of the stone on each timepoint.

Shipping tests. Brazilian-grown 'Tommy Atkins' and 'Kent' mango fruit with dry matter (DM) content of >13% ('Tommy Atkins') or 15% ('Kent') were selected at AgroBras, Petrolina, and imported by Amazon Produce, Vineland, NJ, while Peruvian-grown 'Kent' mango fruit with dry matter content of >15% were selected at Sunshine Export, Piura, and imported by Amazon Produce, Oxnard, CA. The DM values were measured using a NIR spectrometer (F-750, Felix Instruments, Camas, WA) utilizing a previously developed algorithm (Tonetto de Freitas et al., 2022).

The following treatments were applied:

T1 = Air control - No Breatheway™ and no It's Fresh™ filters (Control)

T2 = Breatheway™ + No It's Fresh™ filter (BW)

T3 = Breatheway™ + It's Fresh™ filter (BW+IF)

Mangos in 4.5-kg cartons containing 7 or 8 fruit were collected from the packinglines at AgroBras and Sunshine and either used as is or re-packed with the Breatheway™ bags used as a carton liner, either with or without an It's Fresh™ filter inserted, sealed by twisting the opening, folding over, and affixing a rubber band. The Brazilian fruit were transported by ship at 9 °C to Philadelphia, PA/Vineland, NJ, for 22 days, and from there by truck for 2 days at 9 °C to UF, Gainesville, FL, and held at 9 °C and 85% RH. The Peruvian fruit were transported by ship at 9 °C to Port Hueneme/Oxnard, CA, for 20 days, held in cold storage for 4 days at 9 °C, and from there taken by air-conditioned car to Hazel Technology, Fresno, CA, and held at 9 °C and 85% RH. In both New Jersey and California, the MAP gas concentrations were measured at the importer upon arrival from the port using a portable MAP Gas Analyzer (Bridge Analyzers, Inc., Bedford Heights, OH). When 28 days since the start of each experiment had elapsed, all fruit were transferred to 20 °C and 95% RH. Fruit evaluations were done as previously described after 28 days at 9 °C and after 2 and 4 days at 20 °C. An additional sensory evaluation of the Brazilian fruit was conducted after 10 days of shelf-life at 20 °C, when the T2 (Breatheway™ + No It's Fresh™ filter) and T3 (Breatheway™ + It's Fresh™ filter) fruit were ripe/ready to eat (Day 10). The bags were kept sealed during the initial 4-day shelf-life period since the Breatheway™ MAP adjusts its permeability to accommodate temperature changes. The bags were then opened for the next 6 days before the final sensory evaluation. A completely randomized design was used for this study. There were 21 or 24 replicates (three, 7-, or 8-fruit cartons) for each treatment at all timepoints, and each fruit was considered as one replicate.

Results and Discussion

Laboratory tests. The results (Figs. 1 and 2) showed that tree-ripe mangos (stage 3.5–4) can be stored at low temperature [7 to 8 °C (45 to 46 °F)] for 4 weeks without showing CI symptoms and deterioration of fruit internal quality. Air control fruit were generally unmarketable by the second day of shelf life at 20 °C. With respect to O_2 tolerance, tree-ripe 'Tommy Atkins' did perform better in CA at 6% O_2, developing a small amount of internal discoloration in CA at 4% O_2, while 'Kent' and 'Keitt' mangos both benefited more from 4% O_2. With respect to CO_2 tolerance, tree-ripe 'Tommy Atkins', 'Kent' and 'Keitt' mangos all benefited from CA with both 5 and 10% CO_2, however, CA with 15% CO_2 had deleterious effects on fruit quality. Based on fruit marketability, 'Tommy Atkins' had more marketable fruit when stored in CA with 6% O_2 + 5 or 10% CO_2 while 'Kent' and 'Keitt' did better when stored in CA with 4% O_2 + 5 or 10% CO_2. The mangos in Breatheway™ MAP developed atmospheres at 7 °C of about 3 to 5% O_2 + 8 to 12% CO_2; the atmospheres became quite extreme during the shelf life period at 20 °C, with the O_2 reaching <2% and the CO_2 reaching >30%; however, the fruit did not show any indication of injury from those extreme atmospheres within the short time frame of the shelf-life testing. Including an It's Fresh™ ethylene filter in the MAP bags reduced ethylene concentrations by about 75% to almost 100%, with the A6 filter being more effective than the A5. The CA and MAP treatments reduced lenticel discoloration, CI and decay, while retaining firmer fruit with less color development (i.e., less ripe). Scrubbing ethylene in Breatheway™ MAP slightly further reduced softening and color changes (ripening).

Overall, our results lead us to recommend atmosphere concentrations of 4 to 6% O_2 plus 5 to 10% CO_2 for long distance (4 week) shipping of tree-ripe mangos, with or without ethylene scrubbing.

Shipping tests. The results (Fig. 3) showed that tree-ripe mangos (stage 3.5–4) can be transported or stored at low temperature [9 °C (48 °F)] for 4 weeks without showing CI symptoms and deterioration of fruit internal quality. The results suggest that Breatheway™ MAP protects fruit from stress as expressed by lower lenticel discoloration scoring and may also protect from CI. Fruit in Breatheway™ MAP treatments, especially with It's Fresh™, retained their firmness compared to the non-MAP control fruit by slowing ripening related fruit softening. Results from colorimetric measurements of flesh color showed that these areas were more green and less yellow in Breatheway™ MAP treatments compared to the non-MAP control. The combination of Breatheway™ MAP and It's Fresh™ delayed the ripening processes that consume acids and sugars compared to the other 2 treatments. Fruit sensory quality panelists on shelf life day 4 rated the Brazilian 'Tommy Atkins' and 'Kent' mangos from the T2 (Breatheway™ + No It's Fresh™ filter) and T3 (Breatheway™ + It's Fresh™ filter) treatments as sweeter, less acidic, and with an overall higher eating quality; on shelf life day 10, the T2

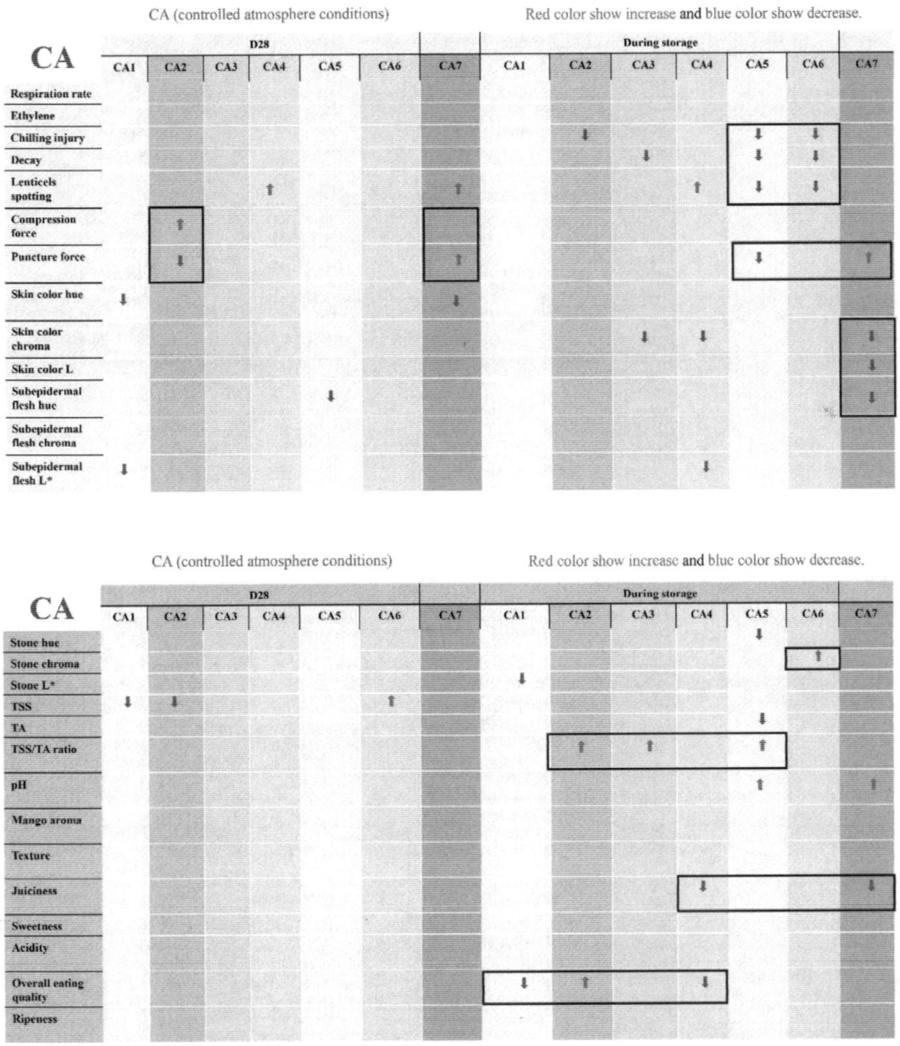

Fig. 1. Summary of laboratory controlled atmosphere (CA) test results showing the effects of treatments relative to the air control for tree-ripe 'Tommy Atkins', 'Kent', and 'Keitt' mangos. The treatments were: CA1 = Air control; CA2 = 4 kPa O_2 + 5 kPa CO_2; CA3 = 4 kPa O_2 + 10 kPa CO_2; CA4 = 4 kPa O_2 + 15 kPa CO_2; CA5 = 6 kPa O_2 + 5 kPa CO_2; CA6 = 6 kPa O_2 + 10 kPa CO_2; and CA7 = 6 kPa O_2 + 15 kPa CO_2. Red, upward-pointing arrows indicate increases and blue, downward-pointing arrows indicate decreases of a given factor relative to the air control treatment.

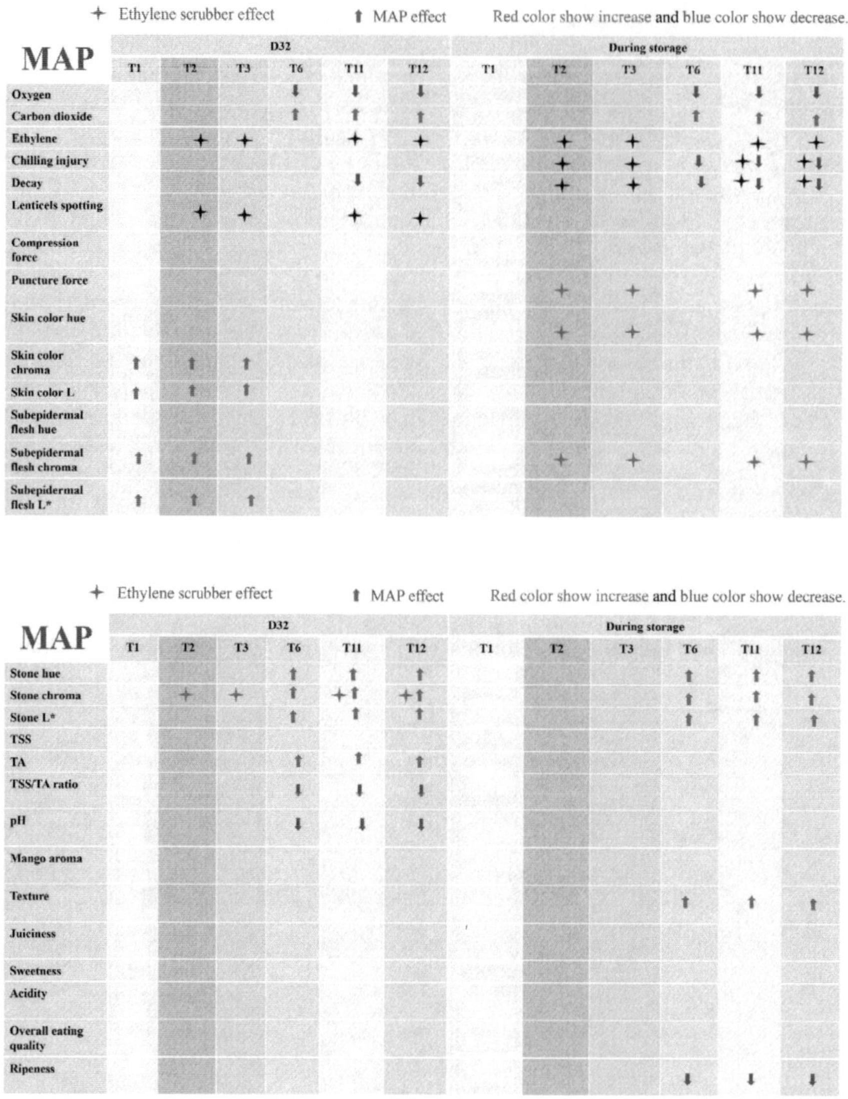

Fig. 2. Summary of laboratory modified atmosphere packaging (MAP) and ethylene scrubbing test results showing the effects of treatments relative to the air control for tree-ripe 'Tommy Atkins', 'Kent' and 'Keitt' mangos. The treatments were: T1 = Air control + No It's Fresh™; T2 = Air control + A5 It's Fresh™; T3 = Air control + A6 It's Fresh™; T6 = Breatheway™ MAP + No It's Fresh™ ; T11 = Breatheway™ MAP + A5 It's Fresh™; T12 = Breatheway™ MAP + A6 It's Fresh™. Red stars or upward-pointing arrows indicate increases and blue stars or downward-pointing arrows indicate decreases of a given factor relative to the air control treatment with stars indicating a response to ethylene scrubbing and arrows a response to MAP.

(BW, Breatheway™ + No It's Fresh™ filter) and T3 (BW/IF, Breatheway™ + It's Fresh™ filter) Brazilian fruit were rated as sweet and juicy, but there were no fruit from the control treatment available at that time for comparison.

Overall, our results lead us to recommend Breatheway™ MAP with It's Fresh™ filters for long distance (4-week) shipping if the goal is to export/import tree-ripe mangos.

Summary and Recommendations

Select fruit at about Stage 3–4 flesh color and/or >15% DM content. Use a Breatheway™ MAP system that establishes and maintains 4 to 6% O_2 + 5 to 10% CO_2. It's Fresh™ filter (ethylene scrubbing) adds additional protection over Breatheway-™ MAP alone for delaying ripening, particularly fruit softening. The mangos will maintain about a week of shelf life after transfer to a ripening-conducive temperature. We did not make a comparison between low maturity versus high maturity fruit; however, we know from experience that there is a large quality difference between mangos harvested at Stage 1–2 versus 3–4 after hot water treatment and long-term cold storage.

Future Research Needs

- Pallet scale Breatheway™ MAP would be more economical than the carton-scale Breatheway™ MAP system used here and is worth testing.
- The benefits of It's Fresh™ filters that effectively remove ethylene should be evaluated as a standalone treatment.
- Hazel Technology 1-MCP release sachets placed within a Breatheway™ MAP is also worth testing.

Red color shows increase and blue color shows decrease.

Shipping Trials	D28			During shelf life		
	Control	Breatheway™	Breatheway™ + It's Fresh™	Control	Breatheway™	Breatheway™ + It's Fresh™
Oxygen		↓	↓		↓	↓
Carbon dioxide		↑	↑		↑	↑
Ethylene			↓			↓
Chilling injury		↓	↓		↓	↓
Decay						
Lenticel spotting		↓	↓		↓	↓
Skin browning						
Compression force						
Puncture force			↑		↑	↑
Skin color hue						
Skin color chroma						
Skin color L						
Subepidermal flesh hue		↑	↑		↑	↑
Subepidermal flesh chroma						
Subepidermal flesh L*						

Red color shows increase and blue color shows decrease.

Shipping Trials	D28			During shelf life		
	Control	Breatheway™	Breatheway™ + It's Fresh™	Control	Breatheway™	Breatheway™ + It's Fresh™
Stone hue		↑	↑		↑	↑
Stone chroma						
Stone L*		↑	↑		↑	↑
TSS		↓	↓		↓	↓
TA		↓	↓		↓	↓
TSS/TA ratio			↑		↑	↑
pH		↑	↑			
Mango aroma						
Texture						
Juiciness					↑	↑
Sweetness		↑	↑		↑	↑
Acidity		↓	↓			
Overall eating quality		↑	↑			
Ripeness						

Fig. 3. Summary of shipping test results for tree-ripe 'Tommy Atkins' and 'Kent' mangos shipped by ocean from Brazil to New Jersey then by refrigerated truck to Florida, and tree-ripe 'Kent' mangos shipped by ocean from Peru to Oxnard, California then by air-conditioned car to Fresno, CA. The fruit were at 9 °C for 28 days, then at 20 °C for 4 days; for the Brazil–New Jersey tests, an additional 10-day shelf life evaluation was conducted [modified atmosphere packaging (MAP) and MAP plus ethylene filter treatments only]. The treatments were: 1) Control (unpackaged); 2) Breatheway™ MAP; and 3) Breatheway™ MAP plus It's Fresh™ filter. Red upward-pointing arrows indicate increases and blue downward-pointing arrows indicate decreases of a given factor relative to the air control treatment.

Literature Cited

Bender R.J., Brecht J.K., Baldwin E.A., & Malundo T.M.M. 2000a. Aroma volatiles of mature-green and tree-ripe 'Tommy Atkins' mangoes after controlled atmosphere vs. air storage. *HortScience*. 35:684–686.

Bender R.J., Brecht J.K., Sargent S.A., & Huber D.J. 2000b. Mango tolerance to reduced O_2 levels in controlled atmosphere storage. *J. Amer. Soc. Hortic. Sci.* 125:707–713.

Brecht J.K. 2018. Mango. p. 443–466. In: S. Pareek and S. Tonetto de Freitas (eds.). Postharvest Physiological Disorders in Fruits and Vegetables. CRC Press, Boca Raton, FL.

Brecht J.K., Sargent S.A., Kader A.A., Mitcham E.A., Maul F., Brecht P.E., & Menocal O. 2020. Mango postharvest best management practices manual. Brecht J.K. (ed.), HS1185. National Mango Board and UF/IFAS, Gainesville, FL. (revised).

Brecht J.K., Sargent S.A., Ristow N.C, Tonetto de Freitas S., Osuna García J.A., MacLeod R., & Singh S. 2016, Demonstrating modified atmosphere packaging (MAP) feasibility for mango export. Final Research Report. 14 p. https://www.mango.org/wp-content/uploads/2017/10/Demonstrating_the_Feasibility_of_MAP_Final_Report_Eng.pdf. Accessed December 12, 2023.

Brecht J.K. & Yahia E.M. 2009. Postharvest physiology. p. 484–528. In: RE. Litz (ed.). The mango, botany, production and uses, 2nd ed., CABI, Wallingford, U.K.

Brecht P.E., Dohring S., Brecht J.K., & Benson W. 2009. Transport technology and applications. p. 51–71. In: E.M. Yahia (ed.). Modified and controlled atmospheres for the storage, transportation, and packaging of horticultural commodities. Taylor & Francis, London.

Kelany A.E., Sohar M.A., Hafeez A.A. & Osman M.T. 2010. Using modified atmosphere and different temperature for storing 'Kent' mango fruit. *J. Hort. Sci. Ornamental Plants*. 2:46–56.

Medlicott A.P., Sigrist J.M., & Sy O. 1990. Ripening of mangoes following low temperature storage. *J. Amer. Soc. Hortic. Sci.* 115:430–434.

Mohammed M. & Brecht J.K. 2002. Reduction of chilling injury in 'Tommy Atkins' mangoes during ripening. *Sci. Hort*. 95:297–308.

O'Hare T.J. & Prasad A. 1993. The effect of temperature and carbon dioxide on chilling symptoms in mango. *Acta Hortic*. 343:244–250.

Oliveira Jr. L.F.G., Coelho E.M., & Coelho F.C. 2005. Utilização de atmosfera modificada na conservação do mamão (*Carica papaya* L.) Golden sob refrigeração. *Revista Brasileira de Armazenamento, Viçosa*. 30(1):73–77.

Sivakumar D., Jiang Y., & Yahia E.M. 2011. Maintaining mango (*Mangifera indica* L.) fruit quality during the export chain. *Food Research International* 44:1254–1263.

Tasneem A. 2004. Postharvest treatments to reduce chilling injury symptoms in stored mangoes. M.Sc. Thesis, Department of Bioresource Engineering, Macdonald Campus of McGill University, Ste-Annede-Bellevue, Q.C. Canada, 105.

Tonetto de Freitas S., Tavares Guimarães I., Vilvert J.C., Amaral M.H, Brecht J.K., & Telles Biasoto Marques A. 2022. Mango dry matter content at harvest to achieve high consumer quality of different cultivars in different growing seasons. *Postharvest Biol. Technol*. 189:111917. https://doi.org/10.1016/j.postharvbio.2022.111917

Thompson A.K. 2010. Controlled atmosphere storage of fruits and vegetables. CABI, Wallingford, U.K.

Yahia E.M. 2009. Introduction. p. 1–16. In: Modified and controlled atmosphere for the storage, transportion and packaging of horticultural commodities. Taylor and Francis, Boca Raton, FL.

Proc. Fla. State Hort. Soc. 136:61–63. 2023.

The Mango Volume and Inventory System (MaVIS)

Rolff V.V. Mitton*, M. Michel, and Leonardo Ortega

National Mango Board, USA

Keywords. MaVIS, inventory system, data uniformity, confidentiality, consumption

The National Mango Board (NMB) as a national promotion and research organization implements strategically tools that support its mission, which is to increase the consumption of mangos in the United State (US). Collecting all the volume information provided by the different industry members (producers, packers, and importers) and making them available in one place will allow for a complete picture of the mango movement daily. In doing so, the industry is able to prevent apparent discrepancies that could have arisen otherwise. This is precisely why the NMB has developed an online information system (database), to record and report detailed and accumulated information on mango production and import volumes and inventories in the US. This inventory system: Mango Volume and Inventory System (MaVIS) is intended to: provide service to stakeholders (suppliers and marketers) that will allow them to improve mango market response times by reacting faster and more accurately to changing market conditions; improve supply chain efficiencies and value for the industry by delivering more timely information; and allow for better planning of promotions for the NMB marketing team. Although the system has been developed by the NMB, Importers and suppliers will need to participate voluntarily to ensure both efficiency and success. As this system will require information on the part of the participants that can be seen as sensitive, measures have already been taken to ensure confidentiality and security. The information provided by the participants will be preserved and protected by the NMB and will not be published, disclosed, or made available to the public. Those participants will be able to see the aggregate data, provided that at least three (3) will have provided data for the specific item. This inventory system has the potential to create a platform that will contribute to the unification of the mango industry through data homogeneity. Its proper use will also allow for more informed decision-making processes for all in the industry and certainly contributes to the per capita consumption increase of mango.

The National Mango Board (NMB) is composed of a diverse group of individuals that represent many sectors within the mango industry. They are growers, exporters, importers and together with staff members, researchers, consultants and other industry partners who work tirelessly to increase the overall consumption of mango in the United States. The synergy with which this important goal is carried out allows each industry member, regardless of their role and function, to actively contribute while being connected and informed on the development and growth of the industry year after year.

The United States (US) predominantly imports mango from Mexico, Peru, Ecuador, Brazil, Guatemala, and Haiti (combined for approximately 98% of the total available volume) (Osuna, 2021). Since its inception in 2005, the NMB has adopted, as part of its main strategy, bringing together and connecting the Foreign Mango Organizations (FMOs) belonging to the main exporting countries. By working together, the NMB together with these FMOs, have managed to impact the overall consumption of mango directly and positively in the United States. Consequently, from 2005 to 2022, the per capita consumption of mango in the US has had a 100% increase (from 1.88 to 3.76 lb). In addition, the import volumes have also increased from 62 million to 139 million, representing an increase of more than 124% (NMB 3-year trend). Furthermore, the combined supply of mangoes to the US from the main exporting countries for the last three years (2020–22) has been, on average, 135 million boxes (4-kg) .

These numbers reflect the successful collaboration between the NMB, the FMO's and other members of the mango industry through an effective and efficient communication process and a fluid data sharing system. The NMB receives the data from the FMOs regarding shipments (total volume, varieties, sizes, vessels and projections) and combining those with the US Department of Agriculture–Agricultural Marketing Service (USDA–AMS) data on arrivals volumes and FOB prices given at the various ports of entry in the US, to create a consolidation report that is sent out on a weekly basis to every industry member registered in the NMB contact database. This report (Crop Report) allows the members to have a complete picture of the current and future state of the mango market in the US.

This report, although robust in terms of information presented, does not however, fully capture certain aspects of the supply chain and, because of the frequency with which it is sent out, it sometimes does not allow for proactivity when it comes to addressing certain situations that may arise. For example, disparity can be noticed between data provided by a specific FMO and that reported by the USDA–AMS, or the units of measure used by the FMO are different from the ones presented in the report, or that the FMO received the data daily while the NMB reports weekly and vice versa. The heterogeneity of the data as far as origin and presentation can been seen as a potential disunity for the industry as it can become challenging to create a consolidated report. (Prokhorov et. al., 2018)

As an alternative, the NMB has developed an online information system (database), to record and report detailed and accumulated information on mango production and import volumes and inventories in the US. This inventory system can complement the information found in the weekly crop report and provide a holistic view of what the mango market looks like, on a day-to-day basis.

*Corresponding author. Email: rmitton@mango.org

This paper intends to provide details on the inventory system that the NMB has developed for the benefit of the mango industry. What are the security, accessibility, and confidentiality aspects of this system? What industry members can join currently? How is the platform managed and by whom? And of course, why should any member join? An attempt will be made to provide suitable answers so as to inform and educate. For any subsequent questions and/or inquiries, reaching out to The NMB staff will therefore be highly recommended.

What Are the Security, Accessibility, and Confidentiality Aspects of This System?

The Mango Volume and Inventory System (MaVIS) is stored and supported by Oracle Cloud (Oracle Corporation, Austin TX), the most robust and secure database platform currently on the market and as such, the channel through which the information travels has a configuration that prevents the data from being intercepted or modified by hackers and intruders. Nagalakshmi and Devi (2014) agreed that "Oracle provides support for cloud computing, and it has been updated periodically in order to meet the customers current needs after getting feedback and usage statistics from millions of customers". Furthermore, it is their conclusion that Oracle cloud provides a robust set of security features and capabilities starting with secure framework level communication to a secure user model (Fig. 1).

Participation in MaVIS is voluntary as it is up to industry members to decide whether they want to share their information. For the time being only importers can participate in MaVIS. Each participant gets access to the platform through personalized logon information and authentication factors. MaVIS uses digital certificates to verify the identity of users, guaranteeing that only verified users can access their information and avoiding impersonations. The data can then be uploaded directly onto the platform. Once this is done, the filters contained within can be applied to view the information in the way that best fits the need of the entity viewing the information. Data in the MaVIS platform is protected by encryption (conversion) to special characters, which makes the data unreadable to non-verified users. The aggregate information of all the participants can also be seen but with one particular caveat: At least three (3) must have submitted the information for it to be seen on the aggregate section. The reason behind that is to maintain the confidentiality and security of the information provided by each participant. Therefore, the individual information of a participant cannot be seen by other participants.

How Is the Platform Managed and by Whom?

To ensure confidentiality on the part of MaVIS participants, the NMB has adopted the following policies: The individual information of the participants, producers and marketers, can only be seen by the MaVIS Administrator, the Director of Research and Industry Relations and the Executive Director of the NMB. On a daily basis, the participants must provide information on arrivals (volume and sizes they imported), inventory (volume they have stored at their facilities) and shipments (where they are distributing the volumes that they imported and placed in inventory). The objective is to have the information updated daily before noon. The participants must also adhere to the coding format under which the platform was built, meaning that the files they submit (in .csv format) must contain the data in a format that will be accepted by MaVIS. This created challenges for the first couple of participants as each company has its unique way of capturing the data and the likelihood that a particular method or strategy would fit MaVIS was highly low. Therefore, it was necessary to both pre-built reports that would be destined for the platform and create a converter tool that would take the pre-built reports and upload them into the platform in the correct format. There three ways for participant to upload their files into the platform: 1) send the files to the converter tool to upload; 2) upload the files directly onto the platform in the proper format; or 3) send the files to the Administrator to have them uploaded.

Though all three have been used by the participants, the NMB recommends that participants get as much interaction as possible

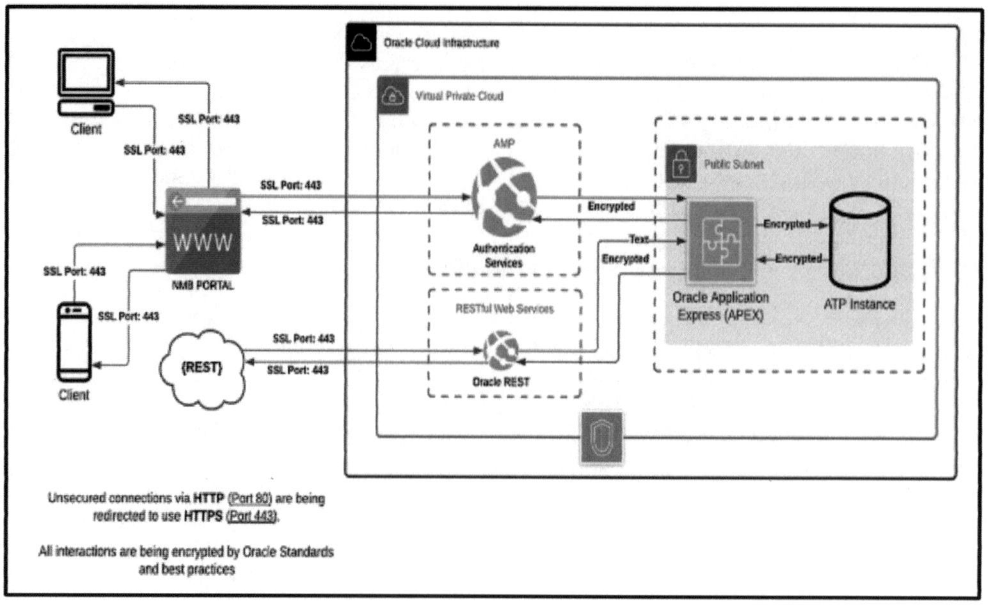

Fig. 1. The Mango Volume and Inventory System (MaVIS) architecture design.

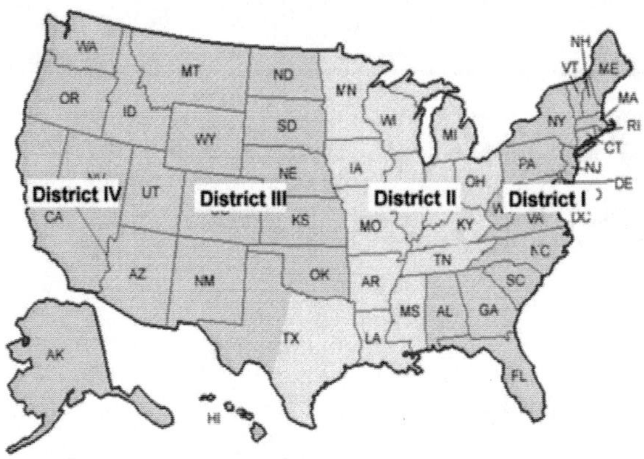

Fig. 2. U.S. district locations.

with the platform by getting involved in uploading directly onto it.

Uploaded data files go to a temporary database until the NMB MaVIS administrator runs a validation process to add data to the final database. Once the files are checked and validated, then the information is available on the MaVIS platform.

Each company registered in MaVIS is allowed to have multiple users, and each user is assigned specific levels of access (administrator, manager, and data entry) to the information available for each company. Each company, through its users, can manage its data. If for some reason technical issues arise, the NMB administrator will be available to explore said issues and work with the company to solve them.

MaVIS has the potential to become a tool that will allow the participants to become more proactive in the way they formulate their responses to shifting conditions in the market. This is a potentiality because the total number of participants need to account for about 70 to 75% of the volume imported in the US. A 360 of the mango industry: volume, countries in season, varieties, sizes, box types, volume by phytosanitary treatment, district location for arrivals, inventory, and shipments (Fig. 2).

MaVIS is intended to: provide a service to stakeholders (suppliers and marketers) that will allow them to improve mango market response times by reacting faster and more accurately to changing market conditions; improve supply chain efficiencies and value for the industry by delivering more timely information; and allow for better planning of promotions.

Why Should Any Member Join?

MaVIS was developed specifically for the mango industry to provide valuable online, up-to-date information on inventory balances and volumes, by importer and consolidated for all importers; including volumes by region within the US. The avocado industry has had success with implementing a similar volume and inventory system (AVIS) which has provided valuable information to its associates, facilitating better decision making. Another commodity implementing such a system is blueberries and, just like the avocado industry, they have found this system to be valuable in their members decision making process. These existing systems have demonstrated valuable performance in terms of reliability, timeliness, and security.

Conclusions

MaVIS was developed specifically for the mango industry to provide valuable online, up-to-date information on inventory balances and volumes, by importers and consolidated for all importers.

Participation to MaVIS is voluntary as it must be up to industry members to decide whether they want to share their information.

MaVIS uses digital certificates to verify the identity of users, guaranteeing that only verified users can access their information and avoiding impersonations.

MaVIS is stored and supported by Oracle Cloud, the most robust and secure database platform currently on the market and as such, the channel through which the information travels has a configuration that prevents the data from being intercepted or modified by hackers and intruders.

Literature Cited

Nagalakshmi V. & Devi V. 2014. Improve security with RSA and Cloud Oracle 10g. p. 497–503. In: Satapathy S., Avadhani P., Udgata S., & Lakshminarayana S. (eds). ICT and critical infrastructure: Proceedings of the 48th annual convention of Computer Society of India–Vol I. Advances in intelligent systems and computing, vol. 248. Springer, Cham. https://doi.org/10.1007/978-3-319-03107-1_55

Osuna J.A.G. 2021. Quality and shelf life of Tommy Atkins mango size 12's and Ataulfo size 20-22 packed in breathable bags. INIFAP/National Mango Board. Accessed April 4, 2023. https://www.mango.org/wp-content/uploads/2022/02/Mango_Packaging_Phase-I.pdf

Prokhorov I., & Kolesnik N. 2018. Development of a master data consolidation system model (on the example of the banking sector). *Procedia Comp Sc*. 145:412–417.

Proc. Fla. State Hort. Soc. 136:64. 2023.

—Scientific Note—

Integrating Cover Crops and Weed Management Practices in Mature HLB-affected Citrus Trees

GABRIEL PUGINA, SARAH L. STRAUSS, RAMDAS KANISSERY, AND UTE ALBRECHT*

Southwest Florida Research & Education Center, University of Florida, FAS, Immokalee, FL

Keywords. *Citrus sinensis*, glyphosate, glufosinate, herbicides, 'Valencia', weed pressure

Huanglongbing (HLB), a devastating citrus disease associated with the bacterium *Candidatus* Liberibacter asiaticus (*C*Las), has significantly impacted the citrus industry in Florida. One of the many challenges of managing HLB-affected citrus trees is the loss of fibrous roots, diminishing the trees' ability to absorb water and nutrients effectively (Graham et al., 2013). The high prevalence of weeds in many Florida production environments exacerbates the competition with citrus fibrous roots. Glyphosate is a broad-spectrum post-emergent herbicide widely used for weed management in Florida citrus groves. However, herbicide use and cultural practices can be combined as integrated weed management to provide a more sustainable approach. Alternative herbicides with different modes of action can reduce the selection pressure, control for possible glyphosate-resistant species, and delay the evolution of herbicide-resistant weed species (Busi et al., 2020). In integration with chemical control, the use of cover crops as a cultural practice can provide benefits such as optimizing nutrient cycling and weed suppression (Castellano-Hinojosa and Strauss, 2020; Osipitan et al., 2019). This study aimed to evaluate the effect of combining cover crops and herbicide treatment on the performance of mature HLB-affected citrus trees.

The large-scale experiment was established on 20 acres of commercial citrus in central Florida in a split-plot design with the cover crop as the main plot factor and weed management as the subplot factor. Each experimental plot comprised three rows of 80 trees, and the middle row, with cover crops on both sides, was evaluated. The trees are 'Valencia' orange (*Citrus sinensis*) grafted on US-942 rootstock (*C. reticulata* × *Poncirus trifoliata*) and were planted in 2016. The cover crops were a mixture of legumes and non-legumes, including daikon radish, sunn hemp, pearl millet, rye, and hairy vetch. Seeds were sown in the row middle of the experimental plots in the summer and fall of each year starting in 2021. No cover crops were planted in the control plots. Herbicides were applied to row middles in all experimental units before establishing each new cycle of cover crops. Weed management under the tree canopy was by glyphosate or glufosinate-ammonium, and herbicides were applied under the canopy two weeks after the cover crop planting. Leaf and soil samples were collected from each plot during summer for nutrient content analysis, and tree canopy measurements were conducted annually. The weed biomass was measured before the cover crop termination using the quadrat sampling method with a 0.5 m^2 frame. Random fruits from 10 trees in each experimental unit were collected for fruit quality analysis to fill one half-bushel fruit bag. In both years, 2021 and 2022, The fruit quality was assessed one week before the harvest in March.

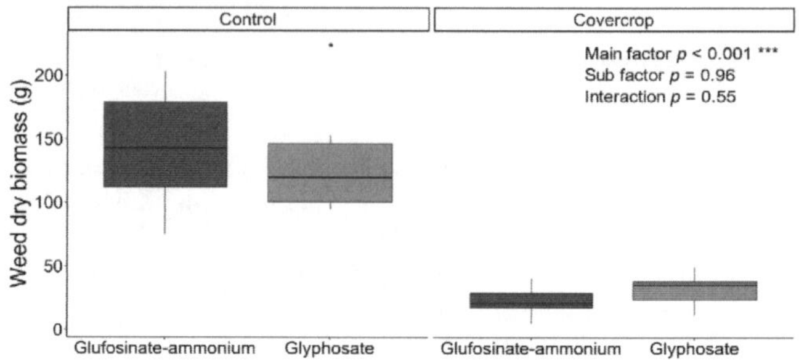

Fig 1. Weed dry biomass in response to glufosinate-ammonium and glyphosate treatment (sub factor) with or without cover crops (main factor) in the row middle of a *Citrus sinensis* 'Valencia' grove.

After two years of treatments, we found that the cover crops reduced weed pressure by more than 80%. The average dry weed biomass for control and cover crop plots were 139 g and 28 g, respectively ($P < 0.001$), but there was no significant difference between the two herbicides (Fig. 1). There were no significant differences in the measured soil physicochemical properties between cover crops or herbicides. In addition, no significant differences in tree size, canopy health, fruit quality, and yield were found. It must be noted that the yield was strongly diminished because of Hurricane Ian, which crossed through this production region in 2022. There was also no significant effect of weed management on any measured variable. These findings suggest that integrating cover crops and weed management practices can be a viable strategy for managing weeds in HLB-affected citrus groves. However, further research is needed to determine the long-term effects of these practices on tree health, yield, and economic feasibility.

Literature Cited

Busi R., Powles S.B., Beckie H.J., & Renton, M. 2020. Rotations and mixtures of soil-applied herbicides delay resitance. *Pest Management Science*. 76(2):487–496. https://doi.org/10.1002/ps.5534.

Castellano-Hinojosa A. & Strauss S.L. 2020. Impact of cover crops on the soil microbiome of tree crops. *Microorganisms*. 8(3):328. https://doi.org/10.3390/microorganisms8030328.

Graham J.H., Johnson E.G., Gottwald T.R., & Irey M.S. 2013. Presyptomatic fibrous root decline in citrus trees caused by huanglongbing and potential interaction with *Phytophthora* spp. *Plant Dis*. 97(9):1195–1199. https://doi.org/10.1094/PDIS-01-13-0024-RE.

Osipitan O.A., Dille J.A., Assefa,Y., Radicetti E., Ayeni A., & Knezevic S.Z. 2019. Impact of cover crop management on level of weed suppression: A meta-analysis. *Crop Science*. 59(3):833–842. https://doi.org/10.2135/cropsci2018.09.0589.

We gratefully acknowledge the U.S. Department of Agriculture for funding: USDA-NIFA ECDRE #2020-70029-33201.

*Corresponding author; email: ualbrecht@ufl.edu

Proc. Fla. State Hort. Soc. 136:65. 2023.

—Scientific Note—

Impacts of Soil Organic Amendments and Weed Control Practices on Nutritional Status and Tree Performance of 'Valencia' Sweet Orange Trees

Ankit Pokhrel, Ramdas Kanissery, Sarah L. Strauss, and Ute Albrecht*

University of Florida, IFAS, Southwest Florida Research and Education Center, Immokalee, FL

Keywords. *Citrus sinensis*, compost, herbicides, nutrient retention, nutrient uptake, weed biomass

Florida's citrus growers are resorting to different management practices to mitigate the impact of the bacterial disease Huanglongbing (HLB) and maintain the productivity of affected citrus trees. Improving soil health to enhance water and nutrient availability may help alleviate some of the disease-induced stress. Florida's sandy soils have little organic matter and a low cation exchange capacity (CEC), making them vulnerable to nutrient leaching and increasing production costs associated with frequent fertilization and irrigation. Incorporating organic matter into the soil through organic amendments can enhance the nutrient and water retention capacity and improve the soil physicochemical properties, creating a favorable environment not only for citrus trees but also for weeds. Therefore, effective weed management strategies are necessary to prevent resource competition. This study examines the effects of organic amendments and weed management on the nutritional status and the performance of young 'Valencia' sweet orange (*Citrus sinensis*) trees grafted on the rootstock US-802 (*Citrus maxima* × *Poncirus trifoliata*) in a commercial southwest Florida production environment.

A large-scale field trial was established in May 2021 in a 3 × 4 split-plot completely randomized block design, replicated five times, with organic amendment (compost, humic acid, and control) as the main plot, and herbicide (post-emergent glyphosate and glufosinate, pre-emergent flumioxazin, and control) as the subplot. Soil and leaf nutrients were determined bi-annually, and tree growth was assessed annually. Weed biomass was assessed in Spring and Summer 2022.

Plots treated with the post-emergent herbicides glyphosate and glufosinate had significantly less weed dry biomass compared to the control and flumioxazin plots in both spring and summer. This suggests that post-emergent herbicides were more effective in controlling weed growth than pre-emergent herbicides. Leaf nutrient analysis revealed that nitrogen and sulfur contents were significantly higher in plots treated with herbicides than in control plots. Glyphosate and glufosinate-treated plots had a significantly higher leaf phosphorus content, while leaf zinc content was significantly higher only in glyphosate-treated plots. There was no significant difference between the main plot treatments for the leaf nutrients, except potassium which was higher in compost-amended than non-amended plots, and boron which was higher in compost plots than humic acid plots. Higher levels of soil potassium, calcium, and boron and a higher cation exchange capacity (CEC) were observed in the compost-amended plots compared to the control and humic acid plots. Conversely, soil copper and manganese contents were significantly lower in compost plots compared to the control and humic acid plots. Post-emergent herbicides significantly improved the tree growth rate in terms of tree height and canopy volume, likely by reducing the competition with weeds. However, none of the organic amendments had any effect on citrus tree growth after 18 months of treatment (Table 1).

The results of this study suggest that organic amendments combined with post-emergent herbicide applications can improve nutrient availability and uptake in Florida soils. However, the visible effects of organic amendments on tree growth and productivity may take several years to manifest. Optimizing the rate of organic amendments for Florida's sandy soils and integrating organic amendments with effective weed control may result in long-term benefits for citrus growers.

Table 1. Change in height and canopy volume (%) of *Citrus sinensis* 'Valencia' trees grafted on 'US-802' rootstock grown in southwest Florida with different organic amendments and herbicides.

Treatment	Change in tree height (%)	Change in canopy volume (%)
Organic amendment		
Humic	18.6	79.6
Compost	17.4	81.2
Control	17.2	73.6
P-value	0.74	0.72
Herbicide		
Glufosinate	20.6 a[z]	81.4ab
Glyphosate	19.0 ab	90.4 a
Flumioxazin	15.8 bc	66.6 b
Control	15.6 c	74.3 ab
P-value	0.0001***	0.014*

[z]Different letters indicate significant differences ($P < 0.05$) within columns according to Tukey's honestly significant difference test.

We gratefully acknowledge the U.S. Department of Agriculture for funding.

*Corresponding author. Email: ualbrecht@ufl.edu

Proc. Fla. State Hort. Soc. 136:66. 2023.

– Scientific Note –

Fruit Size as an Indicator of Drop Likelihood in HLB-affected 'Valencia' Sweet Orange

MARY SUTTON AND TRIPTI VASHISTH*

Citrus Research and Education Center, University of Florida, IFAS, Lake Alfred, FL

Keywords. *Citrus sinensis* (L.), huanglongbing (HLB), preharvest fruit drop

While preharvest fruit drop is a normal physiological occurrence, it is exacerbated in Huanglongbing (HLB)-affected citrus and is a major contributor to the reduced yields seen in HLB-affected citrus. The mechanism behind this drop is still not understood. HLB symptomatic fruit tend to be smaller and more lopsided than healthy fruit. It has been suggested that smaller fruit are more likely to drop during the preharvest period (Tang et al., 2019). To determine if any of the HLB-related fruit characteristics were correlated with drop likelihood, a survey of near-mature fruit was conducted at three timepoints: March 11, 2022 (n = 640), April 15, 2022 (n = 248), and March 22, 2023 (n = 511).

Findings

Fruit were clipped from four HLB-affected trees displaying moderate symptoms. Each fruit was evaluated for equator and peduncle diameters, fruit detachment force (FDF), detachment type (mechanically broken stem or abscission at AZ-C), and lopsidedness. Total soluble solids was determined on a subset of fruit at each timepoint. FDF was lower in abscising fruit compared to mechanically broken fruit at all three timepoints ($P < 0.001$) consistent with the breakdown of cells in the abscission zone during the abscission process leading to a weaker connection between fruit and stem. This also suggests the fruit classified as 'abscission' had already begun the abscission process and would have soon dropped on their own. Average fruit diameters were lower in abscising fruit compared to mechanically broken fruit at 2 timepoints: March 11, 2022 ($P < 0.001$), and March 22, 2023 ($P = 0.0323$). In HLB-affected citrus, preharvest drop begins earlier than in healthy trees. As symptomatic fruit tend to be smaller, it is possible these symptomatic fruit begin dropping earlier in the drop period.

FDF was only lower in lopsided fruit compared to non-lopsided fruit on March 11, 2022 ($P < 0.001$). However, lopsided fruit were also smaller than non-lopsided fruit ($P = 0.006$) at this time point, but not at the other two time points, suggesting the relationship between fruit shape and FDF was due to the differences in fruit size and not fruit shape. No differences in peduncle diameter were found between mechanically broken and abscising fruit either. Lastly, total soluble solids (°Brix) did not differ between mechanically broken and abscising fruit suggesting abscising fruit had attained a similar maturity level as those being retained on the tree. In 'Valencia', fruit stop growing in size several months before preharvest drop begins. Therefore, the smaller fruit on the tree may have ceased growing first and transitioned to fruit maturation (stage 3 of fruit development) while others continued to grow. As these fruit would have begun the maturation process first, it is likely they would have also begun to senesce, leading to fruit abscission. Altogether, this survey demonstrated that in HLB-affected trees, the relatively smaller fruit (due to compromised growth) on the tree were more likely to drop early in the preharvest period while larger fruit began to drop nearer to harvest during the normal senescence-related drop.

Validation

To further confirm this relationship between fruit size and drop, 20 small (< Average fruit size – 1/2 standard deviation) and 20 large (> Average fruit size + 1/2 standard deviation) fruit were tagged on four 'Valencia' trees in mid-December 2021. Fruit were monitored biweekly through harvest. By harvest (April 26, 2022), a larger proportion of the small fruit dropped compared to the large fruit (73% vs. 51%, respectively). This demonstrated that fruit size could be used as an indicator of drop in HLB-related preharvest fruit drop. Future work will focus on why the smaller fruit are more likely to drop.

Literature Cited

Tang L., Chhajed S., & Vashisth T. 2019. Preharvest Fruit Drop in Huanglongbing-affected 'Valencia' Sweet Orange. *J. Amer. Soc. Hort. Sci.* 144(2):107–117. https://doi.org/10.21273/JASHS04625-18

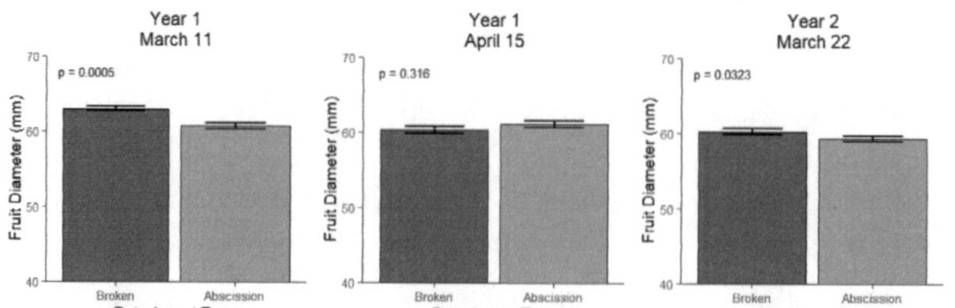

Fig.1. Average fruit diameter in mechanically broken and abscising fruit at three timepoints in HLB-affected *Citrus sinensis* 'Valencia' trees.

*Corresponding author. Email: tvashisth@ufl.edu

Proc. Fla. State Hort. Soc. 136:67. 2023.

Initial Host Response of Sensitive and Tolerant Citrus Varieties to *Candidatus* Liberibacter asiaticus Transmitted by Asian Citrus Psyllid

YUANZHI ZIMMY WANG[1]*, XIAOKANG ZHUO[1], RONALD TAPIA[1], FAHONG YU[2], QIBIN YU[1], PAIGE H. WHITE[1], RICCARDO RUSSO[1], XU WEI[1], AND FREDERICK G. GMITTER JR.[1]

[1]*Citrus Research and Education Center, Institution of Food and Agricultural Sciences, University of Florida, Lake Alfred, FL*

[2]*Interdisciplinary Center for Biotechnology Research, University of Florida, Gainesville, FL*

—abstract—

The citrus industry is one of the economic backbones in the State of Florida. However, citrus Huanglongbing (HLB, or citrus greening disease) has caused over 85% production losses since it was first reported in Florida in 2005. Previous study indicated the immediate transcriptomic-level changes in response to feeding by *Candidatus* Liberibacter asiaticus (*C*Las)-free healthy Asian citrus psyllid (ACP), but responses were delayed against *C*Las-carrying infective ACP, in HLB-sensitive 'Valencia' sweet orange (*Citrus sinensis*; VAL) during a 10-day-post-inoculation (dpi) period. In this study, similar transcriptomic profiling was advanced to a 2-dpi window, specifically at 2-, 12-, 24- and 48-hour-post-inoculation (hpi), between two differently HLB-responding citrus varieties: HLB-sensitive Valencia sweet orange (VAL) and HLB-tolerant LB8-9 Sugar Belle® mandarin (*C. reticulata*; SB). The results reveal distinct patterns of differentially expressed genes (DEGs) in VAL and SB in response to *C*Las: DEGs of SB indicate a gradual and stable increase throughout 2-dpi. By contrast, DEGs of VAL are characterized by a minor fluctuation from 2-hpi to 12-hpi followed by a major burst from 24-hpi to 48-hpi. In response to *C*Las-inoculation, VAL shows a burst of DEGs at 48-hpi, whereas DEGs in SB increase significantly at an earlier 24-hpi. On a transcriptomic level, a delayed response to *C*Las-inoculation is observed in VAL, compared with SB. More down-regulated DEGs are differentiated in VAL than in SB. The initial responsive DEGs in HLB-tolerant variety (SB) are categorized to be related to plant defense response to gram-negative bacterium, kinase activity, and protein-kinase derived complex. In HLB-sensitive variety (VAL), initial responsive DEGS are categorized to be related to phosphatase biosynthesis and dephosphorylation reaction. The transcriptomic study aims at investigating the complete set of RNA transcripts produced by the corresponding genome and identifying differentially expressed genes in response to very early stage of *C*Las infection. It captures as many DEGs as possible and is highly instrumental to thoroughly understand the biological pathways that lead to different tolerance levels in different citrus varieties.

*Corresponding author. Email: ywang8@ufl.edu

Proc. Fla. State Hort. Soc. 136:68. 2023.

—Scientific Note—

Effect of Exogenous Silicon and Potassium in Improving Cold Tolerance in Satsuma Mandarin

Scott Cruz, Mujahid Hussain, Davie Kadyampakeni, Ali Sarkhosh, and Muhammad Adnan Shahid*

Horticultural Science Department, North Florida REC, University of Florida, IFAS, Quincy, FL

Keywords. *Citrus unshiu* Marc., freeze protection

The Satsuma mandarin (*Citrus unshiu* Marc.) has desirable characteristics such as easy to peal, seedlessness, and sweet flavor which are desirable by potential customers. Satsuma mandarin has been grown in North Florida for many decades but freeze events complicated have the development of this industry. The mild winter during the last three decades and emergence of microspinklers as a cold protection technique resulted in the reestablishment of cold hardy citrus industry in north Florida. The use of microsprinklers is beneficial for young plantings but it has some limitations in established orchards, just protecting the lower part inside the canopy (Campbell et al., 2004; Ebel et al., 2004; Nesbitt et al., 2000). This project considers the use exogenous application of silicon (Si) and potassium (K), to provide more protection to the whole tree in established orchards during freeze events.

Two-year-old Satsuma (*Citrus unshiu* 'Owari') grafted on 'Rich 16-6 rootstock were treated with silicon and potassium solutions under controlled conditions in a greenhouse. The treatments were: i) Control (no foliar application); ii) Foliar application SiO_2 (50 ppm); iii) Foliar application SiO_2 (100 ppm); iv) Foliar application K (50 ppm); v) Foliar application (100 ppm); and vi) Foliar application SiO_2 (50 ppm + K 50 ppm). The plants were treated with 60 mL of each treatment every 72 hours. Treatments were applied in a programmed walk-in cold chamber by descending the temperature from 0 to –2 °C, –4 °C, and –6 °C, with 48-hour intervals. Plants were provided with photoperiod of 14 hours light and 10 hours dark.

Findings

Leaf samples were collected before the cold treatment at normal temperature (25 ± 2 °C) and at –6 °C for the biochemical analysis. Overall, exogenous application of Si and K helped to alleviate the negative effect of cold stress by limiting the formation of reactive oxygen species (ROS) formation, lipid peroxidation and electrolyte leakage. Under cold stress conditions (–6 °C) control plants produce 7.02 $\mu mol \cdot g^{-1}$ FW of peroxide (O_2^-) compared with 0.15 $\mu mol \cdot g^{-1}$ FW before cold stress, plants treated with K 50 and 100 ppm under stress, maintained levels of 0.25 and 0.45 $\mu mol \cdot g^{-1}$ FW, respectively. The combined application (Si + K 50 ppm) reduced the level of hydrogen peroxide (H_2O_2) 2.34 $\mu mol \cdot g^{-1}$ FW. Electrolyte leakage was reduced from 71% in control plants under cold stress to 8% and 9% with Si 100 ppm and Si + K 50 ppm respectively. High concentration of Si (100 ppm) was the most effective treatment. Plants treated with Si 100 ppm and Si + K 50 ppm had highest accumulation. Likewise, plants treated with Si and K showed increased enzymatic activities of superoxide dismutase (SOD), peroxide (POD), catalase (CAT), catalase (CAT), and glutathione peroxide (GPX) compared to the control plants. Overall Si 100 ppm and Si + K 50 ppm found to be the best in terms of improved osmolyte concentration and antioxidant activities. (Fig. 1).

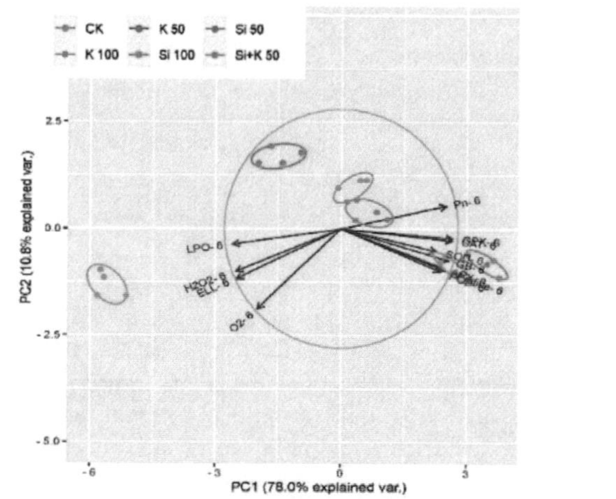

Fig. 1. Principal component analysis of cold protection treatments on two-year-old satsuma (*Citrus unshiu* 'Owari'). CK = control (no foliar application), Si50 = foliar application of 50 ppm SiO_2 , Si100 = foliar application of 100 ppm SiO_2, K50 = foliar application of 50 ppm K, K100 = foliar application of 100 ppm K, and Si+K50 = foliar application of 50 ppm SiO_2 + 50 ppm K.

Suggestions for Future Direction

Future experiments should consider higher levels of exogenous application of silicone and potassium rates and the combination of on large scale farm trials to determine the effect on cold hardiness under commercial setting in open field. There is also a need to investigate the physiological and molecular basis associated with silicon and potassium induced cold hardiness in citrus.

Literature Cited

Campbell B.L., Nelson R.G., Ebel R.C., Dozier W.A., Adrian J.L., & Hockema B.R. 2004. Fruit quality characteristics that affect consumer preferences for satsuma mandarins. *HortScience* 39(7):1664–1669.

Ebel R.C., Dozier W.A., Hockema B., Woods F.M., Thomas R., Wilkins B.S., Nesbitt M., & McDaniel, R. 2004. Fruit quality of satsuma mandarin grown on the northern coast of the Gulf of Mexico. *HortScience* 39(5):979–982.

Nesbitt M.L., McDaniel N.R., Ebel R.C., Dozier W.A., & Himelrick, D.G. 2000. Performance of satsuma mandarin protected from freezing temperatures by microsprinkler irrigation. *HortScience* 35(5):856–859.

*Corresponding author. Email: mshahid@ufl.edu

Proc. Fla. State Hort. Soc. 136:69–72. 2023.

Exploiting OLL Somaclone-derived Nucellar Seedlings to Improve Juice Quality of Early-mid Season Varieties in the Florida Citrus Industry

Maria Brenelli* and Jude W. Grosser

Citrus Research and Education Center, University of Florida, IFAS, Lake Alfred, FL

Keywords. *Citrus sinensis*, early maturity, HLB, somaclone-derived nucellar seedlings, sweet orange

'Hamlin' (*Citrus sinensis*), a primary early sweet orange, is extensively cultivated in Florida. However, it faces challenges in terms of limited juice quality, including color and flavor, which fail to meet the Grade A standards. To overcome this, 'Hamlin' juice is commonly blended with other citrus cultivars like Valencia. Additionally, the Florida citrus industry has been severely affected by Huanglongbing (HLB) disease, which causes reduced fruit quality, lower yields, and economic difficulties. Developing citrus cultivars with improved tolerance to HLB is considered a sustainable solution for the industry. This study evaluates OLL somaclone derived nucellar seedling sweet orange clones for enhanced early-maturity and HLB-tolerance. Fruit quality was assessed using acid content, total Brix, Brix/acid ratio, and juice color analysis in January 2022 and 2023. Selected clones exhibited three consecutive seasons of Brix content equal to or higher than 10.5, with some approaching the desired 12.5 Brix/acid ratio. The selected clones 'RIT26', 'R4T13', and 'R7T35' demonstrated earlier maturity than commercially available OLL clones, and high processing qualities. Furthermore, the juice color assessment confirmed that these selected early-maturity clones met the minimum criteria for Grade A orange juice, reducing the need for extensive blending. These findings highlight the OLL somaclone-derived nucellar seedling population's genetic diversity and desirable characteristics. The early-maturing clones offer potential for citrus production and disease management strategies, particularly in HLB tolerance. Further evaluation through multi-site replicated stage 2 trials is underway to validate the maturity date and assess the HLB tolerance of these newly developed early maturing OLL clones.

The sweet orange (*Citrus sinensis* L. Osbeck) is the most important and widely cultivated citrus species in Florida and globally. In Florida, nearly 90% of the crop and over 30% of the global crop are used for processing (Kimball, 1999; Yu et al., 2021). Efforts have been made to improve the Hamlin cultivar in the Florida citrus industry. 'Hamlin' oranges are known for their high productivity but are lacking in terms of juice quality. The juice obtained from 'Hamlin' oranges tends to have an unsatisfactory color and average flavor and does not meet the Grade A standard. To overcome this limitation, Hamlin juice is often mixed with juices from other cultivars, such as Valencia (Lee and Castle, 2001; Ringblom, 2017).

In recent years, the Florida citrus industry has faced significant challenges due to Huanglongbing (HLB), also known as citrus greening, particularly since 2012 (Gottwald et al., 2007; Graham et al., 2020). HLB-infected citrus trees exhibit symptoms such as smaller, misshapen, and discolored fruits with uneven ripening. These fruits also have reduced sweetness and a more bitter taste (Bové, 2006; Plotto et al., 2010; Bai et al., 2016). As a result, citrus growers face economic challenges due to decreased quantity and quality of harvested fruit, leading to lower income and reduced market demand for affected citrus products (Singerman and Rogers, 2020; Li et al., 2020).

Managing HLB-affected groves involves employing strategies such as insecticide use, additional fertilizers, foliar micronutrient sprays, and careful irrigation practices (Bové, 2006; Gottwald et al., 2012; Atta et al., 2023). However, these practices incur higher production costs, impacting the citrus industry's profitability. The combination of increased expenses, reduced yields, and lower fruit quality in HLB-affected groves presents economic challenges (Singerman, 2019; Singerman and Rogers, 2020; Li et al., 2020). Thus, developing HLB-resistant or tolerant citrus cultivars is a sustainable solution to decrease reliance on costly management practices, enhancing sustainability and profitability for growers (Wang, 2019; Li et al., 2020).

The University of Florida Citrus Research and Education Center breeding program has identified genetic variation in nucellar seedlings derived from somaclones, offering the potential to develop citrus varieties with enhanced processing qualities and improved HLB tolerance. Certain varieties even exhibit early to mid-season maturity, aligning with the harvest dates of 'Hamlin' oranges. The primary objective of this research is to evaluate the fruit quality of a large population of OLL nucellar seedling-derived clones. The goal is to introduce enhanced OLL sweet orange cultivars into commercial production, specifically targeting clones selected for early maturity and improved HLB tolerance.

Materials and Methods

This study was conducted at the Lee Family Groves property, in Saint Cloud, FL. The focus of the study was to evaluate a new population of OLL clones, which were generated from the original OLL somaclone population. Two hundred fifty individual scion clones were produced from nucellar seed obtained from 13 of the original OLL population.

For fruit quality assessment, a random selection of half bushel of fruits was collected from each tree in January 2022 and January 2023. The samples were collected from different positions on the tree to ensure representativeness. Fruit processing was

*Corresponding author. Email: mbrenellidepaiva@ufl.edu

conducted using a state-certified FMC 091B State Test juice extractor located in the pilot plant of the Citrus Research and Education Center (CREC). Various quality variables were evaluated, including acid content, total Brix, and the Brix/acidity ratio. Juice color analysis was performed using a Macbeth Color-Eye Spectrophotometer (Model 1500, 2020+, 3000/3100). Standard methods were applied to assess the selected clones' quality, confirming their compliance with USDA Grade A standards for Brix (10.5), Brix/acid ratio (12.5), and juice color (36 points) when harvested in January, without the need for blending with other varieties (Ringblom, 2015).

Data were analyzed annually for two consecutive growing seasons (2022 and 2023) using the software RStudio version 1.3.1073. Aligned ranks transformation ANOVA (Art in the ARTool package) was used to evaluate the main effects and their interaction. ANOVA was utilized to assess the influence of the somaclone factor on total Brix values. When differences between treatments were significant ($P \leq 0.05$), a Tukey HSD test was used. Clones exhibiting three consecutive seasons of Brix content equal to or higher than the minimum required value of 10.5 Brix were selected. Only those meeting the minimum or close to meeting the 12.5 Brix/acid ratio from January harvested fruit were chosen among the selected clones as early maturity cultivars.

Results and Discussion

During the three evaluation seasons, the total Brix values were significantly influenced by the interaction of the somaclone of origin ($P < 0.001$). Some groups of somaclones demonstrated similar total Brix levels, while others showed significant differences. The group OLL-10 consistently demonstrated the highest significance level throughout the years. In addition to the OLL-10 group, the OLL-21 group also showed a high significance level in 2023 (Fig. 1). The OLL-10 group consistently exhibits a higher significance level in Brix values, indicating that it is relatively easier to identify clones with higher Brix levels within this group. However, clones from other groups can also have high Brix levels and meet the minimum of 10.5 Brix required. Therefore, selecting individuals based not only on their group origin but also considering other factors in evaluating their potential Brix levels accurately is essential. Furthermore, the Brix/acid ratio is a predictive indicator to determine the optimal time for fruit harvesting. Clones that displayed a ratio close to or meeting the minimum required, with consistent results, were selected (Table 1). This selection process ensures that the chosen clones meet the desired standards for both Brix and acid levels, indicating their readiness for harvest in January. The clones selected for early maturity and high processing qualities were 'RIT26', 'R4T13', and 'R7T35'. These 3 selections represent approximately 1% of the total population evaluated, with the remainder exhibiting maturity dates more similar to the original OLL somaclones.

The juice color of the selected trees was assessed in 2022 and 2023 (Table 1). Despite some variation, the values recorded in both years showed that all the selected trees as early maturity varieties still satisfy the minimum criteria for Grade A orange juice. It is worth noting that early-maturity varieties like 'Hamlin' tend to have lighter juice colors compared to other varieties (Ringblom, 2017). Blending early maturity varieties with other varieties possessing deeper juice color is a common method to enhance the color profile and attain optimal juice quality. This blending approach aids in improving the overall juice color and meeting the desired standard for Grade A orange juice. Nevertheless, the blending process can be complicated and labor-intensive (Ringblom, 2017; Kimball, 1999). Including earlier-maturing varieties, like the newly selected OLL clones described in this study, that already display juice colors surpassing the minimum threshold of 36, signifies a substantial improvement.

These findings offer valuable information about the genetic diversity and desirable characteristics of the OLL-derived nucellar seedling population. Three clones were identified that reliably exhibited earlier maturity, namely 'R1T26', 'R4T13 and'R7T35' (Fig. 2). The potential implications of HLB tolerance present compelling opportunities for citrus production and the development of disease management strategies. However, further research is required to investigate the underlying factors contributing to the observed differences and validate these findings on a larger sample size. The newly developed early maturing OLL clones are currently being propagated for further evaluation. This evaluation includes conducting multi-site replicated stage 2 trials to validate their maturity date and assess their tolerance to HLB.

Literature Cited

Atta A.A., Morgan K.T., Ritenour M.A., & Kadyampakeni D.M. 2023. Nutrient management impacts on HLB-affected 'Valencia' citrus tree growth, fruit yield, and postharvest fruit quality. *HortScience*. 58(7):725–732. https://doi.org/10.21273/HORTSCI17110-23.

Bai J., Baldwin E.A., McCollum G., Plotto A., Manthey J.A., Widmer

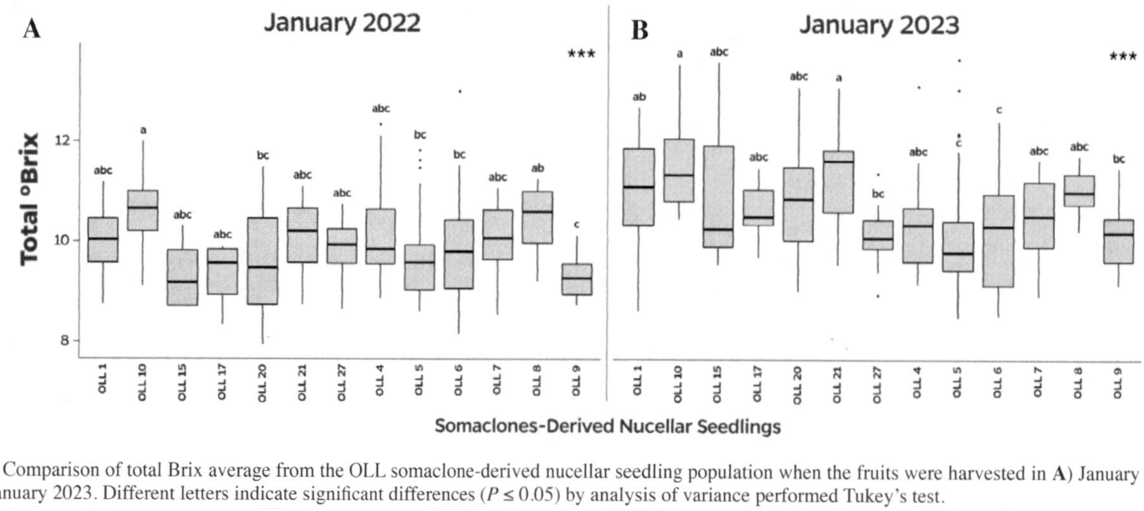

Fig. 1. Comparison of total Brix average from the OLL somaclone-derived nucellar seedling population when the fruits were harvested in **A**) January 2022 and in **B**) January 2023. Different letters indicate significant differences ($P \leq 0.05$) by analysis of variance performed Tukey's test.

Table 1. . Fruit quality assessment (2022 and 2023) of 37 OLL somaclone-derived nucellar seedling clones with superior total Brix.

Tree identification	OLL somaclone of origin (scion)/rootstock	Total Brix 2022	Total Brix 2023	Brix/acid ratio 2022	Brix/acid ratio 2023	Juice color 2022	Juice color 2023
R1T3	OLL 8/ CHANG+50-7	11.26	11.56	8.94	9.25	36.30	36.8
R1T4	OLL 8/ CHANG+50-7	11.25	11.42	7.71	7.77	37.10	36.4
R1T5	OLL 8/ CHANG+50-7	10.89	11.74	7.36	6.83	36.30	35.8
R1T6	OLL 8/ CHANG+50-7	11.05	11.36	8.31	7.94	36.90	36.8
R1T7	OLL 8/ CHANG+50-7	10.70	10.77	9.30	7.75	37.50	37.4
R1T18	OLL 8/ CHANG+50-7	10.51	11.13	9.47	9.59	37.60	36.4
R1T22	OLL 10/ CHANG+50-7	10.97	11.06	8.37	8.44	37.20	36.4
R1T25	OLL 10/ CHANG+50-7	10.76	11.93	9.28	9.39	37.20	37.6
R1T26	OLL 10/ CHANG+50-7	11.68	12.13	11.80	12.51	37.20	38.4
R1T29	OLL 10/ CHANG+50-7	10.66	10.89	9.27	8.64	37.00	36.9
R2T4	OLL 7/ AMB+BENT	10.71	11.51	10.40	9.84	38.10	37.4
R2T5	OLL 7/ AMB+BENT	11.08	11.65	8.94	9.40	36.90	36.2
R2T33	OLL 20/ ORG 3	10.62	13.10	8.17	8.97	37.00	37.8
R2T35	OLL 20/ ORG 3	11.26	10.70	9.01	9.73	37.40	38.7
R3T2	OLL 21/ ORG 3	10.97	10.65	9.46	8.59	37.50	37.0
R3T4	OLL 21/ ORG 3	10.79	11.71	9.38	7.76	37.30	37.1
R3T7	OLL 21/ ORG 3	10.78	11.09	10.78	9.02	37.10	37.0
R3T8	OLL 21/ ORG 3	10.85	12.46	9.52	8.71	37.50	36.8
R3T18	OLL 21/ ORG 3	11.11	13.10	9.42	10.23	37.10	37.5
R4T4	OLL 6/ WHITE 3	11.40	11.86	9.34	8.79	37.50	37.2
R4T5	OLL 6/ WHITE 3	11.18	11.71	11.07	9.44	37.70	36.9
R4T6	OLL 6/ WHITE 3	11.53	11.09	11.77	10.17	37.10	37.1
R4T7	OLL 6/ WHITE 3	11.27	10.67	11.50	9.20	37.70	37.7
R4T9	OLL 6/ WHITE 3	11.36	11.38	11.96	9.41	37.50	37.7
R4T13	OLL 6/ WHITE 3	10.61	10.94	12.06	11.28	36.60	36.2
R5T2	OLL 4/WHITE 1	11.04	10.70	9.12	8.05	37.70	36.7
R5T3	OLL 4/WHITE 1	12.35	11.58	8.88	7.52	38.30	37.0
R5T12	OLL 4/WHITE 1	10.75	13.13	10.14	9.65	36.70	37.4
R5T32	OLL 5/ ORG 13	11.83	13.68	9.94	10.36	36.00	37.3
R6T31	OLL 4/WHITE 1	10.65	11.62	9.42	9.68	37.70	37.5
R7T29	OLL 10/ ORG 13	11.11	13.56	12.08	10.43	37.00	36.9
R7T32	OLL 10/ ORG 13	11.81	12.12	9.84	10.18	37.40	37.5
R7T33	OLL 10/ ORG 13	11.04	10.58	9.86	9.98	37.90	37.1
R7T35	OLL 10/ ORG 13	12.00	10.62	14.12	11.93	37.30	36.6
R10T19	OLL 1/ CHANG+BENT	11.19	12.70	10.46	9.34	37.00	37.5
R10T22	OLL 1/ CHANG+BENT	10.75	11.83	8.96	10.03	37.60	37.5
R10T24	OLL 1/ CHANG+BENT	11.05	11.56	9.21	9.25	36.80	37.9
	Population means values	10.03	10.46	9.64	9.51	37.4	38.1
		n = 245	n = 245	n = 245	n = 245	n = 245	n = 245

Fig. 2. . Fruits of the OLL somaclone-derived nucellar seedling that reliably exhibited earlier maturity A. "R4T13", and B. "R7T35".

W.W., Luzio G., & Cameron R. 2016. Changes in volatile and non-volatile flavor chemicals of 'Valencia' orange juice over the harvest seasons. *Foods*. 5(1):4. https://doi.org/10.3390/foods5010004.

Bové J.M. 2006. Huanglongbing: A destructive, newly-emerging, century-old disease of citrus on JSTOR. Accessed 25 May 2023. https://www.jstor.org/stable/41998278

Gottwald T.R., da Graça J.V. & Bassanezi R.B. 2007. Citrus huanglongbing: The pathogen and its impact. *Plant Health Progress*. 8(1):31. https://doi.org/10.1094/PHP-2007-0906-01-RV

Gottwald T.R., Graham J.H., Irey M.S., McCollum T.G., & Wood, B.W. 2012. Inconsequential effect of nutritional treatments on huanglongbing control, fruit quality, bacterial titer and disease progress. *Crop Protection* .36:73–82. https://doi.org/10.1016/j.cropro.2012.01.004

Graham J., Gottwald T., & Setamou M. 2020. Status of huanglongbing (HLB) outbreaks in Florida, California and Texas. *Trop. Plant Pathol.* 45(3):265–278. https://doi.org/10.1007/s40858-020-00335-y

Kimball D.A. 1999. Citrus-processing management, p. 43–72. In:

Kimball, D.A. (ed.). Citrus processing: A complete guide. Springer US, Boston, MA. https://doi.org/10.1007/978-1-4615-4973-4_3.

Lee H.S. & Castle W.S. 2001. Seasonal changes of carotenoid pigments and color in Hamlin, Earlygold, and Budd Blood orange juices. *J. Agric. Food Chem.* 49(2):877–882. https://doi.org/10.1021/jf000654r.

Li S., Wu F., Duan Y., Singerman A., & Guan Z. 2020. Citrus greening: Management strategies and their economic impact. *HortScience*. 55(5):604–612. https://doi.org/10.21273/HORTSCI14696-19.

Plotto A., Baldwin E., McCollum G., Manthey J., Narciso J., & Irey M. 2010. Effect of Liberibacter infection (huanglongbing or "greening" disease) of citrus on orange juice flavor quality by sensory evaluation. *Journal of Food Science*. 75(4):S220–S230. https://doi.org/10.1111/j.1750-3841.2010.01580.x

Ringblom U. 2017. The orange fruit and its products. Orange Book. Accessed 24 May 2023. https://orangebook.tetrapak.com/chapter/orange-fruit-and-its-products

Singerman A. & Rogers M.E.. 2020. The economic challenges of dealing with citrus greening: The case of Florida. *Journal of Integrated Pest Management* 11(1):3. https://doi.org/10.1093/jipm/pmz037

Wang N. 2019. The citrus huanglongbing crisis and potential solutions. *Molecular Plant* 12(5):607–609. https://doi.org/10.1016/j.molp.2019.03.008

Yu W., Cui J., Zhao S., Feng L., Wang Y., Liu J., & Zheng J. 2021. Effects of high-pressure homogenization on pectin structure and cloud stability of not-from-concentrate orange juice. *Front Nutr.* 8:1–13. https://doi.org/10.3389/fnut.2021.647748

Proc. Fla. State Hort. Soc. 136:73. 2023.

—Scientific Note—

Interactions of Four Citrus Scion/Rootstock Combinations in Response to Salinity Stress

NEUS ALCON BOU[1], SANDRA GUZMAN[1], EDZARD VAN SANTEN[2], AND LORENZO ROSSI[1]*

[1]*Indian River Research and Education Center, University of Florida, IFAS, Fort Pierce, FL*

[2]*Agronomy Department, University of Florida, Gainesville, FL*

Keywords. *Citrus* ×*paradisi*, salt intrusion, stress physiology

Citrus is a salt-sensitive crop. The distressing effects of diseases such as Huanglongbing (HLB), makes the citrus trees more prone to have lower yields and more vulnerability to other biotic stresses. Recently, cultivar selection has been focused on releasing HLB-tolerant rootstocks. However, few studies evaluated the performance of these new rootstocks under salinity stress. In addition, although citrus is a well-studied crop, most of the research has been carried out on non-grafted plants.

The objectives of the study were to: 1) carry out a first assessment of two grafted HLB-tolerant rootstocks under salinity stress and 2) assess the interaction of scion and rootstocks in response to an increase in salt concentrations.

One-year-old seedlings of 'Jackson' and 'Star Ruby' grapefruits (*Citrus* ×*paradisi*) grafted on 'US-802' and 'US-942' were grown in a greenhouse irrigated under three different salinity concentrations: 0 mM, 50 mM, and 100 mM NaCl. Chlorophyll content, dry weight, nutrient content, leaf fluorescence, and root barriers were measured after 6 weeks.

Chlorophyll content and leaf dry mass were lower for 'Star Ruby' and for the plants grafted on 'US-802'. The scion seemed to have a major influence on the concentration of sodium in leaves than the rootstock, being 'Star Ruby' the one that presented higher values. Fluorescence decreased more notoriously for those scions grafted on 'US-802'.

New HLB-tolerant rootstocks seemed to present different tolerance thresholds to salinity and the results suggested that both, scion and rootstock, influenced the response to the salinity stress. Comprehensive long-term field studies are needed to validate and expand upon our results.

*Corresponding author. Email: l.rossi@ufl.edu

Proc. Fla. State Hort. Soc. 136:74–77. 2023.

Impact of Silicon on Citrus Performance: A Review

Jose Prieto Fajardo[1], Muhammad Shahid[2], and Davie Kadyampakeni[1]*

[1]*University of Florida, IFAS, Citrus Research and Education Center, Lake Alfred, FL*

[2]*University of Florida, IFAS, North Florida Research and Education Center, Quincy, FL*

Keywords. citrus, plant immunity, silicon, silicon nanoparticles

The use of silicon as a fertilizer has shown promising results in improving the growth and immunity of various crops. However, its effects on citrus trees are not well documented. This review aims to shed some light on the impact of silicon fertilization on citrus performance, focusing on its ability to improve plant growth and resistance to pests and diseases. The review examines the current literature on the topic and discusses the mechanisms through which silicon can affect citrus trees. Additionally, the review discusses the optimal application rates and methods of silicon fertilization for citrus trees depending on soil and climate conditions. Overall, the review provides valuable insights into the potential benefits of silicon fertilization for citrus growers, particularly those affected by the devastating impact of citrus greening and highlights the need for further research in this area.

Silicon Status

Silicon (Si) is one of the most abundant elements in earth's lithosphere, just below oxygen (Sommer et al., 2006; Epstein, 1999). Silicon refers to the element Si in the periodic table of the elements below carbon (C) and has four valence electrons. Both Si and C can have an oxidation state of $^+2$ and $^+4$ and are non-metals, with a similar density. Silicon shares many properties with C, such as having four electrons in its outer shell (Emsley, 1998). This allows it to form very stable compounds. Si is mostly found in the form of silica [silicon dioxide (SiO_2)] and is the main component of certain types of sand. Silicates are compounds that Si forms in association with the cations Ca^{+2}, K^+, Mg^+, and Na^+. They form the compounds $CaSiO_3$, K_2SiO_3, $NaSiO_3$ and $MgSiO_3$, commonly known crystalline or amorphous depending on their lattice structure. Crystalline silicates with lower impurities tend to release more silicic acid than amorphous silicates with higher concentrations of contaminants (Dixon and Schulze, 2002). Silicic acid, (known as mono-silicic acid or ortho-silicic acid) refers to the soluble form of Si that is readily available for absorption by plant roots (Epstein, 1999). It is usually present in the soil at a pH between 4 and 9. At a pH >10 it forms polysilicic acid, a silicic acid polymer that is not available for plant use (Tubaña & Heckman, 2015; McKeague & Cline, 1963; Datnoff et al., 2001)

Silicon: Plant Essential Nutrient?

Takashi et al. (1990) classified plants based on the mechanism through which they accumulate Si, defining high, intermediate or not accumulating. High accumulators have a highly efficient Si uptake while low accumulators may have an inefficient or passive uptake. Depending on whether they are high or intermediate accumulators, plants can have Si concentrations comparable to some micronutrients, and in some cases even macronutrients (Epstein, 1999). Nutrients are designated as a macro or micronutrients depending on whether the concentration is above or below 0.1%, respectively (Epstein & Bloom, 2005). At the low end of the range (0.1%), Si concentrations may correspond on a dry matter basis with macronutrients such as Ca, Mg, P, and S. At the upper end of 10%, the tissue concentrations will exceed mineral nutrients such as N and K (Epstein, 1994). One of the properties of an essential nutrient is its direct role in plant metabolism and production. There is currently no evidence of the direct involvement of Si in plant metabolism (Ma et al., 2001; Knight & Kinrade, 2001).

Sources of Silicon

There are several solid and liquid sources of Si that have been used as soil conditioners, amendments, or as fertilizers in the field and in greenhouses. For them to be useful, they must have a relatively high soluble Si content, ease of storage, ease of application, and a negligible amount of soil contaminants such as heavy metals (Gascho, 2001). Liquid sources of Si include silica gel, monosilicic acid, and calcium, potassium, and sodium silicates that are used as foliar spray. Liquid sources have also been used as contact biopesticides to suppress plant diseases and insect pests (Rodrigues et al., 2015). More recently, silica nanoparticles have been shown to promote plant growth and enhance disease resistance (El-Shetehy et al., 2020; Suriyaprabha et al., 2012). Silica nanoparticles are a potential source of Si for future use in agriculture, but further studies need to be conducted (Elmer and White, 2018).

Silicon in the Soil

Soil solutions can contain Si concentrations between 2.8 and 17.1 $mg \cdot L^{-1}$, but lower and higher values may be encountered (Epstein, 1994). Concentrations of this magnitude are typical for several other inorganic nutrients such as Ca, K, S, and are sometimes more than phosphorus concentrations in the soil solution. However, the dissolution of Si from soil minerals is slow, and repeated cropping may reduce its levels to the point that supplementation is required for maximum crop yield (Savant et al., 1997). Besides agriculture, there exists other actions that can

*Corresponding author. Email: dkadyampakeni@ufl.edu

Table 1. Silicon (Si) inputs and outputs from a soil system (adapted from Tubaña & Heckman, 2015).

Increase Si concentration in soil	Decrease Si concentration in soil	Remove Si from soil system	Make Si unavailable
Crop residues, animal manure, compost	Absorption of Si by plants	Leaching	Formation of silica polymorphs
Microorganism remains	Absorption of Si by microorganisms	Runoff	
Dissolution of secondary minerals	Precipitation of secondary minerals	Soil erosion	
Iron, aluminum oxides/hydroxides and clay exchange sites desorbing Si	Adsorption of Si		
Dissolution of biogenic silica			
Dissolution of primary minerals			
Dissolution of suspended silicon particles			
Soil deposition, ash, rain, irrigation, fertilization, organosilicon adjuvants in agrochemicals			

increase or decrease Si levels in the soil, while other activities can make Si unavailable for plant absorption or completely remove it from the soil system (Table 1).

Low-Si soils are typically highly weathered, leached, acidic, low in base saturation, and include large amounts of iron and aluminum sesquioxides. Soils such as Oxisols and Ultisols can be relatively low in soluble Si. Histosols, being low in minerals, may also contain little Si. Soils comprised mainly of quartz sand, for example sandy Entisols, may be low in soluble Si despite having a great amount of insoluble Si. Alfisols, Inceptisols, and Spodosols were found to be limiting in soluble Si (Tubaña & Heckman, 2015). Soilless mixes can also be limited in Si.

Silicon Translocation and Accumulation in the Plant

All plants absorb measurable concentrations of Si in the range of 0.005 to 11.3% from the soil (Hodson et al., 2005), and 42 angiosperm orders, representing over 100 orders or families, have been reported to do so. Sugarcane (*Saccharum* spp.), rice, wheat (*Triticum* spp.) and some other grasses are among plants that highly accumulate Si in their tissue. Grasses will tend to accumulate more Si in their tissue when compared to dicots. In general, wetland grasses will average between 4.6 to 6.9%, dryland grasses from 0.5 to 1.4%, and dicots less than 0.23% on a dry matter basis (Epstein, 1994; Hodson et al., 2005). Based on their Si concentration in leaf tissues, plant species were initially classified as low, intermediate, and high accumulators, but without any evidence supporting such classification (Handreck & Jones 1967). This classification was later improved by categorizing high, intermediate, and low accumulator plant species with active, passive, or rejective mechanisms for Si uptake, respectively (Takahashi et al., 1990). When determining whether a plant accumulates Si, past studies focused on measuring Si in the foliage and not in other organs such as the roots (Hodson et al. 2005). Some plant species amended with Si, such as tomatoes, peppers, and coffee, have root Si concentrations that are equal to or exceed Si levels in their shoots (Carré-Missio et al., 2009; French-Monar et al., 2010; Huang et al., 2011). Root application provides an efficient means of Si absorption and distribution in plant tissues. Foliar application of Si has also been used for disease control. When the foliar application of Si has been compared to root (soil) application, plants respond better to the latter. This comparison has been successfully demonstrated in the following different host-pathogen interactions: cucumber/powdery mildew, melon/powdery mildew, wheat/powdery mildew, rice/blast, rice/brown spot, and soybean/Asian soybean-rust (Tubaña, 2015).

There have been no major disagreements among the studies carried out on digestion procedure nor quantification of Si for its determination in plant tissues. Most of the research has revolved around validating and improving procedures in terms of efficiency and safety. Between the gravimetric method, solubilization, and digestion procedures, the selection is made based on precision, turn-around time, cost, and required instrumentation (Tubaña et al, 2015).

Silicon Application Rates

Research reports depict that Si application rates are inconsistent, so it is difficult to understand the amount of Si that is applied to each plant. Also, all the studies report positive effects in their experiments after applying Si. Most studies focused on how it impacted productivity indicators (growth, yield), but did not delve too deep on the specifics of how it impacted plant physiology or the soil (Table 2).

Role of Silicon in Disease Resistance

According to Rodrigues et al. (2015), the mechanism by which Si affects pathogen infection is dependent on whether applied via the roots or foliage. When Si is absorbed by the roots, it moves through the xylem vessels and is distributed throughout the plant tissues, activating several defense responses within the host plant (Epstein, 1999). On the other hand, foliar application of Si will cause Si to polymerize on the leaf surface to form a barrier. This may change the pH and the osmotic potential of the phyllosphere, which could explain the reductions in the development of some foliar diseases (Tubaña et al. 2015)

Research Focused on Silicon Fertilization

The content of Si in leaves of citrus was recorded as 0.04-0.20% of the dry weight (Matichenkov & Calvert, 2001). The Si content in ash ranged from 0.30 to 0.50% in citrus fruit, 1.19 to 1.49% in leaves, 0.61 to 1.45% in wood and 0.84 to 3.17% in roots (Wutscher, 1989). Wutscher (1989) compared the growth of young orange trees (*Citrus sinensis*) in solution culture without Si addition with deliberate inclusion of Si. Results indicated that Si would be required in nonsignificant amounts in citrus nutrition.

In a study investigating the effects of silicon dioxide nanoparticles (SiO_2 NPs) on banana plant growth, it was found that the SiO_2 NPs significantly increased the number and length of roots, plant height, leaf number, and fresh and dry weight of both roots

Table 2. Fruit crop studies that experiment with silicon fertilizer.

Author	Year	Crop	Location	Silicon applied	Amount applied	Goal
Lalithya	2014	Sapota	India	$CaSiO_3$	2.5 kg/tree	Improve fruit characters
More	2015	Mango	India	Silixol (Silicic acid)	0.069 mg/L	Increase yield
Matichenkov	2001	Grapefruit and Orange	Indian River REC	MgO_3Si	8 t/ha	Mitigate NaCl and Al induced stress
Kaluwa	2010	Avocado	South Africa	K_2SiO_3	2940 mg/L	Suppressing ethylene production (fruit)
El-Enien	2017	Valencia orange/ volkamer lemon rootstock	Egypt	K_2SiO_3	10 g / seedling	Improve growth (nursery)
El-Enien	2019	Valencia orange/ volkamer lemon rootstock	Egypt	SiO_2	2 g/L	Improve growth (nursery)

and plantlets in banana plants. These effects were most pronounced at a concentration of 150 mg·L^{-1}. The study suggests that the use of SiO_2 NPs may be a promising strategy for improving the growth of banana plants (El-Kady et al., 2017). In another study investigating the effects of silicon dioxide nanoparticles (SiO_2 NPs) on banana plant growth under water deficit stress, it was found that the SiO_2 NPs improved the number and length of shoots, chlorophyll content, and cell membrane stability, and decreased electrolyte leakage in banana plants under water deficit stress. These effects were most pronounced at concentrations of 100 and 150 mg/L. The study suggests that the use of SiO_2 NPs may be a promising strategy for mitigating the negative effects of water deficit stress on banana plant growth (Mahmoud et al., 2020).

In a study conducted by Mahmoud et al. (2022), it was found that silicon dioxide nanoparticles (SiO_2 NPs) can alleviate salt stress in 'Valencia' sweet orange (*Citrus sinensis*) plants. Treatment with SiO_2 NPs significantly increased root growth, chlorophyll b content, total chlorophyll content, and leaf relative water content in plants subjected to salt stress, indicating a decrease in osmotic stress. The SiO_2 NPs also decreased electrolyte leakage and malondialdehyde content and increased total antioxidant content and total soluble protein content. These effects may be due to SiO_2NP-mediated modulation of antioxidant enzymes and nonenzymatic antioxidants, as well as regulation of protein, lipid and carbohydrate metabolism, metal ion binding and transportation and cell wall synthesis. However, SiO_2 NP treatment also resulted in a decrease in total phenol content, which may be due to the formation of complexes between Si and lignin. Overall, the results suggest that SiO_2 NP treatment may be a promising strategy for improving the tolerance of citrus plants to salt stress (Mahmoud et al., 2022).

Summary

Silicon (Si) is a highly abundant element with chemical properties similar to carbon (C), and while its direct role in plant metabolism is uncertain, certain plants efficiently accumulate silicon in their tissues at levels comparable to macronutrients. Different plant species vary in their silicon accumulation, with grasses generally accumulating more than dicots. Soil solutions can vary in silicon concentrations, and repeated cropping may deplete silicon levels, requiring supplementation for optimal crop yield. There are multiple sources of silicon for agriculture, including nanoparticles that show promise in enhancing plant growth and disease resistance, but further research is required. Silicon can be applied to plants through roots or foliage, activating defense responses and forming a protective barrier against foliar diseases. Studies have demonstrated the benefits of silicon fertilization for citrus plants and the positive effects of silicon dioxide nanoparticles on growth and increased tolerance to water deficit stress and salt stress.

Literature Cited

Carré-Missio V., Rodrigues F.A., Schurt D.A., Resende R.S., Souza N.F.A., Rezende D.C., & Zambolim L. 2014. Effect of foliar-applied potassium silicate on coffee leaf infection by *Hemileia vastatrix*. *Annals of Applied Biology*. 164(3):396–403.

Datnoff L.E., Snyder G.H. & Korndörfer G.H. (eds.). 2001. Silicon in agriculture. Elsevier, Amsterdam, The Netherlands.

Dixon J.B. & Schulze D.G. 2002. Soil mineralogy with environmental applications. Soil Science Society of America Inc., Madison WI.

El-Kady M.E., El-Boray M.S., Shalan A.M., & Mohamed L.M. 2017. Effect of silicon dioxide nanoparticles on growth improvement of banana shoots in vitro within rooting stage. *Journal of Plant Production*. 8(9):913–916.

El-Enien M., Abo El-Kassim A., El-Azaze A., & El-Sayed F. 2017. Effect of Silicon, Potassium and Calcium compounds on growth and increase the efficiency of citrus seedlings to resist citrus leafminer (*Phyllocnistis citrella*). *Journal of Productivity and Development*, 22(3):729–749.

El-Enien M.M.S., Moursi E.A., & El-Rouby W.M. 2019. Effect of some drip irrigation and nano-silicon treatments on growth, yield and water relations of 'Washington Navel' orange trees grown in new reclaimed soils. *Journal of Plant Production*. 10(7):529–537.

El-Shetehy, M., Moradi, A., Maceroni, M., Reinhardt, D., Petri-Fink, A., Rothen-Rutishauser, B., Mauch, F. & Schwab, F. 2021. Silica nanoparticles enhance disease resistance in Arabidopsis plants. *Nature Nanotechnology*. 16(3):344–353.

Elmer W., Ma C., & White J. 2018. Nanoparticles for plant disease management. *Current Opinion in Environmental Science & Health*. 6:66–70.

Emsley J. 1998. The elements (3rd ed.). Clarendon Press, Oxford, United Kingdom.

Epstein E. 1994. The anomaly of silicon in plant biology. *Proceedings of the National Academy of Sciences*. 91(1):11–17.

Epstein E. 1999. Silicon. Ann. Rev. Plant Physiol. *Plant Mol. Biol*. 50:641–664.

Epstein E. & Bloom A. 2005. Mineral nutrition of plants: principles and perspectives. (2nd ed.). Sinauer Associates, Sunderland, MA.

French-Monar, R.D., Rodrigues, F.A., Korndörfer, G.H. & Datnoff, L.E. 2010. Silicon suppresses *Phytophthora* blight development on bell pepper. *Journal of Phytopathology*. 158(7–8):554–560.

Gascho, G.J. 2001. Silicon sources for agriculture. p. 197–207. In: Datnoff L.E., Snyder G.H., & Korndörfer G.H. (eds.). Silicon in agriculture, Vol 8. Studies in plant science. Elsevier, Amsterdam, The Netherlands.

Hodson M.J., White P.J., & Mead A. 2005. Phylogenetic variation in the silicon composition of plants. *Ann. Bot*. 96:1027–1046

Huang, C.H., Roberts, P.D. & Datnoff, L.E. 2011. Silicon suppresses fusarium crown and root rot of tomato. *Journal of Phytopathology*. 159(7–8):546–554.

Kaluwa K. 2010. Effect of postharvest silicon application on'hass' avocado (*Persea americana* Mill.) fruit quality. (Ph.D. dissertation). University of KwaZulu-Natal Pietermaritzburg, South Africa.

Lalithya K.A., Hipparagi K., & Thippeshappa G.N. 2013. Effect of silicon and micronutrients on growth and yield attributes of sapota cv. Kalipatti under hill zone. *Crop Research*. 46(1–3):146–149.

Mahmoud L.M., Dutt M., Shalan A.M., El-Kady M.E., El-Boray M.S., Shabana Y.M., & Grosser J.W. 2020. Silicon nanoparticles mitigate oxidative stress of in vitro-derived banana (*Musa acuminata* 'Grand Nain') under simulated water deficit or salinity stress. *South African Journal of Botany*. 132:155–163.

Mahmoud L.M., Shalan A.M., El-Boray M.S., Vincent C.I., El-Kady M.E., Grosser J.W., & Dutt M. 2022. Application of silicon nanoparticles enhances oxidative stress tolerance in salt stressed 'Valencia' sweet orange plants. *Scientia Hortic*. 295:110856.

Ma J.F., Goto S., & Tamai K. 2001. Role of root hairs and lateral roots in silicon uptake by rice. *Plant Physiol*. 127:1773–1780

Knight C.T.G. & Kinrade, S.D. 2001. A primer on the aqueous chemistry of silicon (Chapter 4). In: Datnoff L.E., Snyder G.H., & Korndörfer G.H. (eds.). 2001. Silicon in agriculture. Elsevier, Amsterdam, The Netherlands.

Matichenkov V., Bocharnikova E., & Calvert D. 2001. Response of citrus to silicon soil amendments. *Proc. Fla. State Hort. Soc*. 114:94–97.

McKeague J.A. & Cline M.G. 1963. Silica in soil solutions: II. The adsorption of monosilicic acid by soil and by other substances. Canadian *Journal of Soil Science*. 43(1):83–96.

More S.S., Gokhale N.B., Shinde S.E., & Korake G.N. 2015. Effect of different sources of silica on nutrient content of leaves and fruit at different stages of Alphonso mango (*Mangifera indica* L.) in lateritic soil. *Journal of Progressive Agriculture*. 6(2):84–88.

Rodrigues F.A., Dallagnol L.J., Duarte H.S.S., & Datnoff L.E. 2015. Silicon control of foliar diseases in monocots and dicots. In: Rodrigues F. & Datnoff, L. (eds) Silicon and plant diseases. Springer, Cham. https://doi.org/10.1007/978-3-319-22930-0_4

Sommer M., Kaczorek D., & Kuzyakov, Y. 2006. Silicon pools and fluxes in soils and landscapes a review. *J Plant Nutr. Soil Sci*. 169:310–329.

Suriyaprabha R., Karunakaran G., Yuvakkumar R., Prabu P., Rajendran V., & Kannan, N. 2012. Growth and physiological responses of maize (*Zea mays* L.) to porous silica nanoparticles in soil. *Journal of Nanoparticle Research* .14(12):1–14.

Takahasi E., Ma J.F., & Miyake Y. 1990. The possibility of silicon as an essential element for higher plants. *Comments Agric. Food Chem*. 2:99–122.

Tubaña B.S. & Heckman J.R. 2015. Silicon in soils and plants. In: Rodrigues F. & Datnoff L. (eds.). Silicon and plant diseases. Springer. Cham. https://doi.org/10.1007/978-3-319-22930-0_2

Wutscher H.K., 1989. Growth and mineral nutrition of young orange trees grown with high levels of silicon. *HortScience*. 24:275–77.

Proc. Fla. State Hort. Soc. 136:78. 2023.

—Scientific Note—

Does Brassinosteroid Application Advance Internal Fruit Quality in Mature Citrus Trees?

Divya Aryal, Saoussen Ben Abdallah, and Fernando Alferez*

Southwest Florida Research and Education Center, University of Florida, Institute of Food and Agricultural Sciences, Immokalee, FL

Keywords. Brassinosteroids (Brs), Brix, huanglongbing, titrable acidity

Huanglongbing (HLB), the most destructive disease in citrus worldwide at the current time, has led to significant harm to the citrus industry in Florida since first detected in 2005. Main effects of the disease include exacerbated fruit drop, reduced fruit quality and yield and shortened tree productive lifespan. Fruit attributes are altered, resulting in changes in sugars and acid composition and content. Treatments to alleviate these effects are desperately needed. Brassinosteroids (Brs), a group of polyhydroxylated plant steroid hormones can improve fruit quality in blood oranges and advance maturation in strawberries (Habibi et al., 2021; Chai et al., 2013). The present study was designed to identify the ideal maturation stage for applying Brs to advance maturation in sweet oranges (*Citrus sinensis* 'Hamlin'), so fruit quality is improved, and maturation is advanced, in order to avoid preharvest fruit drop.

The application of Brs on *C. sinensis* 'Hamlin' trees in a commercial farm in southwest Florida, started on September 2022. One foliar application of Br (at a concentration of 186 mL/100 gallons) was performed every other week in a completely randomized block design in different tree rows (one single application per row). Fruit quality parameters like fruit size, color, Brix:acid ratio were determined every two weeks after each application.

Findings

There was an increase in Brix in juice from fruits treated during mid-November, early December and mid-December as compared to non-treated control fruits. Moreover, Brix:acid ratio was also higher in these fruits indicating improvement in the fruit quality. Interestingly, the Brix:acid ratio observed in non-treated control fruit during harvest (early January) was similar to the ratio in Br-treated fruits in early December. This indicates that Br treatment promotes advancement of fruit maturation by almost one month. At harvest stage, fruits treated with Brs in mid-December showed significantly higher sucrose and fructose contents as compared to non-treated control fruits. Furthermore, the peel color significantly changed to orange more in Br-treated than in non-treated control fruit changing more in the Brs treated group than controlled groups.

Suggestion for Future

Our findings highlight the potential of Brs treatment to improve the fruit quality. The study provides a foundation for further studies to understand the underlying mechanisms of the effects we observed. Treatments on different citrus varieties will be performed to assess if the observed affects can be variety specific. Furthermore, additional research is required to determine the frequency of Br application to achieve maximum benefits. Implementing Br treatments at the appropriate fruit maturation stage is a promising tool to mitigate the detrimental effects of HLB on citrus production and improve fruit quality, and will help to ensure sustainability of the citrus industry.

Literature Cited

Chai Y., Zhang Q., Tian L., Li C-L., Xing Y., Qin L., & Shen Y-Y. 2013. Brassinosteroid is involved in strawberry fruit ripening. *Plant Growth Regul.* 69:62–69.

Habibi F., Serrano M., Zacarías L., Valero D., & Guillén F. 2021. Postharvest application of 24-epibrassinolide reduces chilling injury symptoms and enhances bioactive compounds content and antioxidant activity of Blood Orange fruit. *Front. Plant Sci.* 12:629733.

*Corresponding author. Email: alferez@ufl.edu

Proc. Fla. State Hort. Soc. 136:79. 2023.

Optimise 2,4-D Applications and Investigate Alternatives for Citrus Calyx Retention

Susan Mostert[1], Paul Cronje[2], Wilma du Plooy[3], and Fernando M. Alferez[4]*

[1]*Department of Horticultural Science, University of Stellenbosch*

[2]*Citrus Research International (CRI), Department of Horticultural Science, Stellenbosch University*

[3]*Citrus Research International (CRI), Nelspruit SA*

[4]*University of Florida, IFAS, SWFREC Immokalee, FL*

—abstract—

Abscission is a naturally occurring physiological process that enables plant organ separation in response to various developmental, hormonal and/or environmental stimuli. Synthetic auxins, such as 2,4-dichlorophenoxy acetic acid (2,4-D), are incorporated into postharvest practices to impede citrus calyx deterioration. However, restrictions on 2,4-D use require alternative chemicals and practices. The aim was to assess the efficacy of alternative agro-chemicals, re-evaluate the role of 2,4-D in the packline and determine the impact of postharvest practices, such as degreening and storage, on the calyx condition. Citrus type and cultivar influenced the calyx's susceptibility to senescence, abscission, and response to plant growth regulators. For mandarin and navel, orange 3,5,6-trichloro-2-pyridyloxyacetic acid (3,5,6-TPA), amimoethoxyvinyl-glycine (AVG) and 2-(4-amino-3,5-dichloro-6-fluoropyridin-2yl) oxyacetic acid (Fluroxypyr) showed similar effects on the calyx to 2,4-D and s-ethyl-4-chloro-O-tolilooxytioacetate (MCPA) for lemon. At shorter storage periods (< 20 days), less 2,4-D can be used, while citrus stored for longer periods (> 30 days) should receive the standard 2,4-D dose. Degreening adversely affected fruit quality when not countered with a preceding 2,4-D drench. Prolonging degreening increased calyx deterioration, especially when 48 hours was exceeded, suggesting a possible ethylene tolerability threshold. Fruit storage at –0.5 °C, 2 °C and 4 °C maintained calyx condition and minimized weight loss, although 2 °C and 4 °C showed greater calyx conservation. Temperature regulation became increasingly important at longer storage durations. Macroscopic calyx evaluations showed that citrus type and cultivar had a degree of influence over calyx abscission sensitivity and progression. Reduced reliance on 2,4-D is possible if degreening could be avoided. However, when color development is inadequate and degreening is required in the early season, this auxin is recommended.

The full publication can be found at: https://doi.org/10.1016/j.postharvbio.2023.112629

*Corresponding author. Email: alferez@ufl.edu

Proc. Fla. State Hort. Soc. 136:80. 2023.

—Scientific Note—

Effect of Rootstock on the Cold Tolerance Potential of UF950 Mandarin Citrus

MUJAHID HUSSAIN AND MUHAMMAD ADNAN SHAHID*

University of Florida, IFAS, North Florida Research and Education Center, Horticulture Sciences Department, Quincy FL

Keywords. *Citrus reticulata*,UF950, freezing tolerance, rootstock, sour orange, UFR5

North Florida gets more freezes during the winters as compared to the South Florida which requires to grow freeze tolerant varieties in the North Florida. At the current time, citrus growers in North Florida, are planting Satsuma (*Citrus unshiu* 'Owari') due to its ability to tolerate freezing temperatures. However, due to the saturation of only one cultivar with a shorter harvest window, the growers are making less money. For diversification of citrus cultivars in North Florida, the trials are ongoing to test the new freezing tolerant cultivars and rootstocks. The objective of the current study was to test the best rootstock that can enhance the freezing tolerance of the Mandarin (*C. reticulata* UF950).

The one-to two-year-old *C. reticulata* UF950 plants grafted with eight different rootstocks (UFR5, Bitters, Blue1, C-146, UFR07TC, UFR09TC, US942, and Sour Orange) were kept in cold room at 18 °C for 24 hours and the temperature was gradually decreased to –6 °C for the next 48 hours. A 10-h light and 14-h dark period was maintained in the cold room. The first leaf sampling was done at 18 °C, 24 h after the start of the experiment and the second sampling was done at –6 °C. The leaf samples were analyzed for different biochemical indicators of cold stress. The Tukey–Kramer honestly significant difference test was performed using JUMP software version 2017 for the statistical analysis.

Findings

Although we used eight different rootstocks, we will only be discussing the rootstock showing best and the poorest results. At –6 °C, the amount of reactive oxygen species (ROS) like superoxide and catalase were higher in UF950 plants grafted with sour orange and were lower in UFR5. Similarly, the activities of antioxidant system were significantly different for these two rootstocks at –6 °C as the antioxidant activities in the leaves of UFR5 was highest as compared to the plants grafted with other rootstocks. The lower activities of antioxidant system can cause membrane damage which is quantified through electron leakage (EL) and the rate of lipid peroxidase (LPO) enzyme (Karami-Moalem et al., 2018). The EL and LPO activity were significantly higher in sour orange and lower in UFR5. At freezing temperatures, the osmolytes like proline and glycine betaine and the activities of nitrogen metabolism related enzymes (nitrate reductase and nitrite reductase) were lower in sour orange while the highest in UFR5. Under stress, the plant produces more hormones to cope with the stress and it is well established that the plants

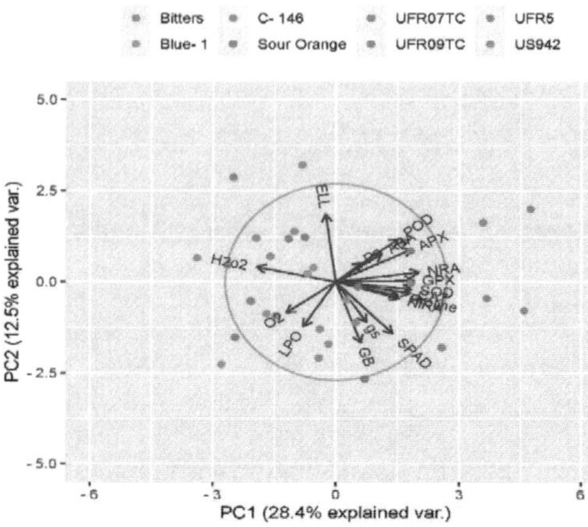

Fig. 1. Principal component analysis (PCA) based on different stress markers show that the UFR5 rootstock (supporting *Citrus reticulata* UF950 grafted onto it) is a potentially better rootstock as compared to the other rootstocks for the freezing winters of North Florida.

produce abscisic acid (ABA) under freezing stress (Shinkawa et al., 2013). Our results also showed that high amount of ABA was produced in UFR5 grafted plants and lower amount in sour orange grafted plants. Overall, the principal component analysis (PCA) also indicates (Fig. 1) that the UF950 plants grafted onto UFR5 performed well under freezing stress as compared to the plants grafted on other rootstocks. However, according to our data, the sour orange grafted plants had the lowest cold hardiness. The he results of a field trial, at the North Florida Research and Education Center, Quincy, were also consistent with this study. During the freezing event of December 2022, the UFR5 grafted plants performed better as compared to the sour orange grafted plants grown in open field.

Suggestions for Future Direction

Based on the findings of this study, further lab experiments and field trials should be conducted for the deeper studies on genomics.

Literature Cited

Karami-Moalem S., Maali-Amiri R., & Kazemi-Shahandashti S.S. 2018. Effect of cold stress on oxidative damage and mitochondrial respiratory properties in chickpea. *Plant Physiology and Biochemistry*.122. https://doi.org/10.1016/j.plaphy.2017.11.011

Shinkawa R., Morishita A., Amikura K., Machida R., Murakawa H., Kuchitsu K., & Ishikawa M. 2013. Abscisic acid induced freezing tolerance in chilling-sensitive suspension cultures and seedlings of rice. *BMC Research Notes*. 6(1). https://doi.org/10.1186/1756-0500-6-351

*Corresponding author. Email: mshahid@ufl.edu

Proc. Fla. State Hort. Soc. 136:81–82. 2023.

—Scientific Note—

The Physiological Responses of Citrus Tree Roots to Soil Acidification

Duplicate Sambani[1], Tripti Vashisth[2], and Davie M. Kadyampakeni[1]*

[1]*University of Florida, IFAS, Soil, Water and Ecosystems Sciences Department, Citrus Research and Education Center, 700 Experiment Station Rd, Lake Alfred, FL*

[2]*University of Florida, IFAS, Horticultural Sciences Department, Citrus Research and Education Center, 700 Experiment Station Rd, Lake Alfred, FL*

Keywords. apoplastic pH, *Citrus sinensis*, leachate, nutrient uptake, soil pH

The citrus greening disease, also called Huanglongbing (HLB) is the most serious threat to the Florida citrus industry and has caused a severe decline in citrus production. Some of the signs of HLB include mottled leaves, deformed fruit, aborted seeds, premature fruit drop, and root loss. The HLB causal pathogen, called *Candidatus* Liberibacter asiaticus, spreads to the roots before the disease shows symptoms in other parts of the tree canopy (Johnson et al., 2014). Therefore, there is a need to better understand the overall architecture of the citrus root system and determine how effectively a citrus plant absorbs water and nutrients from the soil. This study aims to investigate the physiological responses of citrus tree roots to soil acidification, focusing on the impact of varying soil pH on root morphology, nutrient uptake, and overall root health.

The objectives of the study were: to characterize and quantify root morphology variations; to identify the optimum pH range for citrus root functioning and; to understand the significance of apoplastic pH on plant water and nutrition uptake efficiency.

A greenhouse study was conducted at the Citrus Research and Education Center (CREC) in Lake Alfred, FL. Forty young 'Valencia' [*Citrus sinensis* (l.) Osbeck] citrus plants on Swingle (*Citrus paradisi Poncirus trifoliata*) rootstocks were planted in rhizotrons whose dimensions were 54-cm tall, 40-cm wide, and 3-cm thick, made of clear acrylic for transparency. Each rhizotron was filled with Candler fine sand on top and black aquarium stones on the bottom for proper drainage without excessive loss of sand. These trees were irrigated three days a week with four different pH-level treatments including 5.5, 6.5, 7.5, and 8.5 named Treatments 1, 2, 3, and 4, respectively (Fig. 1). A protocol to obtain irrigation water at different pH levels was prepared in the laboratory using sulfuric acid for pH levels 5.5 and 6.5, sodium hydroxide for pH level 8.5, and tap water was used for pH level 7.5. Each side of the rhizotron was irrigated with 500 mL of water for sufficient leachate collection. Soil acidity and alkalinity were routinely monitored with pH probe sticks (Extech PH100, Extech Instruments, China). Roots were observed through weekly root images to assess root condition whether live or dead (Fig. 2). Once the soil was stabilized, feeder root samples were taken for apoplastic and phloem pH determinations.

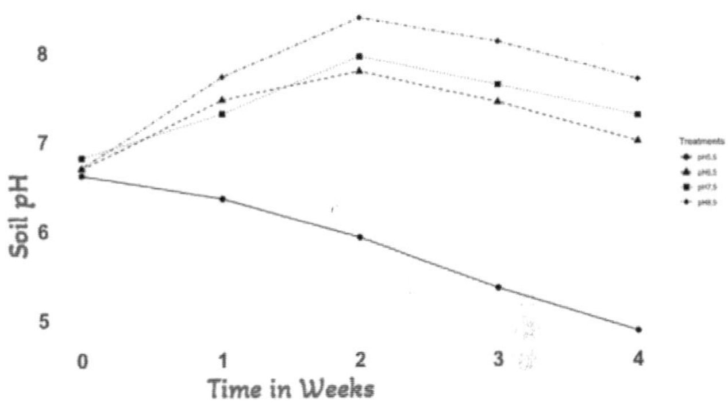

Fig. 1. Changes in soil pH over time with irrigation treatments at varying water pH levels: pH 5.5, 6.5, 7.5, and 8.5.

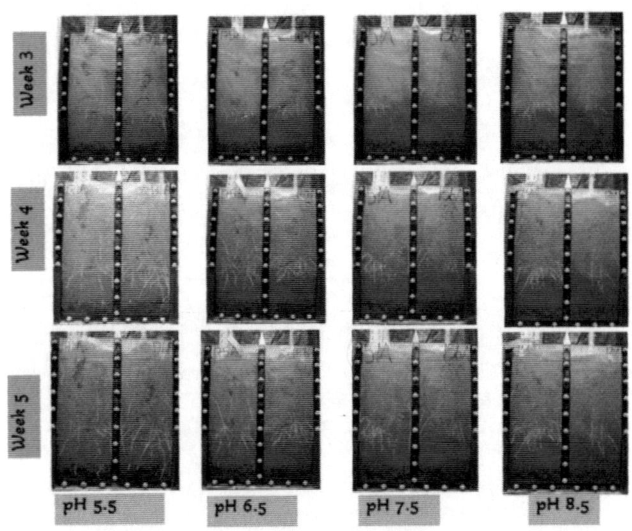

Fig. 2. Visual differences in root morphology variations of *Citrus sinensis* 'Valencia' as affected by irrigation water pH values. Photo credit: Duplicate Sambani.

*Corresponding author. Email: dkadyampakeni@ufl.edu

Early Results

We observed that irrigation water pH levels 7.5 and 8.5 resulted in comparable visual characteristics on the citrus plants; from weeks 3 to 5, showing less root growth in length and width compared to pH level 5.5, which demonstrated more significant visual root growth to the extent of almost growing out of the rhizotron. These are just notable visual differences that are inconclusive since we have yet to do statistical analysis to quantify the root morphology for the different treatments at the end of the study.

After application, soil pH for pH level 5.5 decreased drastically, and pH levels 6.5 and 7.5 constantly maintained near-neutral levels, gradually decreasing over time, possibly due to soil buffering capacity. In Fig. 3, some spotting is noticed by having a small quantity of dye within some cell boundaries in week 0. We are still evaluating why some cells have labeling, and some do not show any labels. We speculate that the roots might use an internal pH compensatory mechanism, making the plants not show any dye in week 4.

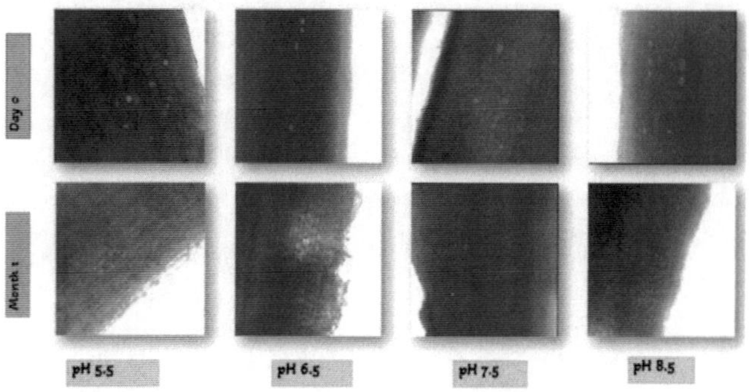

Fig. 3. Apoplastic pH changes over time on *Citrus sinensis* 'Valencia' as affected by irrigation water pH values. Photo Credit: Duplicate Sambani.

Suggestions for Future Direction

Since this is an on-going study, we expect more results by the end of the 12 weeks. The soil, water, and tissue sample results will highlight effects of soil pH on soil and plant nutrition. Even though we have no expected changes from the apoplast pH over time, we are looking to measure the fluorescence intensity by the end of the experiment and compare the apoplastic pH results obtained from this study with root samples obtained from mature HLB-affected affected trees.

Literature Cited

Johnson E.G., Wu J., Bright D.B., & Graham J.H. 2014. Association of '*Candidatus* Liberibacter asiaticus' root infection, but not phloem plugging with root loss on Huanglongbing-affected trees prior to appearance of foliar symptoms. *Plant Pathol.* 63:290–298. https://doi.org/10.1111/ppa.12109 .

Proc. Fla. State Hort. Soc. 136:83. 2023.

—Scientific Note—

Cover Crop Studies in Citrus Groves: A Renewed Interest in a Century Old Soil Management Practice

Lukas M. Hallman and Lorenzo Rossi*

University of Florida, IFAS, Indian River Research and Education Center, Fort Pierce, FL

Keywords. legumes, soil health, sweet orange

Cover crops have been successfully used for centuries to improve soil chemical, physical, and biological properties in agricultural systems throughout the world. The history of cover cropping in Florida's citrus industry dates to the late 19th century. Early citrus growers utilized cowpea, beggarweed, buckwheat, rye, oats, and rape as cover crops. H. Harold Hume, former University of Florida Provost of Agriculture and interim University President (1947), in his 1911 edition of *Citrus Fruits and Their Culture*, reported that "Clean culture cannot produce good results indefinitely, and the sooner cover crops are given attention the better for the future of the citrus industry" (Hume, 1911).

In the 1920s, the introduction of the crotalarias represented a first big gamechanger. These new cover crops were well adapted to the southern climates and offered new research opportunities to soil chemists at the University of Florida. Soon after long-term field studies on summer cover crops and citrus groves were established in Gainesville and Lake Alfred. Dr. Stokes presented his results at Florida State Horticultural Society meetings (Stokes, 1930).

During the 1940s and the 1950s citrus production expanded south, and new agricultural research centers were established at Fort Pierce and Immokalee. The search for winter cover crops began and field studies were conducted in the Indian River area, but most of the findings were inconclusive. Following the end of the World War II and the advent of synthetic fertilizers the use of cover crops greatly declined in the mid-20th century.

With the arrival and subsequent spread of Huanglongbing (HLB, citrus greening) in the early 2000s, there is a renewed interest in cover cropping for HLB-affected trees. With no cure available, growers rely on management methods that reduce disease vector populations and improve tree health. While nutritional management has been documented to improve the yield and quality of HLB-affected trees, increased fertilizers make up a significant portion of grove management costs. Implementing cover crops into an HLB-endemic system may improve soil conditions, particularly soil fertility, possibly allowing for a reduction in costly fertilizer applications while sustaining tree health.

Literature Cited

Hume H.H. 1911. Citrus fruits and their culture. Orange Judd Co. New York, NY.

Stokes W.E. 1930. Cover crops for citrus groves. *Proc. Florida State Hortic Soc.* 43:12–15.

*Corresponding author. Email: l.rossi@ufl.edu

Proc. Fla. State Hort. Soc. 136:84–86. 2023.

Florida Growers' Perceptions of Adopting Cover Crops in Citrus Production

SHOURISH CHAKRAVARTY[1,2] AND TARA WADE[1,2*]

[1]*University of Florida, IFAS, Food and Resource Economics Department, 1103 McCarty Hall B, Gainesville, FL*

[2]*University of Florida, IFAS, Southwest Florida Research and Education Center, 2685 State Rd. 29 North, Immokalee, FL*

Keywords. adoption, citrus, cover crops, perception, willing to pay (WTP)

Florida is a leading producer of citrus (*Citrus* spp.). However, the state's citrus industry has been devastated by the citrus greening disease [Huanglongbing (HLB)], and production and acreage have almost halved since 2005–06. Cover crops are non-cash crops that are used extensively in row crop production to improve soil microbial activity and organic matter, for nitrogen fixing, and for reducing soil erosion. Given that Florida has sandy soils with low organic matter, citrus growers could benefit from adopting cover crops as healthier soils could improve yields and fruit quality of citrus trees. Cover crop adoption, however, would depend on various factors including potential increases in production costs, profitability, actual tree improvement, and growers' attitudes toward the practice. In this study we develop a survey that gathers information on Florida citrus growers' perceptions and attitudes towards cover crop adoption. We collect information on demographics, production practices, conservation practices, and perceptions of cover crops benefits. We use this information to compare perceptions of cover crop benefits and costs, grove characteristics, and the production challenges for adopters and nonadopters. The survey offers insights into uncertainties related to cover crop adoption in perennial fruit production systems that could outweigh the perceived benefits of cover crops among growers. Understanding the factors that affect adoption will provide important insights which could be used by growers, researchers, and policymakers to promote and incentivize adoption of cover crops or similar conservation practices. Our study is the first of its kind to inform about the factors that affect cover crop adoption among citrus growers.

Cover crops, or non-cash crops, are used primarily in row crop agriculture. Citrus growers who adopt cover crops could benefit from improved soil health, reduced weed growth, improved soil organic matter, and improve the overall growing conditions that positively impact fruit quality and yield. Cover crop adoption, however, would depend on various factors including potential increases in production costs, profitability, actual tree improvement, and growers' attitudes towards the practice. In this study we developed a survey that gathered information on Florida citrus growers' perceptions and attitudes towards cover crop adoption. We analyzed the responses and found the factors that influence cover crop adoption. Few studies have analyzed the costs associated with cover crop adoption and the factors that affect adoption. Chakravarty and Wade (2023) found that it would cost an average of $210/acre per year for growers to adopt cover crops. While Cosgray et al. (2023) found that growers were willing to pay (WTP) $509.48/acre per year for adopting cover crops in citrus production in Florida. The primary motivation of our survey was to understand citrus growers' cognitive and socio-economic factors that determine their adoption of a conservation practice such as cover crops.

This work is partially funded by USDA-NIFA Multistate Hatch project W4133, "Costs and Benefits of Natural Resources on Public and Private Lands: Management, Economic Valuation, and Integrated Decision-Making" and the Citrus Research Development Foundation project 18-059C, "Citrus Row Middle Management to Improve Soil and Root Health."

*Corresponding author. Email: tara.wade@ufl.edu

Survey Design and Methodology

We developed a survey of citrus growers to understand their perceptions on adopting cover crops in citrus production, and to estimate their willingness to pay (WTP) for the same. Thus, our survey consisted of three parts. In the first part, we asked the growers about their demographics, types of citrus crops that they produce, their cover crop practices, enrollment in federal or state conservation programs, irrigation practices used, and other questions related to their production practices and grove characteristics. In the second part, we asked questions to understand growers' perceptions of using cover crops. These included the respondents' levels of agreement or disagreement on the benefits and drawbacks of using cover crops in citrus, their rankings of the benefits associated with cover crops, their assessment of their grove soil quality, and the number of years that they were willing to wait to see the benefits from cover crops . Finally, we used a double-bounded dichotomous choice model to measure respondents' Willingness to Pay (WTP) for cover crops. To eliminate duplication and allow for follow-up questions, we asked the respondents to share their contact information and promised confidentiality. We present the results from the first and the second parts of our survey.

We developed the survey for citrus growers from California, Florida, and Texas, and respondents were recruited through various channels beginning in early 2021: Qualtrics, University of Florida Citrus Extension Agents' email listservs, the Indian River Citrus League email listserv, and the California Citrus

Table 1. Summary statistics of key variables in a survey on cover crops of Florida citrus growers.

Statistic	Own no till drill (%)	Citrus acres	Cover crop awareness (%)	Years citrus experience	Crop types (no.)	Rented land (%)	Whether organic grown (%)
Average	0.36	3034.20	0.80	22.71	5.66	15.81	0.27
Maximum	1	33000	1	50	18	100	1
Minimum	0	1	0	1	1	0	0

Research Board e-newsletter. We also called groves and nurseries and conducted in-person interviews at the 2021 Citrus Show in Fort Pierce, FL. Our data collection started in early 2021 and ended on November 2, 2021. The response rate was 35%, and we received responses from 359 growers from Florida, Texas, California, and Georgia, out of which 173 identified themselves as citrus growers. Since the samples from California and Georgia were not representative of commercial citrus producers, we excluded them from our study. After removing inconsistent and incomplete observations, we had 79 observations with 59 from Florida and 20 from Texas. For this analysis, we present results for growers from Florida with cumulative grove acreage of 179,018.5 acres. This represents approximately 48% of the total of the total orange acreage in 2021. The average orange acreage for a respondent was 3,034.2 acres (Table 1) and the median acreage was 80 acres.

Results and Discussion

In our sample, 80% of the growers are aware of cover crops (Table 1), 27% grow organic citrus, and on average a citrus grower who responded had 22.7 years of experience of growing citrus crops. In terms of the use of cover crops, 41% of the respondents used cover crops in their groves (Table 2). Among those that also grew organic citrus, 75% used cover crops, while only 28% of those who did not grow organic citrus used cover crops in their groves. Respondents were enrolled in various federal and state conservation programs. The two most popular programs are Best Management Practices cost share through Florida Department of Agriculture & Consumer Services (FDACS-BMP) with 44% of respondents enrolled, and the Environmental Quality Incentives Program (EQIP) federal program with 27% enrolled (Table 3).

Growers were asked about the soil health of their groves (Table 4). Many (47%) reported soil health was "Fair", 1 out of 16 organic growers reported "Fair" soil health, while 63% of the non-organic citrus growers reported the same. A greater proportion of organic growers reported "Good" or "Great" soil health than non-organic growers—94% versus 23%. The most common technology for irrigation was sprinkler, followed by drip/localized irrigation (Fig. 1). Growers who had surface or flood irrigation in their groves were most likely to report "Good" or "Great" soil health. Organic growers were more likely to use surface or flood irrigation with 50% reporting the same, while

Table 2. Florida citrus growers responses by cover crop use and status of organic citrus production.

	Organic citrus		
Cover crop use	Yes	No	Total
Yes	12	12	24
No	4	31	35
Total	16	43	59

Table 3. Florida citrus growers responses by cover crop use and enrollment in federal and state conservation programs.

	EQIP		FDACS-BMP		
Cover crop use	Yes	No	Yes	No	Total
Yes	8	16	9	15	24
No	8	27	17	18	35
Total	16	43	26	33	59

Table 4. Florida citrus growers responses by organic production status and self-reported soil quality.

Soil health	Organic grove	Not organic grove	Total
Poor	0	6	6
Fair	1	27	28
Good	9	9	18
Great	6	1	7
Total	16	43	59

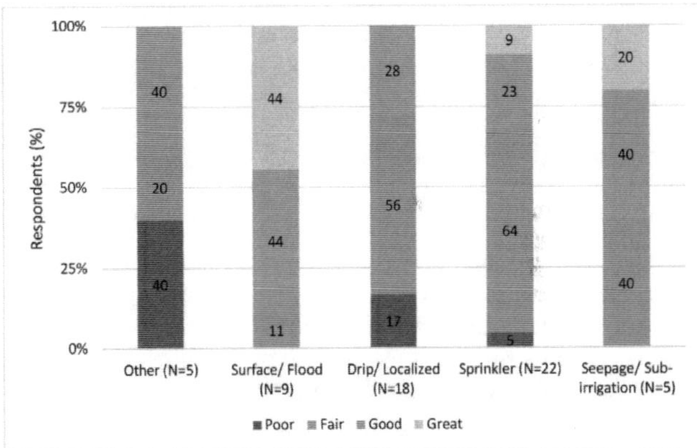

Fig. 1. Irrigation type used by Florida citrus growers by grove soil quality (in percentages).

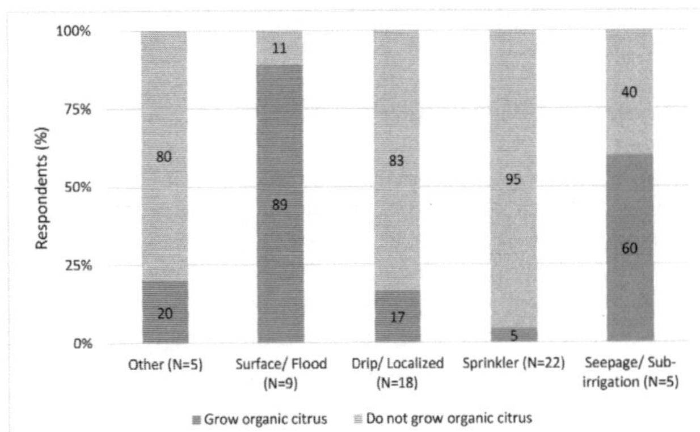

Fig. 2. Irrigation type used by Florida citrus growers respondents by organic production status (in percentages).

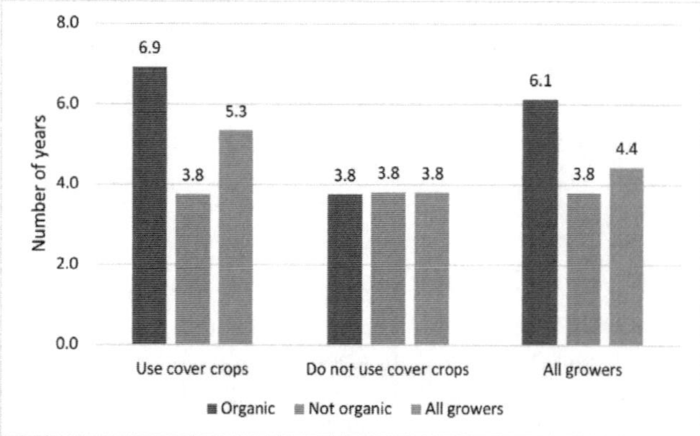

Fig. 3. Average number of years Florida citrus growers/respondents are willing to wait to see changes in soil health from cover crops—by organic production status and cover crop adoption status.

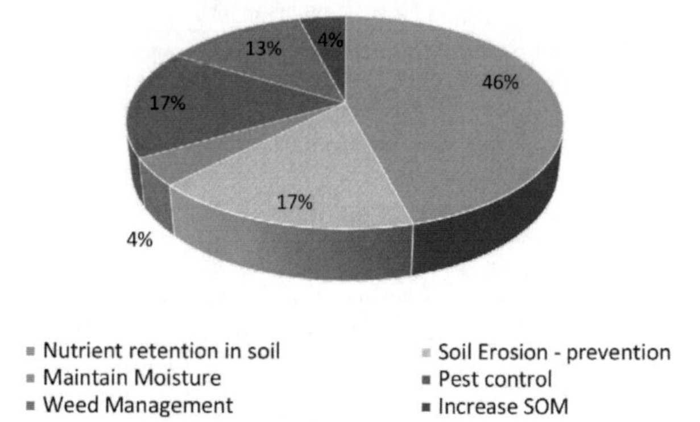

Fig. 5. Perceived number 1 ranked benefit of cover crops by Florida citrus growers who adopted cover crops (in percentage).

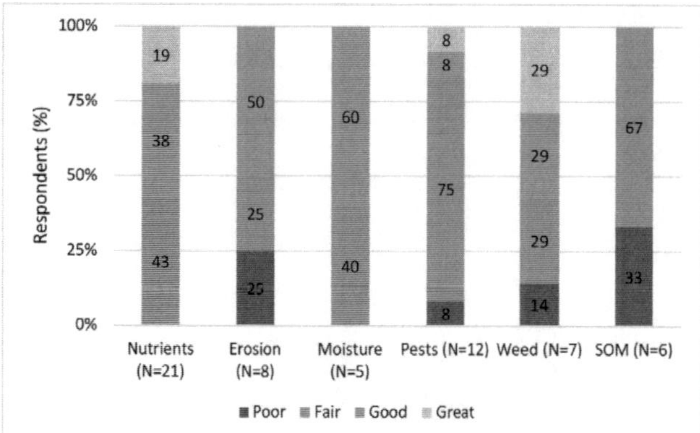

Fig. 4. Florida citrus growers respondents' number 1 reported benefit of cover crops and their self-reported soil quality (in percentages).

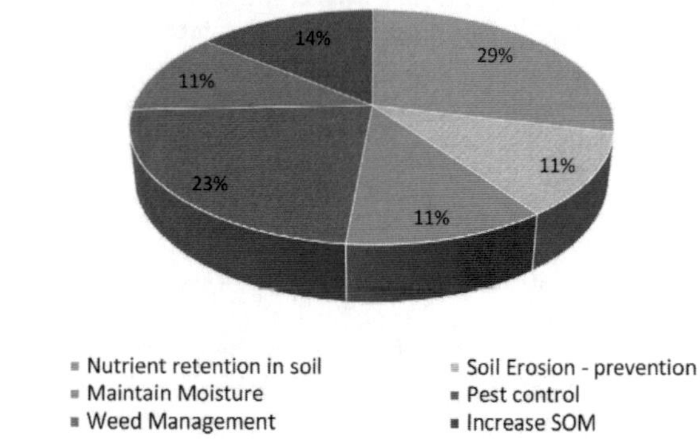

Fig. 6. Perceived number 1 ranked benefit of cover crops by Florida citrus growers who did not adopt cover crops (in percentage)..

non-organic growers mostly used sprinkler and drip or localized irrigation with 35% and 49% using the two respective technologies. Eighty nine percent of the surface or flood irrigation users were organic growers while 95% and 83% or the sprinkler and drip/localized irrigation users, respectively, were non-organic growers (Fig. 2). Organic growers who were using cover crops reported an average wait time of 6.9 years to see changes in soil health from cover crops, while all growers on average reported 4.4 years (Fig. 3).

In terms of the perceived benefits of using cover crops, more respondents ranked nutrient benefits as the number 1 benefit of cover crops compared to other benefits (Fig. 4). Moreover, among these respondents 57% reported "Good" or "Great" soil health. Among those who reported pest control as the number 1 benefit, 75% reported that their soil health was 'Fair'. Most of the cover crop adopters reported nutrient retention as the number 1 benefit of cover crops. The second most number 1 ranked benefits were soil erosion prevention and pest control (Fig. 5). Among non-adopters of cover crops, majority ranked nutrient retention as the number 1 benefit and the second most number 1 ranked benefit was pest control (Fig. 6).

Conclusion

We found that organic growers were more likely to adopt cover crops than non-organic citrus growers. Moreover, a greater proportion of growers growing organic crops reported "Good" or "Great" soil health than those not growing organic citrus. In terms of growers' perceptions, we find that 36% of the growers ranked nutrient benefits as the number 1 benefit from adopting cover crops, while 20% reported pest control as the number 1 benefit. However, these responses depend on the soil health of the respondents' groves. We found that factors affecting cover crop adoption are diverse and depend on growers' perceptions of the benefits of cover crops in addition to the physical characteristics of the grove and weather.

Literature Cited

Chakravarty S. & Wade T. 2023. Cost analysis of using cover crops in citrus production, *HortTechnology* 33(3):278–285. https://doi.org/10.21273/HORTTECH05126-22

Cosgray C., Chakravarty S., Wade T., & Gao Z. 2023. Citrus growers' willingness to pay and perceptions of cover crops. *Agricultural and Resource Economics Review*. 1–22. https://doi.org/10.1017/age.2023.15

Proc. Fla. State Hort. Soc. 136:87. 2023.

Impact of Cover Cropping on Temporal Nutrient Distribution and Availability in the Soil

Miurel Brewer[1], Sarah Strauss[2], Ramdas Kanissery[2], and Davie Kadyampakeni[1]*

[1]University of Florida, IFAS, Citrus Research and Education Center, Lake Alfred, FL

[2] University of Florida, IFAS, Southwest Florida Research and Education Center, Immokalee, FL

—abstract—

Cover cropping is a best management practice that can improve soil quality by reducing soil erosion, building soil organic matter (SOM), and improving soil nutrient availability. Southwest (SW) Florida citrus growers have the challenge of growing citrus in sandy soils characterized by low organic matter (< 2%), extremely low water and nutrient-holding capacities, and high sand content (> 90%), and therefore are looking for methods to improve SOM and nutrient retention and availability in sandy soils. A trial of two cover crop (CC) mixtures planted in the row middles (RM) of Huanglongbing-affected citrus 'Valencia' [*Citrus sinensis* (L.) Osbeck] orchards in sandy soils in SW Florida was conducted. This study explored how incorporating CCs in the RM of the orchards could affect soil ammonium (NH_4^+), soil nitrate (NO_3^-), exchangeable macronutrients, and SOM temporal availability. These parameters were measured under the tree canopy (UC) and within RM of two orchards: South Grove (SG) and North Grove (NG), both located in SW Florida. The two seeded CC mixtures were legume + non-legume (LG+NL) and non-legume (NL) and were compared to a control no-CC grower standard (GSC). Phosphorus, calcium, magnesium, and NH_4^+ were not statistically significantly different among treatments in either of the two sampling positions (UC and RM). Cover cropping significantly ($P < 0.05$) increased NO_3^--N concentrations in the RM area of the citrus orchards after seven consecutive seasons (brassicas, legumes, and grasses) by 31% in the LG + NL and 29% in the NL with reference to the GSC. In addition to the significant increase in NO_3^-N, SOM significantly ($P < 0.05$) increased in the RM in the NG site only in both CCs treatments by 17% and 16% for LG + NL and NL treatments, respectively, compared with GSC.

The full paper published on 22 October 2023 in *Horticulturae*.
https://doi.org/10.3390/horticulturae9101160

*Corresponding author. Email: dkadyampakeni@ufl.edu

Proc. Fla. State Hort. Soc. 136:88. 2023.

—Scientific Note—

Oak Mulch to Increase Citrus Groves Soil Health: Conclusions from a Three-year Study

LUKAS M. HALLMAN[1], JOHN M. SANTIAGO[1], JOHN-PAUL FOX[1], ROBERT G. SHATTERS, JR.[2], AND LORENZO ROSSI[1]*

[1]*University of Florida, IFAS, Indian River Research and Education Center, Fort Pierce, FL*

[2]*Horticultural Research Laboratory, U.S. Department of Agriculture, Agricultural Research Services, Fort Pierce, FL*

Keywords. *Citrus sinensis*, huanglongbing, soil amendment

Florida's citrus industry has experienced a massive decline since the arrival of Huanglongbing (HLB). This disease is caused by the phloem limited bacterium *Candidatus* Liberibacter asiaticus (*C*Las) and is vectored by the Asian citrus psyllid, *Diaphorina citri*. With nearly all commercial citrus groves affected and no cure on the horizon, growers and researchers have been searching for alternative strategies to manage diseased trees. After reports showed reduced HLB symptoms in citrus trees growing in close proximity to oak trees, many growers began incorporating oak mulch into their production systems.

No studies have investigated the impact of oak mulch on HLB-affected citrus trees or the impact of oak mulch on soil characteristics. Therefore, this study was created to i) study the capability of oak mulch to contain and suppress *C*Las and ii) measure the effect of oak mulch on soil characteristics and nutrient uptake in HLB-affected trees.

Twenty-four field grown 'Valencia' sweet orange (*Citrus sinensis*) trees were assigned to a completely randomized design replicated 3 times. A 3-inch (7.6 cm) thickness layer of oak mulch was applied every September for a three-year period and control trees did not have mulch. Soil, root, and leaf nutrient analysis, leaf bacteria titer, and soil moisture levels were measured monthly.

Data showed higher levels of phosphorus (P), potassium (K), and magnesium (Mg) in oak mulch treated soils in certain seasons. No significant differences were observed in leaf bacteria titer and leaf nutrient levels. Differences were observed in soil moisture levels in certain months between the control and mulch treatments.

The results of this study show that applying oak mulch may not suppress *C*Las but can lead to higher nutrient levels in the soil. Higher soil nutrient levels may increase the productivity of HLB-affected citrus trees and reduce costly fertilizer inputs for growers. Longer studies are needed to better understand the dynamics of nutrients in the soils and the ability of trees to uptake them.

*Corresponding author. Email: l.rossi@ufl.edu

Proc. Fla. State Hort. Soc. 136:89. 2023.

—Scientific Note—

Effect of Nitrogen and Phosphorus Rates on the Citrus Leaf, Soil Nutrient Concentrations, Fruit Yield, and Postharvest Quality of HLB-affected Sweet Orange

Alisheikh Atta[1]*, Davie M. Kadyampakeni[1], Samuel Kwakye[1], and Kelly Morgan[2]

[1]*University of Florida, IFAS, Soil, Water and Ecosystem Sciences Department, Citrus Research and Education Center, 700 Experiment Station Road, Lake Alfred, FL*

[2]*University of Florida, IFAS, Soil, Water and Ecosystem Sciences, 2181 McCarty Hall, Gainesville, FL*

Keywords. best management practices (BMPs), *Candidatus* Liberibacter asiaticus, *Citrus sinensis*, huanglongbing, leaf nutrient concentrations

Water and nutrient movement are challenging on Florida sandy soils linked with severely impacted huanglongbing (HLB)-affected citrus tree root systems. Nutrient and irrigation management based on crop demand has been the emphasis of the study for benign ecosystems recently. The objective of this study is to investigate the effect of different fertilization rates of nitrogen (N) and phosphorus (P) on soil, leaf nutrient concentrations, fruit yield, and juice quality of HLB-affected citrus trees per best management practice (BMP) modalities in Florida. N (100, 150, 200, 250, and 300 80 lb/acre) and P (0, 10, 20, 40, and 80 lb/acre) rates will be applied on 'Valencia' and 'Hamlin' *Citrus sinensis* trees in three splits each year following the tree leaf flushes. Soil samples will be collected at depths of 0–6, 6–12, and 12–24 inches every six months. Canopy volume, trunk diameter, leaf area index, chlorophyll content, and leaf N and P concentrations will be collected in response to the fertilization rate. Fruit yield and postharvest fruit quality will also be collected and evaluated each year. Baseline data of soil, leaf samples, canopy volume, trunk diameter, fruit yield, and postharvest fruit quality parameters have been collected. At the end of the study, targeted rates of N and P as BMPs will be produced, a benchmark to recommend for grower use, and improved Florida citrus production.

Water and nutrient dynamics are challenging on Florida sandy soils associated with severely impacted huanglongbing (HLB)-affected citrus tree root systems. Nutrient and irrigation management based on crop demand have been the emphasis of the study for benign ecosystems recently (Obreza et al., 2020). This study aims to investigate the effect of different fertilization rates of nitrogen (N) and phosphorus (P) on soil and leaf nutrient concentrations, fruit yield, and juice quality of HLB-affected citrus trees to revise the BMPs for Florida citrus (Fig. 1).

Based on the preliminary research findings, N, and calcium (Ca) were deficient in the three study sites in Florida—Clewiston, Arcadia, and Lake Wales—whose soil orders fall under the Histosols, Spodosols, and Entisols, respectively. These soil orders commonly share sandy soil characteristics with low water and nutrient-holding capacities. According to "Liebig's law of minimum", the yield attainable is dictated by the most limiting nutrient. Not only N but also Ca, manganese (Mn), and zinc (Zn) were deficient in the Clewiston and Arcadia sites. Meanwhile, iron (Fe) was also deficient in the Lake Wales and Clewiston sites (data not shown). Potassium (K) was also deficient in Lake Wales and Clewiston sites. Results of canopy volume and trunk cross-sectional area of 'Hamlin' and 'Valencia' citrus trees did not show variation among each of the study plots.

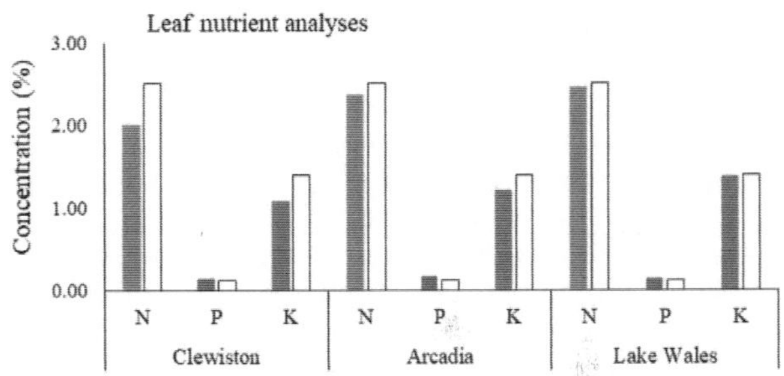

Fig. 1. Leaf nutrient concentrations in *Citrus sinensis* as compared with the optimum nutrient ranges (open bars) at three Florida study sites.

Monitoring tree growth and leaf and nutrient concentrations at each study site and variations between cultivars, if any, that may arise in the future could be linked because of treatment effects. The different N and P rates applied to the plots could also be used to identify the best N and P rates recommended for citrus growers.

Literature Cited

Obreza T.A., Boman B.J., Zekri M., & Futch S.H. 2020. Methods of fertilizer application. Chapter 7. In: Morgan K.T. and Kadyampakeni D.M. (eds.). Nutrition of Florida citrus trees, 3rd Ed. Extension publication SL 461, University of Florida, Gainesville, FL.

*Corresponding author. Email: aatta@ufl.edu

Proc. Fla. State Hort. Soc. 136:90. 2023.

—Scientific Note—

Plant Sap Analysis as a Tool to Optimize Fertilizer Application for Sustainable Citrus Production

Guilherme Locatelli, John-Paul Fox, Neus Alcon Bou, John M. Santiago, Ashley E. Boyle, Lukas M. Hallman, and Lorenzo Rossi*

University of Florida, IFAS, Indian River Research and Education Center, Fort Pierce, FL

Keywords. *Citrus* ×*paradisi*, Huanglongbing, nutritional status

Over the past decade, Florida's citrus industry has been devastated by a disease named citrus greening (HLB; Huanglongbing). Although no cure is available, growers have found some success in managing the disease, and research has shown increased nutrient applications have extended the producing life of affected trees. Using leaf nutrient analysis to determine the nutritional requirements of the plant has become a standard practice for citrus growers in the United States. However, it is difficult to detect nutritional deficiencies in the early phases; this is because traditional leaf nutrient analysis may show the nutrient accumulation from the entire season, reducing the overall precision of fertilizer management. One popular management tool gaining attention is the use of sap analysis. This approach can increase the sustainability of citrus operations by improving the precision in tree nutrient status determination and fertilizer application. For this reason, sap analysis has been considered by many authors as real-time analysis, where indicates the concentration of nutrients that is available for the plants during the season. This method analyzes the nutrient concentration of the fluids from the xylem and phloem in the leaf (sap).

In this study, the treatments tested were one citrus variety 'Ray Ruby' grapefruit (*Citrus* ×*paradisi*) and two analysis methods (standard plant leaf tissue analysis, and plant sap analysis) arranged in a complete randomized design with four replications. Leaf samples were taken during the most import phenological stages (i.e., flowering, fruit set, fruit fill, and pre/early harvest) between the hours of 8:00 am to 11:00 am, including both new and old growth. Similarly, standard soil and plant leaf tissue analysis has been conducted. The N, P, K, Ca, Mg, S, B, Cu, Fe, Mn, and Zn results were compared to the standard leaf nutrient concentration ranges currently used for citrus production in the area.

The final aim of the study is to consolidate the sampling strategy for citrus in Florida (time of the day, leaf age, leaf part, and number of leaves), sap testing (equipment, sample storage, and reading time frame), testing options (commercial minilabs for in-house testing and outsource analysis in commercial lab), and establish the initial guidelines for N, P, K, Ca, Mg. S, B, Cu, Fe, Mn, and Zn ranges in HLB-affected grapefruit trees.

The data collected showed great variability when compared and contrasted among different sap analysis and traditional laboratories, making their interpretation quite complicated.

An additional second year of data collection is planned for the 2023–24 season.

*Corresponding author. Email: l.rossi@ufl.edu

Proc. Fla. State Hort. Soc. 136:91. 2023.

Combining Individual Protective Covers (IPCs) and Brassinosteroids to Protect Young Citrus Trees Against *Diaphorina citri* and *Candidatus* Liberibacter asiaticus

Saoussen Ben Abdallah, Ute Albrecht, Ozgur Batuman, Jawwad Qureshi, and Fernando Alferez*

Southwest Florida Research and Education Center, University of Florida, IFAS, Immokalee, FL

—abstract—

The Asian citrus psyllid, *Diaphorina citri* Kuwayama (Hemiptera: Liviidae), is an invasive citrus pest that vectors the phloem-dwelling bacterium, *Candidatus* Liberibacter asiaticus (*C*Las), which is a putative causal agent of citrus greening disease known as Huanglongbing (HLB). In Florida, the *D. citri*–*C*Las pathosystem complex is at present considered the greatest threat the citrus industry must manage. Brassinosteroids (Brs), a relatively new class of plant hormones, are approved for commercial use in Florida citrus and are believed to delay HLB progression. Individual protective covers (IPCs), a type of psyllid exclusion mesh bags, are increasingly being adopted to efficiently protect newly planted citrus trees from HLB infection. However, IPCs typically must be removed after 2–3 years due to tree growth. With the aim of prolonging tree health after IPC removal, we are investigating the efficacy Brs in protecting and/or preventing trees from ACP and *C*Las infection once they are left exposed. Upon IPC removal from trees covered for 3 years, monthly Br treatments started in trees from three mandarin cultivars (Early Pride, SugarBelle and Tango) on two different rootstocks, sour orange and US-942. The IPCs effectively excluded psyllids, prevented *C*Las infection, and clearly improved most of the horticultural traits measured. Although this is an on going experiment, we have already found that treatment with Brs reduced *C*Las bacterial titer by 10-fold on average two months after IPC removal. The long-term efficacy of the treatments still needs to be assessed and will be discussed.

*Corresponding author. Email: alferez@ufl.edu

Proc. Fla. State Hort. Soc. 136:92. 2023.

Efficacy of Gibberellic Acid Treatment on HLB-affected 'Hamlin' Sweet Orange

Taylor Livingston and Tripti Vashisth*

Citrus Research and Education Center, University of Florida, IFAS, Lake Alfred, FL

—abstract—

Huanglongbing (HLB) is a severely destructive disease of citrus. Gibberellic acid (GA), a plant growth regulator, has garnered a lot of attention for the possibility it may improve the health of HLB trees as past research has shown that GA has potential in mitigating some symptoms such as preharvest fruit drop and increasing canopy density by promoting vegetative growth in 'Valencia' sweet orange. Gibberellic acid also inhibits flowering, which could prove beneficial in saving resources for increased productivity as flowering is a resource-costly event for a stressed HLB-affected tree. However, its effect on Hamlin is unknown. Grower collaborated field trials evaluating the use of GA on HLB-affected 'Hamlin' were initiated at two locations in August of 2021 and repeated in 2022. Gibberellic acid was applied 3 times from August to November at 45-day intervals. Yield results from year 1 showed that GA-treated trees retained on average 15.6 kg more fruit than control trees. In year 2, after Hurricane Ian (category 3) devastated Hardee County, the GA trees still retained on average 22.7 kg more fruit than the control. In year 1, fruit drop was significantly reduced by GA treatments at both sites by 11.1% in Hardee County and 4.7% in Polk County. Significant improvements in canopy density of the GA-treated trees are also evident at both sites. Results from this trial suggest that use of GA on 'Hamlin' can improve canopy health and reduce pre-harvest fruit drop caused by HLB and inclement weather.

*Corresponding author. Email: tvashisth@ufl.edu

Proc. Fla. State Hort. Soc. 136:93–94. 2023.

—Scientific Note—

Exploring Circadian Rhythms in Huanglongbing-affected Sweet Orange

Shalini Pareek, Daniel L. Stanton, and Tripti Vashisth*

University of Florida, IFAS, Citrus Research and Education Center, Lake Alfred, FL

Keywords. circadian rhythms, citrus, *Citrus sinensis*, huanglongbing, 'Valencia'

To anticipate and coordinate with diurnal environmental changes living organisms have an internal time-keeping mechanism called circadian clock/rhythms (~24 hours). This mechanism consists of three interconnected feedback loops, viz., the morning, central, and evening loops. The central loop consists of two morning expressed MYB-related transcription factors namely, CIRCADIAN CLOCK ASSOCIATED 1 (CCA1)/ LATE ELONGATED HYPOCOTYL (LHY) and another evening expressed TIMING OF CAB EXPRESSION (TOC1). In plants, several physiological processes like morphogenesis, flowering, photosynthesis, respiration, senescence, hormone signaling, and defense responses are regulated by circadian rhythms. Biotic and abiotic stresses can disturb plant circadian rhythms resulting in shorter or longer period length, phase advancement or delay, reduced amplitude, or cause arrhythmicity (Pruneda-Paz & Kay, 2010).

Huanglongbing (HLB), a citrus disease, presumably caused by *Candidatus* Liberibacter asiaticus (*C*Las), share similarities with symptoms of affected disturbed circadian rhythms in *Arabidopsis*, maize, wheat, poplar (Graf et al., 2010; Ko et al., 2016; Edwards et al., 2018; Gong et al., 2022). These symptoms are reduction in chlorophyll content, plant growth, fruit/grain size, lateral root density; disturbance in starch metabolism and stomatal aperture regulation; increasing leaf drop and reactive oxygen species, etc. Bacterial infection (*Pseudomonas syringe*) systemically lengthens the period, and reduces the amplitude of circadian genes in *Arabidopsis thaliana*, and cause disease symptoms (Li et al., 2018). There are eight RNA-seq studies showing differential regulation of circadian clock genes in HLB affected citrus.

Therefore, the aim of the present study was to investigate whether the HLB symptoms is the direct result of *C*Las infection or that *C*Las infection disturbs the circadian clock in citrus which show disease symptoms. This study will help to gain insight into the circadian clock of citrus and the changes in circadian rhythms because of HLB disease. The study will open avenues for future studies aiming at small adjustments in clock for increasing plant performance after pathogen infection.

For this study, leaves were collected from greenhouse growing three healthy (HLY) and three HLB affected sweet orange *Citrus sinensis* 'Valencia' trees at the Spring Equinox time (12 hours light/12 hours dark) at a time gap of 6 hours for 48 hours. Analysis of *CCA1* gene expression at 9 time points showed a continuous rhythmic pattern of 12 h light/12 h dark. Its expression peaked at 7:30 AM and was lowest at 7:30 PM. A significant difference in *CCA1* gene expression was observed between 7:30 AM and all other time points ($P < 0.001$) (Fig. 1). HLB affected *C. sinensis* plants showed significant downregulation in *CCA1* gene expression ($P < 0.05$) compared to HLY plants at time point 1.30 AM, but no phase change was observed (Fig. 2).

Future work will include the study of other circadian clock genes patterns in HLB-affected sweet oranges. The connection

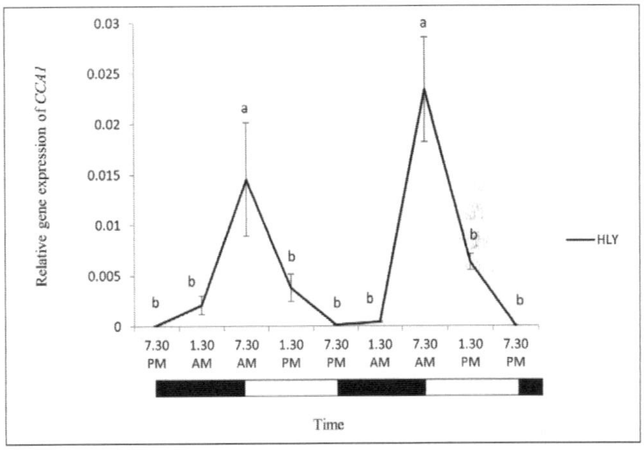

Fig 1. Circadian pattern of *CCA1* gene expression in HLY sweet orange (*Citrus sinensis*) leaves not infested with Huanglongbing (HLB), a citrus disease.

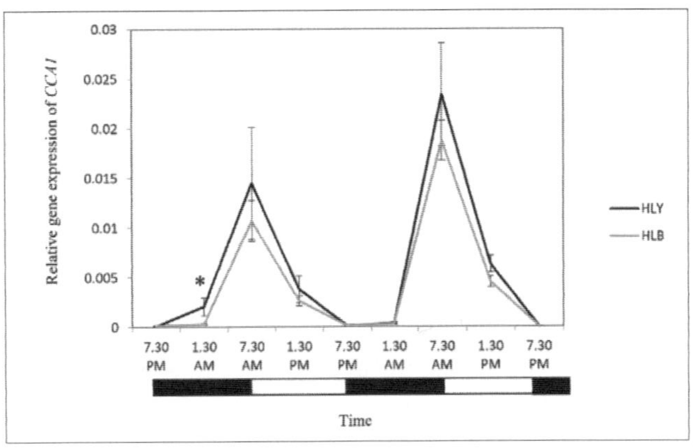

Fig 2. Comparative circadian pattern of *CCA1* gene expression in HLY (not infested with Huanglongbing (HLB), a citrus disease), and HLB affected sweet orange (*Citrus sinensis*) leaves.

*Corresponding author. Email: tvashisth@ufl.edu

between disrupted regulation of circadian genes in HLB and carbohydrate metabolism in citrus will also be investigated.

Literature Cited

Edwards K.D., Takata N., & Johansson M. 2018. Circadian clock components control daily growth activities by modulating cytokinin levels and cell division-associated gene expression in *Populus* trees. *Plant Cell Environ*. 41(6):1468–1482.

Gong J., Tang Y., & Liu Y. 2022. The central circadian clock protein TaCCA1 regulates seedling growth and spike development in wheat (*Triticum aestivum* L.). *Front. Plant Sci*. 13:946213.

Graf A., Schlereth A., Stitt M. & Smith A.M. 2010. Circadian control of carbohydrate availability for growth in *Arabidopsis* plants at night. *Proc. Natl. Acad. Sci. USA*. 107(20):9458–63.

Ko D.K., Rohozinski D., & Chen Z.J. 2016. Temporal shift of circadian-mediated gene expression and carbon fixation contributes to biomass heterosis in maize hybrids. *PLoS Genet*. 12(7):e1006197.

Li Z., Bonaldi K., Uribe F., & Pruneda-Paz J.L. 2018. A localized *Pseudomonas syringae* infection triggers systemic clock responses in *Arabidopsis*. *Curr. Biol*. 28(4):630–639.e4

Pruneda-Paz J.L. & Kay S.A. 2010. An expanding universe of circadian networks in higher plants. *Trends Plant Sci*. 15(5):259–65.

Proc. Fla. State Hort. Soc. 136:95–96. 2023.

—Scientific Note—

Biochemical Profiling of HLB-symptomatic Sweet Orange Fruit

Meryam Manzoor*, Mary Sutton, Taylor Livingston, and Tripti Vashisth*

University of Florida, IFAS, Citrus Research and Education Center, Lake Alfred, FL

Keywords. *Citrus sinensis,* citrus greening, fruit development, fruit quality, huanglongbing, lopsided fruit

Citrus fruits are highly valued for their sensory and nutritional properties, making them an important crop. However, the Florida citrus industry has suffered significant economic losses due to the citrus disease, Huanglongbing (HLB), which affects both the citrus trees' health and the fruit's development and quality. HLB-affected trees yield small, poor quality, misshapen, and lopsided fruit (the central core may be curved) with aborted seeds. HLB-affected fruit have poor juice quality and taste and the trees suffer more preharvest mature fruit drop, a major industry concern. Although previous research has compared asymptomatic and symptomatic fruit, no studies have investigated the cause of developmental defects in HLB-symptomatic fruit, making it an important gap in our understanding. Therefore, this study aimed to understand the mechanism leading to the development of lopsided fruit in HLB-affected sweet orange *Citrus sinensis* Osbeck 'Valencia'. These findings could lead to the development of new strategies to mitigate this kind of developmental defect for improving citrus fruit quality and benefiting the citrus industry.

This study collected visually asymptomatic and symptomatic mature fruit from HLB-affected sweet orange trees grown in field conditions. The asymptomatic (AS) fruit were divided into equal halves, while the symptomatic-lopsided fruit were further divided into two sub-sections: small side (S-S) and large side (S-L) of symptomatic-lopsided fruit (Fig. 1). Fruit samples were analysed and compared for fruit growth characteristics (weight, diameter, size, peel thickness ratio, and fruit central core curve), total soluble solids content, and acidity. The enzymatic profile of these fruits' peels was also analysed, including superoxide dismutase, peroxidase dismutase, catalase, total soluble protein, and proline content.

Results showed that symptomatic fruit were smaller (less weight, diameter, and size), and more acidic than asymptomatic fruit. The small side of symptomatic fruit showed slow growth and developmental pattern with poor juice quality than the large side within the same fruit (Fig. 2 A, B). Correlation analysis indicated that large sized fruit had less lopsidedness, less peel thickness, and more total soluble solids (data is not shown). Regarding the enzymatic profile of fruits' peels, more peroxidase and catalase enzymatic activity was found on the small side of the symptomatic fruit (Fig. 2 C, D).

Our preliminary findings reveal distinct differences in growth and development patterns within symptomatic fruit, including variations in fruit weight, diameter, size, and shape. Notably, we observed a higher incidence of developmental defects and

Fig. 1. Physical attributes and internal growth morphology of HLB-affected (asymptomatic and symptomatic) mature *Citrus sinensis* Osbeck 'Valencia' fruit.

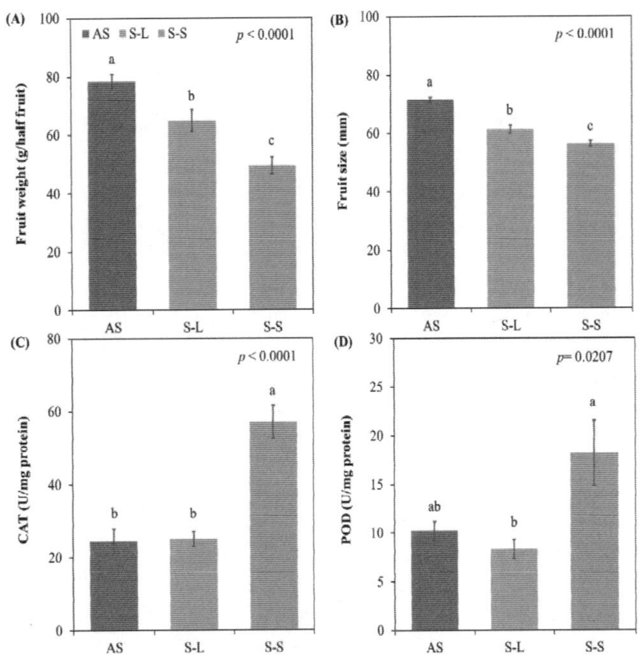

Fig. 2. Average fruit weight (**A**), fruit size (**B**), catalase (**C**), and peroxidase (**D**) activity of asymptomatic (AS), symptomatic large-side (S-L), and symptomatic small-side (S-S) of HLB-affected mature sweet orange (*Citrus sinensis* Osbeck 'Valencia') fruit. Different lowercase letters in each graph indicate significant differences at $P < 0.05$ among samples

*Corresponding authors. Email: lmanzoor@ufl.edu; tvashisth@ufl.edu

indicators of oxidative stress on the smaller side of the symptomatic fruit. These findings strongly suggest the involvement of a stress-induced mechanism in the development of symptomatic-lopsided fruit.

Using the finding of this study as a guide, the profiling of other enzymatic activities and differentially regulated fruit growth and quality attributes during the fruit growth and development phase is needed. A comprehensive new study focusing on the growth and development pattern of HLB-affected symptomatic fruit has been initiated. Fruit will be collected at monthly intervals (June to March) for their growth and quality attributes, and other biochemical and enzymatic profiles. So far, the result of the June study showed that early in June, symptomatic lopsided fruit were present on the tree and the stress induction and ceased growth of peel is at its early stages, and with the passage of time, more enzymes could be activated.

Proc. Fla. State Hort. Soc. 136:97. 2023.

—Scientific Note—

Cold Hardy Citrus in North Florida: Future Prospects and Challenges

Muhammad A. Shahid*

University of Florida, IFAS, Horticultural Sciences Department, North Florida Research and Education Center, Quincy, FL

Keywords. *Citrus unshiu*, freezing tolerance, rootstock, Satsuma, sour orange, UFR5, UF950

Cultivation of cold hardy citrus (CHC) in north Florida has a long history. There were about 23,000 acres of citrus in 1920s in north Florida, but this industry was eliminated by climatic factors such as hurricanes and freezes, and lack of labor because of World War II. The weather in north Florida has been characterized by hot summers and mild winters for the last three decades resulting in increased interest in cold-hardy citrus production for the fresh market. Satsumas are the most grown citrus for fresh markets in north Florida. Since CHC is a newly emerging industry, most of the citrus groves are young (5–8 years), and growers coming to citrus production have limited fruit growing, particularly citrus growing background. Currently, growers are depending on the information on nutrient and water management, disease, and pest control, weed management, pre- and postharvest plant care etc., developed for other conventional citrus growing regions, such as central and south Florida. Therefore, an integrated research-driven extension program focusing on different aspects of cold hardy citrus production.

Findings

North Florida has been characterized by hot summers and mild winters for the last three decades. This characterization has increased interest in cold-hardy citrus production for the fresh market. Cultivation of cold hardy citrus (CHC) in north Florida has a long history. In 1565, the Spanish first introduced different citrus cultivars in St. Augustine, and then in 1700 Satsuma (*Citrus unshiu*) was introduced. In the late 1800s, Owari, a cultivar of satsuma, was introduced in the lower Gulf Coast States. However, the area in which the satsuma had been grown was severely reduced in the north Gulf Coast states by severe freeze events from 1894 to 1899. The fruit was marketed as 'Sugar Sweet' in northern and midwestern regions of the United States. The satsuma industry was again eliminated due to climatic factors such as hurricanes and freezes, and lack of labor because of World War II in the Gulf Coast States until the 1990s.

The development of freeze protection techniques in the form of microsprinklers, and mild winters in the early 1990s led to a resurgence of CHC production in north Florida. Satsuma cultivars are most commonly grown for fresh markets in north Florida. The citrus acreage in north Florida increases each year with major cultivation in Gadsden, Jackson, Jefferson, Taylor, Madison, and Suwannee counties. According to the north Florida Cold Hardy Citrus Association, there are 2000 acres approximately under CHC for the fresh market in north Florida, and these acreages are continuously increasing. Likewise, there are approximately 600 and 2500 acres of cold-hardy fresh citrus in the south of Alabama and south of Georgia, respectively. The Cold Hardy Citrus Association, made up of growers, packers, processors, industry suppliers, nurseries, consultants, homeowners, and academia, is very active and organizes annual CHC meetings to discuss challenges and opportunities for CHC production. Since CHC is a newly emerging industry, most of the citrus groves are young (5–8 years), and growers beginning citrus production have limited citrus growing background. Currently, growers are depending on the information on nutrient and water management, disease, and pest control, weed management, pre- and postharvest plant care etc., developed for the citrus growing regions in central and south Florida. Growers cannot directly benefit from that information because of the different nature of soil and climatic conditions. In addition, some of that information is developed for citrus production for processing while we are dealing with cold hardy citrus for fresh markets in north Florida. Therefore, an integrated research-driven extension program focusing on rootstock/scion evaluation for early fruiting, best management practices for citrus in cold-hardy regions, marketing, fruit storing, freeze and cold protection techniques etc., is needed to develop the recommendation for sustainable citrus production in north Florida.

Suggestions for future direction include the evaluation of scion/rootstock combinations for high cold tolerance and disease resistance, and the development of research-based recommendations for nutrient and irrigation management for citrus production in north Florida.

*Corresponding author. Email: mshahid@ufl.edu

Proc. Fla. State Hort. Soc. 136:98. 2023.

Hamlin N13-32: A Somaclone Under Observation for Huanglongbing Tolerance

John M. Chater*

Citrus Research and Education Center, University of Florida IFAS, Lake Alfred, FL

—abstract—

Huanglongbing (HLB) has devastated citrus production in Florida, with juice orange production down approximately 90% as measured by boxes of fruit per year. Sweet orange (*Citrus sinensis*) scions are highly susceptible to HLB and the disease causes severe problems for tree growth, yield, and fruit quality. There have been major efforts to produce HLB tolerant sweet oranges using biotechnological approaches. Utilizing somaclonal variation of 'Hamlin' sweet orange has been one of these approaches to select for tolerance. During the 2022–23 season, a 'Hamlin' somaclone named 'N13-32' on the UFR-4 rootstock has been noted to display plant health that has been described as indicative of HLB tolerance at multiple sites. 'Hamlin N13-32' is a somaclone (protoclone), one of three selected from 1000 clones under evaluation. This study collected fruit quality and health data from these trees to determine the potential of this clone for crop production in the HLB environment. Data collected included tree height, yield, Brix, titratable acidity, sugar to acid ratio, fruit size, fruit mass, *Candidatus* Liberibacter asiaticus (*C*Las) titer, and health based on hyperspectral drone imagery. Conventional 'Hamlin' trees on an industry standard rootstock were compared to 'N13-32' on the UFR-4 rootstock. There was evidence for statistically significant differences between scion-rootstock combinations for yield, fruit quality, and tree health. Further evidence is needed to determine the level of HLB tolerance of this promising somaclone.

*Corresponding author. Email: jchater@ufl.edu

Proc. Fla. State Hort. Soc. 136:99. 2023.

—Scientific Note—

Identification of a New Natural Bud Sport from the 'Flame' Grapefruit

MANJUL DUTT[1]*, LAMIAA MAHMOUD[1], AND GARY ENGLAND[2]

[1]*University of Florida, IFAS, Citrus Research and Education Center, Lake Alfred, FL*

[2]*University of Florida, IFAS, Volusia County Extension, DeLand, FL*

Keywords. citrus, citrus greening, *Citrus ×paradisi*, grapefruit, natural mutation

Grapefruits (*Citrus ×paradisi*) are a relatively new introduction, having been discovered only during the 17th century from Barbados. A hybrid between the pummelo and sweet orange, all present day grapefruit cultivars have been produced through natural or induced mutations from the original discovery and thus have a narrow genetic base (Spiegel-Roy & Goldschmidt 1996; Barry et al. 2020). Recently a naturally occurring bud sport was identified from a 'Flame' grapefruit tree and this scientific note reports the initial findings on this mutant which has been provisionally named as the 'Lennon Flame'.

The 'Lennon Flame' mutant displays several notable and unique characteristics, including darker leaves, improved canopy density, and later maturing fruit that distinguishes itself from 'Flame' grapefruit (Fig. 1). Despite similar *Candidatus* Liberibacter asiaticus (*Ca*Las) Ct values, enhanced chlorophyll levels were recorded in the mutant's leaves when compared to the parent 'Flame' tree. Fruit color scores were comparable between the two, but we recorded Brix level of 7.5 ± 0.2 from the mutant and which surpassed the parent tree's 6.3 ± 0.4 °Brix. Budwood from the 'Lennon Flame' mutant has been deposited with the Florida Department of Agriculture and Consumer Services Division of Plant Industry's parent tree program for certification and replicated field trials will be conducted once certified budwood becomes available.

Fig. 1. (**A**) *Citrus × paradisi* 'Flame' tree and (**B**) mutant branch identified from one tree in a grapefruit block in Lake County, FL. (**C**) Fruit of the original *Citrus ×paradisi* 'Flame' tree. (**D**) Fruit of the identified branch.

Literature Cited

Barry G.H., Caruso M, & Gmitter Jr., F.G. 2020. Commercial scion varieties. p. 83–104. In: Talon M, Caruso M, & Gmitter Jr. F.G. (eds.). The genus citrus. Elsevier, New York, N.Y.

Spiegel-Roy P. & Goldschmidt E.E. 1996. The biology of citrus. Cambridge University Press, Cambridge, U.K.

*Corresponding author. Email: manjul@ufl.edu

Citrus Section

Proc. Fla. State Hort. Soc. 136:100. 2023.

—Scientific Note—

Morphological and Physiological Changes to 'Valencia' Sweet Orange Scion Grafted onto Australian Lime Derived Citrus Rootstocks

LAMIAA M. MAHMOUD*, JUDE W. GROSSER, AND MANJUL DUTT

Citrus Research and Education Center, University of Florida, Lake Alfred, FL

Keywords. citrus, *Citrus australasica, Citrus australis,* citrus greening, *Citrus sinensis,* finger lime, hybridization

Huanglongbing (HLB), caused by *Candidatus* Liberibacter asiaticus (*C*Las) is currently the most destructive citrus disease, and poses a significant threat to citrus cultivation (Bové, 2006). HLB tolerant citrus species can be used to develop commercially acceptable HLB tolerant citrus cultivars through conventional breeding and genetic engineering strategies. In this study, we utilize the genetics of the finger lime (*Citrus australasica*) and the round lime (*Citrus australis*) to develop HLB tolerant hybrids. A subset of the HLB tolerant germplasm was generated by crossing various citrus cultivars with *C. australasica* or *C. australis*. Selected hybrids were clonally propagated and used as rootstocks. They were divided into two sets: one grafted with *C*Las-free *Citrus sinensis* 'Valencia' budwood and the other with *C*Las-positive 'Valencia' budwood. Genomic DNA was isolated six months post-grafting from 'Valencia' leaves for *C*Las titer diagnosis. Physiological and biochemical changes in infected 'Valencia' leaves grafted onto finger lime hybrids were recorded, with 'Valencia' on Swingle rootstock serving as a control.

High *C*Las Ct values were recorded in 'Valencia' leaves (ranging from 29.11 to 35.00) grafted onto selected Australian lime derived hybrids. This indicated a reduction of the *C*Las titer when compared to the initial Ct (22.25 ± 1.11) that was recorded from the infected budwood used to graft onto the rootstocks. 'Valencia' leaves from trees on the Australian lime hybrids also exhibited enhanced redox control factors, including glutathione-S-transferases (GSTs), catalase, thioredoxin, and growth hormones (Weber et al., 2022). Upregulation of defense-related factors, such as R-proteins, Cys-rich secretory proteins (PR proteins), and hormone signaling-related genes, was observed in finger lime hybrids. 'Valencia' grafted onto R1T98 performed the best with lower *C*Las titer, enhanced growth, and chlorophyll pigments compared to the Swingle control.

Australian lime derived hybrids show promise as HLB-tolerant rootstocks, offering practical advantages over other HLB management methods. The observed physiological and biochemical changes, along with reduced *C*Las titer, suggest the potential utility of some of these hybrids as replacements for commercial rootstocks for 'Valencia' sweet orange cultivation. Further research and field trials are ongoing to confirm these findings and to optimize the integration of Australian lime hybrids into citrus cultivation practices.

Literature Cited

Bové J.M. 2006. Huanglongbing: A destructive, newly-emerging, century-old disease of citrus. *Journal of Plant Pathology*. 88:7–37.

Weber K.C., Mahmoud L.M., Stanton D. Welker S., Qiu W., Grosser J.W., Levy A., & Dutt M. 2022. Insights into the mechanism of Huanglongbing tolerance in the Australian finger lime (*Citrus australasica*). *Frontiers in Plant Science*. 13:1019295.

*Corresponding author. Email: lamiaa.mahmoud@ufl.edu

Proc. Fla. State Hort. Soc. 136:101. 2023.

—Scientific Note—

Huanglongbing Pathogen Levels in Juice Across Citrus Genotypes Do Not Correlate with Sensory and Volatile Profiles

Gabriela Olmedo[1], Kristen Jeffries[1], Wei Zhao[1], Elizabeth Baldwin[1], Weiqi Luo[1], Zhen Fan[2], Xiuxiu Sun[3], Matt Mattia[1], Ed Stover[1], Anne Plotto[1]*, and Jinhe Bai[1]*

[1]*USDA-ARS Horticultural Research Laboratory, Fort Pierce, FL*

[2]*University of Florida, Gulf Coast Research and Education Center, Wimauma FL*

[3]*USDA-ARS Daniel K. Inouye U.S. Pacific Basin Agricultural Research Center, Hilo, HI*

Keywords. *C*Las titer, juice quality

Huanglongbing (HLB), caused by *Candidatus* Liberibacter asiaticus (*C*Las), has caused a 90% decline in Florida citrus production since 2005. Juice extracted from HLB-affected oranges is associated with deteriorated quality, such as low sweetness and high bitterness. Our group has observed a positive correlation between *C*Las titer and sensory quality loss in 'Hamlin' and 'Valencia' oranges (Zhao et al., 2021). In this research, we hypothesized that *C*Las content could be used as an indicator of juice quality across citrus species. Hence, correlations and principal component analyses (PCA) were performed between *C*Las titer, sensory scores and volatile profiles from a large citrus genotype population including sweet orange (*Citrus sinensis*), mandarin (*C. reticulata*), grapefruit (*C. ×paradisi*) and their hybrids with or without poncirus (*Poncirus trifoliata*) in their pedigrees.

Findings

Fruit from 125 harvest/genotype combinations were used. Fruit were collected between 2017 and 2022 from trees grown in different locations in Florida. After juicing, genomic DNA was extracted following a modified CTAB protocol and qPCR was performed using LJ primers. Results were subjected to correlations and principal component analyses (PCA) with data derived from volatile determination by HS-SPME-GC-MS (Bai et al., 2014) and sensory descriptive evaluation (Deterre et al., 2023).

The analysis of *C*Las titer and sensory data resulted in low Pearson's coefficients, indicating no correlation. This outcome was expected as determinations describing "species-specific" flavors were higher for each particular species. Additionally, some negative sensory descriptors, like aftertaste, were strongly correlated with grapefruit flavor and, as a result, often rated higher in grapefruit than in other species. In the PCA, nearly all samples with low *C*Las titer scored close to sweetness and orange flavor eigenvectors and away from negative descriptors eigenvectors (Fig. 1). Finally, a positive correlation was found between *C*Las titer and linalool content across all genotypes. Linalool is a monoterpene alcohol usually associated with fruity and sweet flavors. This is not the first time that increase in this compound is reported in HLB infected fruit versus healthy fruit (Kiefl et al., 2018).

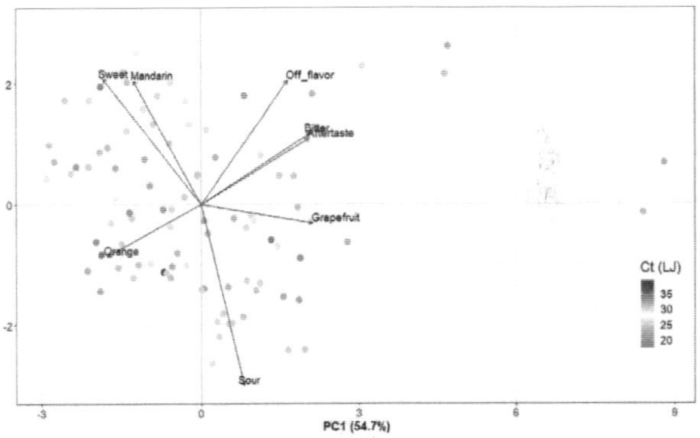

Fig. 1. Principal component analyses (PCA) for sensory results with classification for *Candidatus* Liberibacter asiaticus (*C*Las) titer, sensory scores and volatile profiles from a large citrus genotype population including sweet orange (*Citrus sinensis*), mandarin (C. reticulata), grapefruit (*C. ×paradisi*) and their hybrids with or without poncirus (*Poncirus trifoliata*) in their pedigrees.

Suggestions for Future Direction

Since a clear correlation between *C*Las content and juice quality in two orange cultivars was previously found, we were wondering if a similar trend could be observed in citrus juice in general. Not surprisingly, juice quality was more determined by genotype than degree of *C*Las infection. Further studies should focus on trees within similar genetic group. Furthermore, the correlation between linalool and *C*Las titer was intriguing and should be pursued.

Literature Cited

Bai J., Baldwin E., Hearn J., Driggers R., & Stover, E. 2014. Volatile profile comparison of USDA sweet orange-like hybrids versus 'Hamlin' and 'Ambersweet'. *HortScience*. 49:1262–1267. https://doi.org/10.21273/HORTSCI.49.10.1262

Deterre S., Jeffries K., McCollum G., Stover E., Leclair C., Manthey J., Bai J., Baldwin E., Raithore S., & Plotto A. 2023. Sensory quality of *Citrus* scion hybrids with *Poncirus trifoliata* in their pedigrees. *J. Food. Science*. 88:1684–1699. https://doi.org/10.1111/1750-3841.16499.

Kiefl J., Kohlenberg B., Hartmann A., Obst K., Paetz S., & Krammer G. 2018. Investigation on key molecules of Huanglongbing (HLB)-induced orange juice off-flavor. *J. Agric. Food Chem.* 66:2370–2377. https://doi.org/10.1021/acs.jafc.7b00892.

Zhao W., Baldwin E., Bai J., & Plotto A. 2021. Method for assessing juice/cider quality and/or safety. U.S. Patent 11,046,994 B2. Filed: July 5, 2018. Issued: June 29, 2021.

This work is supported in part by the U.S. Department of Agriculture, National Institute of Food and Agriculture, USDA-NIFA-CDRE project 2018-70016-27387.

*Corresponding authors: anne.plotto@usda.gov; jinhe.bai@usda.gov

Proc. Fla. State Hort. Soc. 136:102. 2023.

Transcriptional Analysis of Carotenoids Accumulation and Metabolism in a Pink-fleshed Lemon Mutant

Giuseppe Lana[1]*, Jaime Zacarias-Garcia[2], Gaetano Distefano[1], Alessandra Gentile[1], María J. Rodrigo[2], and Lorenzo Zacarias[2]

[1]*Department of Agriculture, Food and Environment, University of Catania, 95123 Catania, Italy*

[2]*Food Biotechnology Department, Instituto de Agroquímica y Tecnología de Alimentos, Consejo Superior de Investigaciones Científicas (IATA-CSIC), Paterna, 46980 Valencia, Spain*

Keywords. carotenoids, citrus, *Citrus limon*, fruit quality, gene expression, lycopene, pigments

—abstract—

Pink lemon is a spontaneous bud mutation of lemon (*Citrus limon* L. Burm. f) characterized by the production of pink-fleshed fruits due to an unusual accumulation of lycopene. To elucidate the genetic determinism of the altered pigmentation, comparative carotenoid profiling and transcriptional analysis of both the genes involved in carotenoid precursors and metabolism, and the proteins related to carotenoid-sequestering structures were performed in pink-fleshed lemon and its wild-type. The carotenoid profile of pink lemon pulp is characterized by an increased accumulation of linear carotenoids, such as lycopene, phytoene and phytofluene, from the early stages of development, reaching their maximum in mature green fruits. The distinctive phenotype of pink lemon is associated with an up-regulation and down-regulation of the genes upstream and downstream the lycopene cyclase, respectively. In particular, 9-cis epoxycarotenoid dioxygenase genes were overexpressed in pink lemon compared with the wild-type, suggesting an altered regulation of abscisic acid biosynthesis. Similarly, during early development of the fruits, genes of the carotenoid-associated proteins heat shock protein 21, fibrillin 1 and 2, and orange gene were overexpressed in the pulp of the pink-fleshed lemon compared to the wild-type, indicating its increased capacity for sequestration, stabilization or accumulation of carotenes. Altogether, the results highlighted significant differences at the transcriptomic level between the pink-fleshed lemon and its wild-type, in terms of carotenoid metabolism and the capacity of stabilization in storage structures between the two accessions. Such changes may be either responsible for the altered carotenoid accumulation or in contrast, a metabolic consequence.

The full manuscript can be found at https://doi.org/10.3390/genes11111294

*Corresponding author. Email: giuseppelana@ufl.edu

Proc. Fla. State Hort. Soc. 136:103–104. 2023.

—Scientific Note—

Optical and Thermal Data Integration to Analyze the Impact of the Mealybug Pest on Citrus Trees

FÀTIMA DELLA[1]*, BELEN FRANCH[1,2], DRAŽEN SKOKOVIĆ[1], ALBERTO SAN BAUTISTA[3], CONSTANZA RUBIO[4], ITALO MOLETTO-LOBOS[1], BERTRAN MOLLÀ-BONONAD[1], AND MARÍA JOSÉ SÁNCHEZ-TORRES[1]

[1]*Global Change Unit, Image Processing Laboratory, Parc Científic, Universitat de València (Paterna), 46980, Spain*

[2]*Department of Geographical Sciences, University of Maryland, College Park MD, United States*

[3]*Departament of Production of Plants, Universitat Politècnica de València (Valencia), 46022, Spain*

[4]*Centro de Tecnologías Físicas, Universitat Politècnica de València (València), 46022, Spain*

"Cotonet de les Valls" (*Delotococcus aberiae*) is a mealybug pest that is causing serious economic losses to the Spanish agricultural sector, especially to citrus fruits (Conselleria d'Agricultura, Desenvolupament Rural, Emergència limàtica i Transició Ecològica. 2021). The European Copernicus program has enabled the development of numerous agricultural monitoring tools by exploiting the remote sensing technology using earth orbiting satellites. Our work is based on integrating optical (Sentinel-2 satellite) and thermal (Sentinel-3, S3 and Landsat 8 satellites) data to develop a near real-time monitoring tool for mealybug population levels on citrus fruits.

This study was conducted in the area of Vall d'Uixó (Castellón, Spain), and utilized two citrus parcels that had been classified as healthy and unhealthy, depending on their pest infestation level. The study was conducted between the years 2017 to 2022, with the first signs of mealybug infestation observed in 2018.

Temporal evolution of each Sentinel-2 satellite sensor (Level 2A) (European Space Agency. 2015) band's average reception over each of the 2 citrus parcels has been done. We also conducted an analysis of the trends of the seasonal average 95th percentile representing the maximum surface temperatures (June), and average 5th percentile representing the minimum surface temperatures (December) through the timeseries of healthy and affected citrus parcels. The mean values of Landsat 8 satellite sensor's land surface temperature (LST) (U.S. Geological Survey, 2019) (band 10) have been measured. Temperature differences have been also observed (as healthy LST citrus parcel-insect infested LST citrus parcel). The meteorological dataset ERA 5 (Muñoz, 2019) yearly August, September, and October air temperature mean values were also used.

Finally, the integration of thermal and optical data has been performed through a linear regression between S3 satellite senor (Polehampton, 2023) LST (1km) and the normalized difference vegetation index (NDVI) derived from Sentinel-2 satellite Level 1C (1 km), atmospherically corrected using LaSRC (Vermote, 2016). Assuming that the LST-NDVI relationship would hold regardless of resolution, the LST resolution upscale can be performed, considering the NDVI at 10 m and at 100 m as the independent variable (Mukherjee et al., 2014). For validation (against LST Landsat 9 satellite (U.S. Geological Survey, 2022), LST at 10 m and 100 m has been used.

The results show that the RED and SWIR bands (bands 4 and 11) present good separability during December, especially for the minimum values evolution (Fig. 1). We observed that healthy and affected minimum value trends progressively diverge beginning in 2017. NIR shows good separability during June (maximum temperature values) (Fig. 1). The thermal analysis (Fig. 2) shows temperature differences during 2018, 2020 and 2021 larger than 2K during August, September and October. The years 2019 and 2022 LST differences are low, and this may have been caused by the higher air temperatures. And from the downscaling method validation (Fig. 3) we see that RMSE is too high (above 2K) and is therefore not useful in determining mealybug infestation levels.

The separability found in the optical spectrum shows that remote sensing data can be leveraged to monitor the mealybug infestation levels in a timely, objective and cost-efficient way. We may be able to differentiate stressed and healthy citrus groves during the year using certain spectral ranges: RED and SWIR in December; NIR in June; Thermal in August, September and October. On the other hand, the downscaling method does not show reliable LST differences but L8 displays that affected parcels presents warmer temperatures. We also checked that air temperature may influence the LST results. The need to acquire temperature images with high spatial and temporal resolution would be very beneficial.

Based on these preliminary results, in the future we will seek to introduce more study plots and check if the separability between healthy and affected citrus plots is maintained. We will also try to improve the downscaling process by introducing some intermediate resolution steps or the correction of the regression by means of the residuals, as well as testing the use of some other vegetation index. Finally, we also propose to normalize the air temperature data, as we have seen that it influences the LST and it would be optimal to reduce its impact in order to be able to make interannual comparisons.

*Corresponding author. Email: fatima.della@uv.es

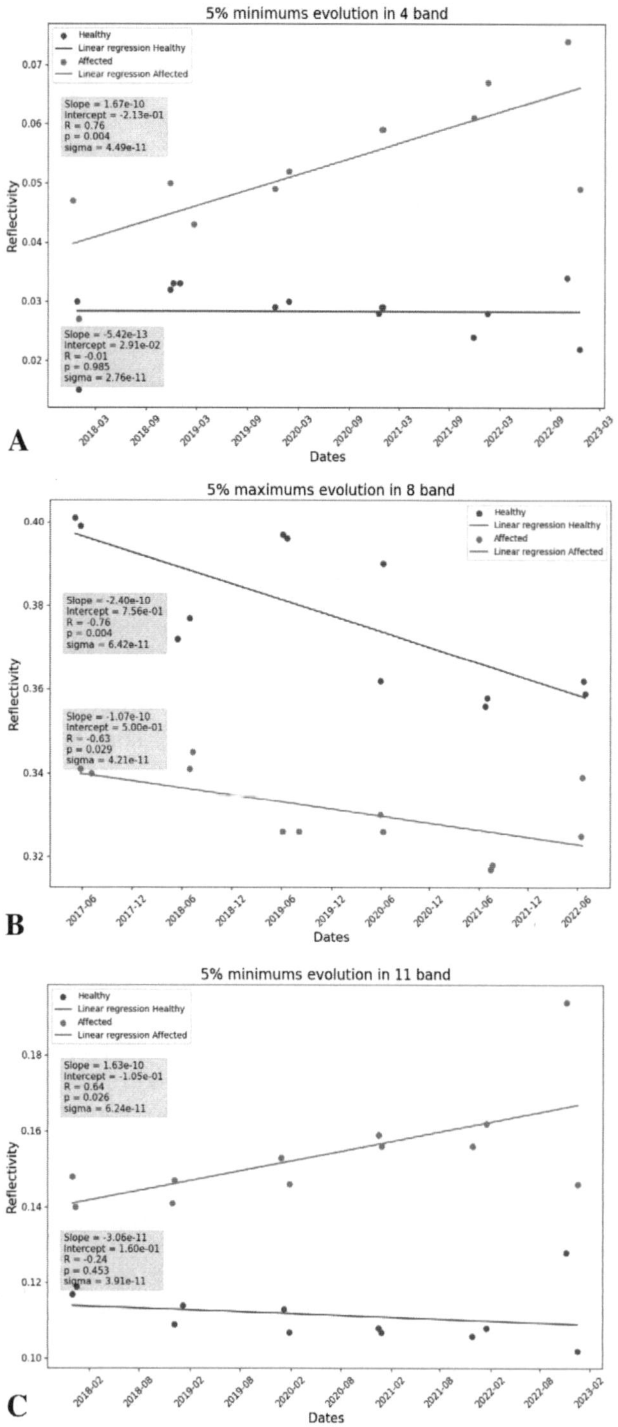

Fig. 1. RED and SWIR bands [4 (**A**) and 11 (**C**)] minimums and NIR [band 8 (**B**)] maximums reflectivity trends from 2017 to 2022. Blue color indicates the reflectivity from healthy fields and red color refers to the affected ones..

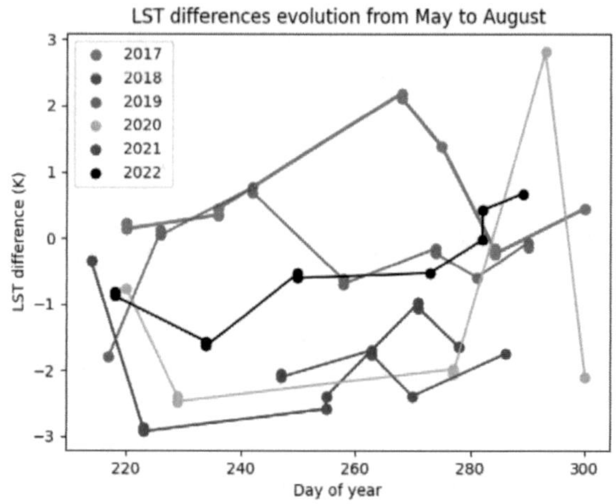

Fig. 2. Temperature differences [land surface temperatures (LST)] evolution from May to August for all years studied.

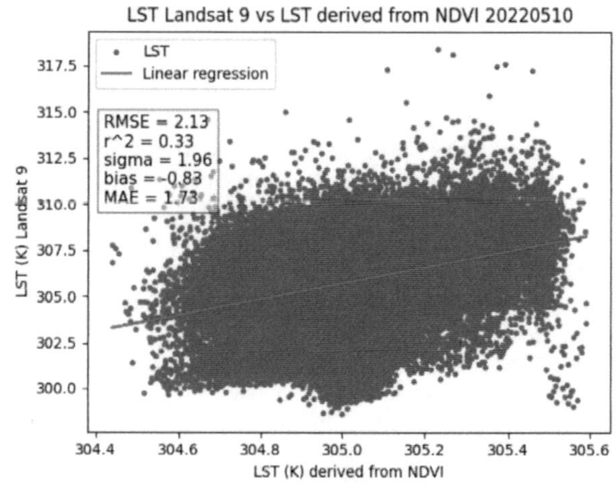

Fig. 3. Validation against land surface temperatures (LST) from Landsat 9 satellite using LST from the downscaling at 10 m.

Literature Cited

Conselleria d'Agricultura, Desenvolupament Rural, Emergència limàtica i Transició Ecològica. Num. 9153 / (2021, 08, 17). 36564 [DO]. Available: https://dogv.gva.es/datos/2021/08/17/pdf/2021_8695.pdf

European Space Agency. 2015. Sentinel-2 user handbook. Available: https://sentinels.copernicus.eu/documents/247904/685211/Sentinel-2_User_Handbook

Mukherjee S., Joshi P.K., & Garg R.D. 2014. A comparison of different regression models for downscaling Landsat and MODIS land surface temperature images over heterogeneous landscape. *Advances in Space Research*. 54(4):665–669. https://doi.org/10.1016/j.asr.2014.04.013

Muñoz Sabater, J. 2019. ERA5-Land monthly averaged data from 1981 to present. Copernicus Climate Change Service (C3S) Climate Data Store (CDS). Acessed 11 June 2023. https://doi.org/10.24381/cds.68d2bb30

Polehampton, E. 2023. Copernicus Sentinel-3 SLSTR land user handbook. European Space Agency. 2023. https://sentinels.copernicus.eu/documents/247904/4598082/Sentinel-3-SLSTRLand-Handbook.pdf

U.S. Geological Survey. 2019. Landsat 8 data users handbook. Available: https://d9-wret.s3.us-west-2.amazonaws.com/assets/palladium/production/s3fs-public/atoms/files/LSDS-1574_L8_Data_Users_Handbook-v5.0.pdf

U.S. Geological Survey. 2022. Landsat 9 data users handbook. Available: https://d9-wret.s3.us-west-2.amazonaws.com/assets/palladium/production/s3fspublic/media/files/LSDS-2082_L9-Data-Users-Handbook_v1.pdf

Vermote E., Justice C., Claverie M., & Franch B. 2016. Preliminary analysis of the performance of the Landsat 8/OLI land surface reflectance product. *Remote Sensing of Environment*. 185:46–56. https://doi.org/10.1016/j.rse.2016.04.008

Proc. Fla. State Hort. Soc. 136:105–106. 2023.

—Scientific Note—

Application of Remote Sensing for Monitoring the Incidence of *Delotococcus aberiae* in Citrus Farms at Castellón (Spain)

Constanza Rubio[1], Belen Franch[2,3], Fran Valls[4], Daniel Tarrazo-Serrano[1], Sergio Castiñeira[1], Antonio Uris[1], David Fita[4], Patricia Arizo-García[4], and Alberto San Bautista[4]*

[1]*Physical Technologies Center, Universitat Politécnica de València, Spain*

[2]*Global Change Unit, Image Processing Laboratory, Universitat de València, Spain*

[3]*Department of Geographical Sciences, University of Maryland, USA*

[4]*Departament of Production of Plants, Universitat Politécnica de València, Spain*

Keywords. citrus industry, cotonet, European Space Agency, near-infrared, neural analyses, reflectance bands, Sentinel-2 earth satellite

The citrus industry of Spain is at a critical moment due to the emergence of a new pest insect known as South African Cotonet or Cotonet de les Valls (*Delotococcus aberiae*), a mealybug that is currently causing substantial damages and whose complex management is putting the Spanish citrus industry in a compromising situation. Remote sensing enables improving mathematical models for pest monitoring and detection (Fletcher, 2005; Reisig and Godfrey, 2006). Earth orbiting satellite sensor reflectance is related to the crop photosynthetic rate and biomass (You et al., 2017). The present work aims to find a methodology to model the presence of the pest to prevent and monitor its incidence and effect on citrus production.

This study was conducted at citrus production locations at Vall d'Uixó, La Llosa and Moncofar (Castellón). From 2018 to 2021, a citrus crop area of 47 ha, containing groves that were both affected and non-affected by the Cotonet pest was analyzed.

For each available Sentinel-2 earth orbiting satellite date (European Space Agency, 2015), the evolution of the B4 (Red) and B8 (NIR) bands and the normalized difference vegetation index (NDVI) for healthy and unhealthy fields were studied. An annual linear model was obtained between the two studied bands at the pixel level. The value of the coefficient of determination (r^2) and mean absolute error (MAE) were used to evaluate the model performance. Finally, to classify plots affected and not affected by Cotonet, different neural analyses were elaborated using all Sentinel-2 satellite sensor bands.

The results show that the NDVI index shows an opposite behavior to the reflectance of the B4 and B8 bands (Fig. 1), varying

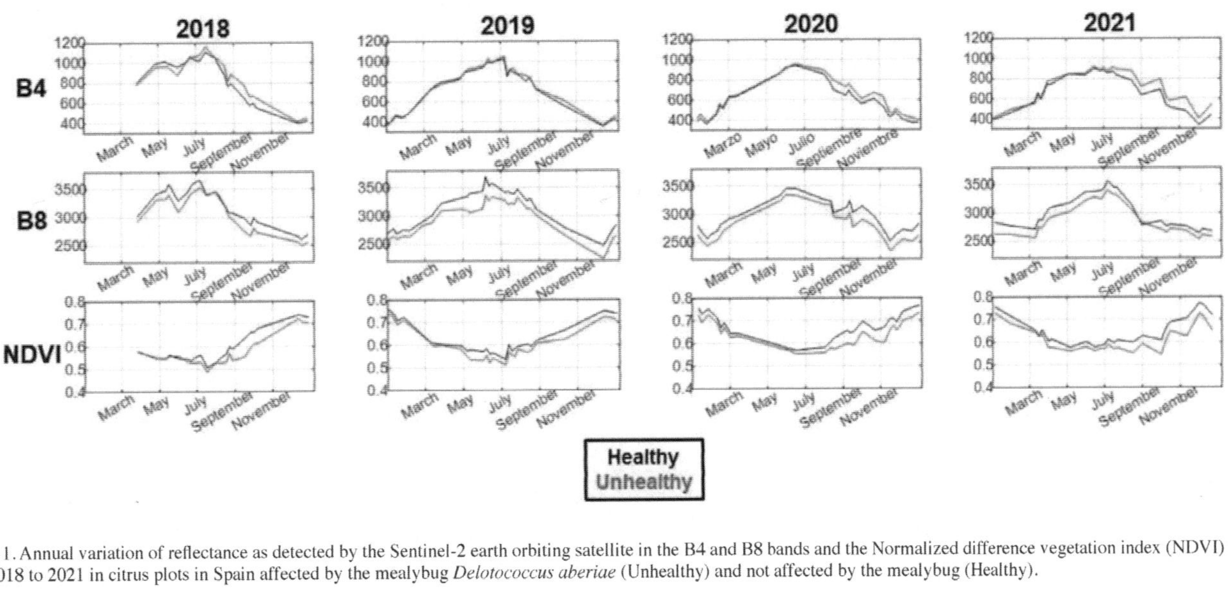

Fig 1. Annual variation of reflectance as detected by the Sentinel-2 earth orbiting satellite in the B4 and B8 bands and the Normalized difference vegetation index (NDVI) from 2018 to 2021 in citrus plots in Spain affected by the mealybug *Delotococcus aberiae* (Unhealthy) and not affected by the mealybug (Healthy).

This research was funded by the SABIOCONTROL Project SCPP2100C008733XVD, Ministry of Science and Innovation, PRTR, State Agency Research.

*Corresponding author; email: asanbau@prv.upv.es

in a significant way the reflectance values between healthy and unhealthy plants, with lower values of NDVI and B8 reflectance observed from the beginning of the cycle in unhealthy crops. On the other hand, differences observed in B4 in unhealthy crops were not statistically significant. However, the year factor showed significant differences in reflectance in B4 and B8, and NDVI, which may indicate the variability of conditions from the year-to-year pest condition. A linear relationship between the reflectance in the red and the NIR was found, clearly differentiating between the two groups of plants. Given the value of the r^2 coefficient, these linear models could be used to classify the plots according to their level of pest insect infestation. The neural analysis has allowed the correct classifying of the pixels affected by the pest in 92.3%, using the model with the reflectance at 20 m resolution in all possible ranges except the bands in the shortwave infrared (SWIR). In conclusion, "Cotonet de les Valls" pest monitoring and detection can be carried out by analyzing the spectral signature of the plots, checking if the mathematical relationship between B4 and B8 fits one of the calculated slopes, and validating the recorded data from the model obtained by neural analysis.

Based on the results, future work will focus on expanding the study area and introducing new locations in different countries to develop models that could be used to propose measures against the pest. Due to this enlargement of the database, it will be necessary to use tools based on artificial intelligence.

Literature Cited

European Space Agency. 2015. Sentinel-2 User handbook. https://sentinels.copernicus.eu/documents/247904/685211/Sentinel-2_User_Handbook

Fletcher R.S. 2005. Evaluating high spatial resolution imagery for detecting citrus orchards affected by sooty mold. *Intl. J. Remote Sensing*.26(3):495–502. https://doi.org/10.1080/01431160512331316414

Reisig D. & Godfrey, L. 2006. Remote sensing for detection of cotton aphid–(homoptera: aphididae) and spider mite–(acari: tetranychidae) infested cotton in the San Joaquin Valley. *Environmental Entomology*. 35(6):1635–1646. https://doi.org/10.1093/ee/35.6.1635

You J., Li X., Low M., Lobell D., & Ermon S. 2017. Deep gaussian process for crop yield prediction based on remote sensing data. *Proceedings of the Thirty-First AAAI Conference on Artificial Intelligence*. 31(1):4559–4565.

Proc. Fla. State Hort. Soc. 136:107. 2023.

—Scientific Note—

Hedging and Thinning to Mitigate Fruit Disorder Incidence in Huanglongbing-affected 'Sugar Belle' Mandarin

Joe H. Volpe and John M. Chater*

University of Florida, IFAS, Citrus Research and Education Center, Lake Alfred, FL

Keywords. *Citrus reticulata*, fruit disorder, hedging, 'Sugar Belle'

'Sugar Belle', a mandarin (*Citrus reticulata*) hybrid released by the University of Florida, is known to have a high level of tolerance to Huanglongbing, a devastating disease affecting citrus production across much of the world. The fruit has potential for use in both the fresh market and the juice stream. However, in the past two to three seasons many growers in Florida have reported increased incidence of fruit disorders, mainly misshapen and/or soft fruit. The objective of this study was to examine if different timing for hedging treatments (early summer, middle summer, and late summer) as well as a manual thinning treatment (removing approximately 50% of the fruit by hand in late summer) will mitigate fruit disorders and increase fruit quality.

The experiment was replicated at two sites (the Citrus Research and Education Center (CREC) in Lake Alfred, FL. and Tamiami Citrus in Avon Park, FL) and included 30 trees each for the early and middle hedge, 10 trees each for the late hedge and thinning treatment and 60 control trees at each site. Hedging was completed using a DeWalt battery powered hedger with the goal of removing approximately 0.5 m of canopy around the whole tree. Fruit growth as well as juice quality data were collected throughout the growing season as well as several fruit and juice quality characteristics at harvest time from mid-December to mid-January. Hurricane Ian in late September severely damaged approximately 15% of the data trees at each site and data was omitted from these trees, however the other trees experienced some effects mainly in terms of increased fruit drop.

Some of the most significant results were from the thinning treatment with increased Brix and increased pounds solids per box compared to the controls for whole tree harvests at the CREC (Fig. 1). The fruit from the thinning treatment also had significantly higher mass than those from all the hedging treatments. Interestingly there were no significant differences among any of the treatment groups or control trees for yield in terms of mass or number of fruit for the whole tree harvests at CREC. There was also significantly more asymptomatic fruit from the control and early hedge trees than later hedging treatments or thinning at both sites.

*Corresponding author. Email: jchater@ufl.edu.

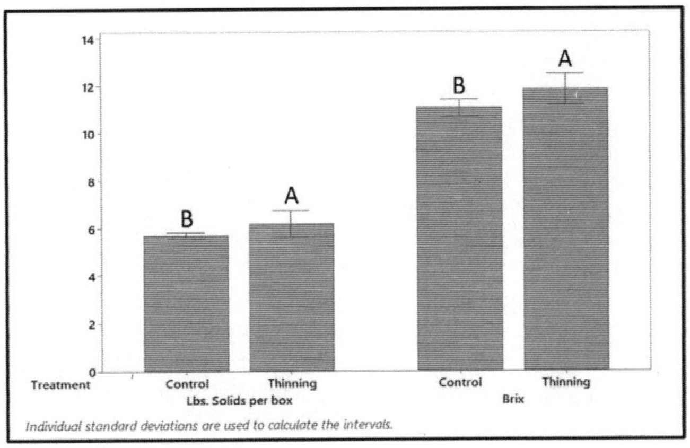

Fig. 1. Mean whole tree harvests (n = 5) of 'Sugar Belle' mandarin (*Citrus reticulata*) hybrid at the Citrus Research and Education Center in Lake Alfred, FL, as affected by pruning. Letters above bars indicate significant differences between treatment groups.

Suggestions for Future Direction

Using the findings from this study as a guide future experiments to further evaluate the effects of hedging and thinning on 'Sugar Belle' trees are needed. Increasing the number of trees in the thinning treatment will bolster the robustness of the experiment and adding a controlled release fertilizer treatment and an oxytetracycline injection treatment would provide further insights on the potential to mitigate fruit disorders in this cultivar.

Additionally, because of the effects from hurricane Ian, it is important to replicate this experiment to evaluate the treatments again during a more normal season in terms of weather with no major hurricane or flooding events.

Literature Cited

Zekri M. 2018. Mechanical pruning of citrus trees. Extension Publication HS 1267. University of Florida, Gainesville, FL.

Rouse, R.E. 2017. Rehabilitation of Huanglongbing-affected citrus treesusing severe pruning and enhanced foliar nutritional treatments. *HortScience*. 52(7):972–978.

Proc. Fla. State Hort. Soc. 136:108. 2023.

—Scientific Note—

High-throughput Phenotyping and Fruit Juice Quality of 'OLL-8' Sweet Orange on Different Rootstocks

EMILY WORBINGTON*, HEAVEN HEFLIN, JOE H. VOLPE, AND JOHN M. CHATER

University of Florida, IFAS, Citrus Research and Education Center, Lake Alfred, FL

Keyword. *Citrus sinensis*

A majority of the Florida citrus industry produces fruit for juice processing and is affected by the devastation that huanglongbing (HLB) has caused. There has been nearly a 90% decline in boxes of fruit production since the identification of HLB in commercial groves in Florida. Due to the high susceptibility of sweet orange (*Citrus sinensis*) cultivars, new treatments or tolerant cultivars must be produced. In this study, the 'OLL-8' sweet orange scion was evaluated on six different rootstocks, including three industry standards and three UFR rootstocks located at a field trial in Lake Alfred, FL. The 'OLL-8' selection was derived from the original unstable 'OLL' tree via somatic embryogenesis. Of the released sweet orange lines from the University of Florida Institute of Food and Agricultural Sciences, 'OLL-8' is one selection that has shown potential for having increased tolerance to HLB and improved fruit and juice quality.

This rootstock experiment was conducted at a field trial in Lake Alfred, FL, and investigated three industry standard rootstocks ('US-942', 'US-897', and 'Swingle') and three UFR rootstocks ('UFR-2', 'UFR-4' and 'UFR-6'). For each rootstock study, twelve trees were evaluated across three blocks at the site. Harvests oc-curred on three dates (late February, mid-March, and early April) with the first two harvest being half bushel fruit samples and for the third harvest, trees were stripped of all remaining fruit. Leaf samples were collected mid-May and sent to the Central Florida Soil Laboratory for nutrient analysis. Vigor (normalized difference vegetation index), health (normalized difference red edge index), and canopy area (sq. ft) were measured via drone flight in December 2022.

Of the rootstocks evaluated, 'UFR-4' trees outperformed two industry standard rootstocks, 'US-942' and 'US-897', as well as 'UFR-6' for total sample and juice weights (Fig. 1). It is important to note that fruit from 'UFR-4' and 'Swingle' had significantly smaller fruit diameters than other rootstocks in this study. Although 'US-942' trees were younger than the other trees evaluated, conversations within the commercial citrus growers indicate that a tree at this age should be higher producing than the results indicated. These trees also performed significantly lower in total soluble solids and pounds solids which are vital characteristics that determine pay out for growers who produce citrus for the juice industry.

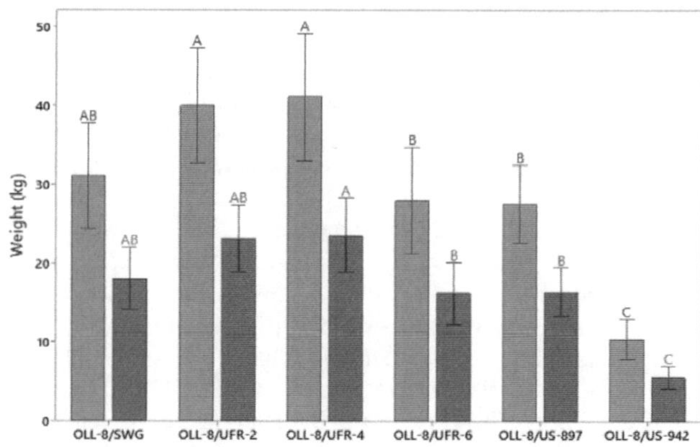

Fig. 1. Average yield and juice weight per Citrus *sinensis* tree. Letters above bars indicate significant differences among treatment groups. Separate anaylses of variance and post hoc tests were run for yield (blue) and juice weight (red).

In regard to this experiment, data collection is needed for at least one more season to further confirm these results. It is also important to assess data from a year in which a major hurricane or flooding event does not occur. Evaluating these six rootstocks at other locations across Florida will account for numerous variables such as differences in watering regimens, psyllid control, and soil nutrition. Increasing the number of trees per evaluated rootstock as well as increasing the number of rootstocks evaluated would provide a fuller understanding as to which rootstock pairs better with the 'OLL-8' selection. Another future direction to expand this experiment would be to evaluate the effects of antimicrobial injections on fruit and juice quality of the 'OLL-8' scion.

As cultivars are created and released, it is important to follow their quality and productivity across multiple seasons and sites to better evaluate tolerance to HLB symptoms.

*Corresponding author. Email: emilyworbington@ufl.edu

Proc. Fla. State Hort. Soc. 136:109. 2023.

—Scientific Note—

'Donaldson': Characterization of a Rediscovered Variety for Use in Orange Juice

MATTHEW MATTIA*, JINHE BAI, ZHEN FAN, KRISTEN JEFFRIES, GABRIELA OLMEDO, AND ANNE PLOTTO

United States Department of Agriculture, Agricultural Research Service, Fort Pierce, FL

Keywords. citrus industry, *Citrus sinensis*, Huanglongbing, orange juice, sweet orange

The rediscovered sweet orange (*Citrus sinensis*) variety, 'Donaldson,' located within the U.S. Department of Agriculture, Agricultural Research Service citrus germplasm collection in Groveland, FL, exhibits promising characteristics for utilization in the orange juice industry. With a documented history dating back to the 1940s, 'Donaldson' has a unique background involving selection and movement throughout Florida. Notably, there is a single tree of 'Donaldson,' which has survived in the field for more than three decades (Fig. 1). During a field display day in December 2021, fruit samples from 'Donaldson' were evaluated by growers, who were impressed by its exceptional fruit quality, overall health in the context of high disease pressure from Huanglongbing (HLB), and high yield potential on a single tree. Subsequently, an amplification project was spurred lead by the Florida Department of Citrus aimed at making 'Donaldson' commercially available to the citrus industry. In addition, the genetic background of the 'Donaldson' rootstock was investigated. Analysis of the rootstock using SSR markers revealed that the rootstock of 'Donaldson' at the Whitmore Farm in Groveland, FL, was 'Swingle'. This information enhances our understanding of the genetic background of the 'Donaldson' rootstock and provides the citrus industry with additional information about this interesting single tree.

Fig. 1. A picture of the single mature *Citrus sinensis* 'Donaldson' tree at the A.H. Whitmore Foundation Farm in Groveland, Florida, taken on May 18, 2023.

To further validate the potential of 'Donaldson' as a contributor to the orange juice stream, extensive characterization of this individual tree and its fruit quality is necessary. More replications of 'Donaldson' and establishing additional field trials are required to confirm its suitability for large-scale citrus production and orange juice processing. Given the devastating impact of diseases like Huanglongbing (HLB) on the citrus industry, it is essential to conduct additional testing to confirm whether 'Donaldson' exhibits tolerance to HLB Moreover, the genetic basis and variability of 'Donaldson' should be thoroughly investigated to assess its potential to contribute positively to the genetic diversity of sweet orange varieties.

*Corresponding author. Email: matthew.mattia@usda.gov

Vegetable Section

Proc. Fla. State Hort. Soc. 136:110. 2023.

—Scientific Note—

Developing Artichoke as a New Winter Crop in Florida through Cultivar Selection and Vernalization Strategies

Shinsuke Agehara* and Weining Wang

Gulf Coast Research and Education Center, University of Florida, IFAS, Wimauma, FL

Keywords. *Cynara cardunculus*, chilling requirement, gibberellic acid, vernalization

Artichoke (*Cynara cardunculus* var. *scolymus* L.) belongs to the sunflower family and is cultivated for its flower buds. This specialty crop requires adequate winter chill to induce the flowering process, generally 200 to 1300 hours of temperatures below 10 °C (Oregon State University, 2012), yet is susceptible to freeze damage (Agehara, 2017). Therefore, artichoke is adapted to Mediterranean climates with mild winter. Consequently, almost 100% of artichoke production in the United States comes from California. In 2021, the US artichoke production was 37784 tons, valued at $57 million (U.S. Department of Agriculture National Agricultural Statistics Service, 2022). Insufficient winter chill is the major limiting factor for artichoke production in subtropical or tropical climates.

Vernalization is the process by which flowering is induced by prolonged exposure to winter chill (Taiz & Zeiger, 2010). Gibberellins (GAs) are phytohormones that play important roles in vernalization in many perennial crops (Taiz and Zeiger, 2010). The major forms of bioactive GAs include GA_1, GA_3, GA_4, and GA_7, among which GA_3 is one of the most active and widely used GAs in agriculture (Rodrigues et al., 2012; Yamaguchi, 2008). Our previous study demonstrated that foliar application of GA_3 is an effective vernalization technique for selected artichoke cultivars (Agehara & Silvasy, 2019). The objective of this study was to examine production strategies for winter artichoke production under Florida's subtropical climatic conditions. Field experiments were conducted in west-central Florida during the 2022–23 growing season. In the first field experiment, we evaluated three planting dates (early and mid-October and early November) for four cultivars ('Green Globe Improved', 'Green Queen', 'Imperial Star', and 'Opal'). All plants were treated with GA_3 (37 g·ha^{-1}) at 7 and 9 weeks after transplanting (WAT). In the second field experiment, we evaluated three GA_3 application timings (37 g·ha^{-1} at 6 and 8, 7 and 9, or 8 and 10 WAT) for three cultivars ('Green Globe Improved', 'Green Queen', and 'Imperial Star'). All plants were transplanted mid-October.

Cumulative winter chill hours were 136 hours. Nonetheless, foliar application of GA_3 successfully induced bud formation in all four tested cultivars. In the first experiment, yields were significantly affected by the cultivar × planting date interaction. The highest yield was recorded for 'Imperial Star' planted in early October (17.7 t·ha^{-1}). Delaying planting date to early November reduced yields of 'Imperial Star' and 'Opal' by up to 71% and 34%, respectively, but it had no significant effect on the other two cultivars. In the second field experiment, the cultivar × GA_3 application timing interaction had no significant on yields. Pooling across the three tested cultivars, marketable yields were very low without GA_3 application (1.7 t·ha^{-1}) and maximized with GA_3 application at 6 + 8 WAT (11.0 t·ha^{-1}). When GA_3 application was delayed from 6 + 8 to 8 and 10 WAT, all three cultivars showed yield reductions by 24% on average. With GA_3-induced vernalization and three planting dates, high-quality buds were harvested from mid-January through late March. The results suggest that, with GA_3-induced vernalization and optimum cultural practices, selected artichoke cultivars are highly adapted to winter production under subtropical climatic conditions.

Literature Cited

Agehara S. 2017. Production guidelines for globe artichoke in Florida. EDIS HS1289.

Agehara S. & Silvasy T. 2019. Development of management practices for winter artichoke production in Florida. *Proc. Fla. State Hortic. Soc.* 132:152.

Oregon State University. 2012. Artichoke, Globe. Accessed 26 Aug. 2022. https://horticulture.oregonstate.edu/oregon-vegetables/artichoke-globe

Rodrigues C., Vandenberghe L.P.D.S., De Oliveira J., & Soccol C.R. 2012. New perspectives of gibberellic acid production: A review. *Crit. Rev. Biotechnol.* 32:263–273.

Taiz L. & Zeiger E. 2010. Plant physiology, 5th ed. Sinauer Associates, Inc., Sunderland, MA.

U.S. Department of Agriculture National Agricultural Statistics Service. 2022. Quick stats. Accessed 26 Aug. 2022. https://quickstats.nass.usda.gov/

Yamaguchi S. 2008. Gibberellin metabolism and its regulation. *Annu. Rev. Plant Biol.* 59:225–251. https://doi.org/10.1146/annurev.arplant.59.032607.092804

*Corresponding author. Email: sagehara@ufl.edu

Vegetable Section

Proc. Fla. State Hort. Soc. 136:111. 2023.

—Scientific Note—

Genetic Mechanisms Underlying Nuclear Male-sterility in Squash

GEOFFREY MERU*, EMILY WANG, YUQING FU, AND VINCENT MICHAEL

Tropical Research and Education Center, University of Florida, IFAS, Homestead, FL

Keywords. *Cucurbita pepo*, F_1 hybrid, *ms-2* gene

Commercial deployment of male sterility is highly valued as a tool for efficient hybrid seed production by eliminating the need for emasculation. In squash (*Cucurbita pepo* L.), genetic male sterility is controlled by a single recessive gene designated *ms-2*. The phenotype conferred by *ms-2* is characterized by shriveling and browning of the androecium. Marker-assisted selection targeting *ms-2* would allow rapid introgression into elite squash inbred lines, but its genomic location is currently unknown. In the current study, an F_2 bi-parental mapping population segregating male sterility was phenotyped in the greenhouse and subsequently used to identify genomic regions linked to the trait. QTL detection was performed using a combination of traditional linkage mapping and QTL-seq based on whole-genome resequencing. Using traditional linkage mapping, a major QTL explaining > 50% of the phenotypic variation was observed on LG4 of the squash genome. The QTL-seq strategy confirmed the loci on LG4, but additionally detected five other QTLs across the genome. A candidate gene encoding a 'Novel INteractor of JAz' (NINJA) protein was identified within the QTL on LG4. Cloning and sequence analysis of this gene identified sequence variants that could be useful for marker-assisted selection in squash.

*Corresponding author. Email: gmeru@ufl.edu

Proc. Fla. State Hort. Soc. 136:112. 2023.

—Scientific Note—

Cauliflower Production During Anticipated Climate Change Conditions

G. Papathanasiou[1], P. Tsouvaltzis[1,2*], and A.S. Siomos[1]

[1]*Vegetable Crops, Department of Horticulture, Aristotle University, Thessaloniki, Greece*

[2]*Southwest Florida Research and Education Center, University of Florida, IFAS, Immokalee, FL*

Keywords. cauliflower, climate change, quality, temperature, yield

In order to study the effects of climate change on cauliflower production, three crop cycles including three commercial cultivars were established on three consecutive planting dates with the latest corresponding to the conventional establishment period, whereas the other two, refer to periods with higher prevailing temperatures, simulating the anticipated conditions induced by climate change. The air temperature, relative air humidity and solar radiation intensity, as well as the amount of water supplied through irrigation and precipitation were recorded daily. Plant height, the total number and weight of leaves, the length and weight of stem, the diameter, weight and quality of the curd at harvest were also recorded on a weekly basis. The mean air temperature and solar radiation intensity increased by 2.4 to 4.1 °C and 52 to 90%, respectively during the two earlier growing seasons, in comparison to the conventional latest one. Consequently, the period between the transplanting and the harvest dates was extended by 15–58 days and the amount of irrigation water increased by up to 54 to 66%. Moreover, the plants' morphology was significantly modified (in terms of the plant height, the length and weight of stems, the total number and the weight of the leaves which all increased), a greater defoliation rate of the basal leaves was also observed (acceleration of plant aging). Despite the slight degradation of curd's quality, the increase in the air temperature and solar radiation intensity only slightly affected the marketable yield of the cauliflower crop.

*Corresponding author. Email: ptsouv@ufl.edu

Proc. Fla. State Hort. Soc. 136:113–115. 2023.

Integrating Cover Crops as a BMP for Improving Soil Health into Commercial Vegetable Production Systems

LAURA JALPA AND RAO MYLAVARAPU*

Soil, Water, and Ecosystem Sciences Department, University of Florida, Gainesville, FL

Keywords. bulk density, saturated hydraulic conductivity, soil health, *Solanum lycopersicum L.*, vegetable production

Alternative sustainable management practices such as incorporation of cover crops into commercial crop production systems have been shown to positively influence physical, chemical, and biological aspects of soil health without negatively affecting the yields. Although the majority of soil health research has been conducted with field crops, benefits such as reduced loss of nitrogen (N) and phosphorus (P), improved nutrient balances, and increased soil carbon levels were also found in rotational systems of cover crops with vegetables as the main crop. Due to limited information on the specificity of the parameters measured for soil health for specific soils, cropping systems, and ecosystems in Florida, a two-year field study has been initiated in a commercial vegetable-cover crop system. Data presented will focus on the physical parameters of soil health and will use saturated hydraulic conductivity and bulk density as indicators of soil pore space and water movement withing the soil profile in a commercial tomato-cover crop system produced on sandy soils.

Florida agriculture is highly specialized, owing to the state's unique ecosystems like wetlands, Everglades, forests, and pastures. It encompasses a wide array of production systems including vegetable, fruit, citrus, row crop, timber, and pasture production (Mylavarapu et al., 2014). Due to competitive markets combined with lower fertility soils, commercial vegetable production systems require intensive management of nutrients and water. To minimize any negative water quality impacts from agricultural fields, nutrient best management practices (BMPs) were developed to decrease the loss of fertilizer inputs and maintain nutrients in the rootzone for crop use (Mylavarapu, 2003). Commercial vegetable nutrient management practices involve the use of tillage, raised beds, plastic mulch, and surface drip irrigation in most of the southeastern United States (Zotarelli et al., 2009; Jalpa et al., 2020, 2021, Jalpa and Mylavarapu, 2021). In these vegetable plasticulture systems, the intent is to maintain water and nutrients in the rootzone for plant use and are therefore typical BMPs for this system. Vegetable nutrient BMPs are in place when a vegetable crop is actively growing, but after the conclusion of the harvest, the field is usually left fallow until the land is prepared again for the next season. During the fallow period, residual nutrients, particularly N, is subjected to deep leaching as the plastic mulch is removed from the system and the ground is exposed to any precipitation. During the removal of the plastic mulch, non-fruit biomass is also cleared from the field, resulting in the depletion of nutrients from the lack of organic matter inputs (Lewis and Johnson, 2007; Jalpa and Mylavarapu, 2023). Such a system, therefore, necessitates the addition of fertilizers season to season.

In order to scavenge the residual soil nutrients [nitrogen (N) in particular] and incorporate this biomass into the system, production and incorporation of cover crops were proposed as a means during intervening periods between two commercial crops. Both soil carbon (C) and nutrients can be returned to the soil enhancing soil health. When adding an alternative sustainable management practice such as the incorporation of cover crops into commercial vegetable production, soil physical, chemical, and biological attributes need to first be carefully assessed before recommending it as BMP for commercial vegetable production. For growers to adopt practices that promote positive soil health attributes, these parameters need to be measured individually. Once several parameters are screened for relevance, ease, and costs, then a soil health index has to be developed and implemented after field validation.to be measured and information disseminated in a clear way that depicts how soil functions change due to soil management practices that are typical for a production system (Bagnall et al., 2020). The assessment of multiple soil health indicators can be quantified into an index (Lehmann et al., 2020) and this project will be testing a suite of soil health indicators in north and south Florida that can be used to help develop a soil health index for Florida. This index should represent the practical soil health indicators that best inform growers on the sustainability of their soil management practices (Tahat et al., 2020).

Cover crop incorporation has been shown to positively influence soil physical, chemical, and biological properties without negatively affecting yields. In the literature, soil health in vegetable systems has mostly focused on the chemical aspects, where some biological, and no reports of soil physical properties have been assessed with cover crops (Norris and Congreves, 2018). This proceeding publication will present preliminary results on the soil physical indicators of saturated hydraulic conductivity and bulk density after a two-year rotation of fall tomato-winter rye-spring tomato-sorghum-sudan.

Materials and Methods

A two-year, tomato-cover crop field experiment was conducted at the University of Florida, Plant Science Research and Education Unit in Citra, FL (29.405400 N; –82.140000 W). The experiment started in Fall 2019 and concluded in Summer 2021. Before the commencement of the experiment, bulk soil samples of the field were taken to obtain background values of soil physical proper-

*Corresponding author. Email: raom@ufl.edu

ties. Raised beds (0.3 m high) on 1.5m centers were created and fumigant Pic-Clor 60 (TriEst Ag. Group Inc, Greenville, NC) was applied at a rate of 336 kg·ha^{-1} to prevent nematode infestation and suppress weeds. During fumigant application, drip tape (12-inch emitter spacing) was laid, and beds were covered in plastic mulch. Black plastic mulch was used for fall tomato production and white plastic mulch was used for spring tomato production. In preparation for transplanting, holes were punched into the plastic about 0.46 m apart. Three weeks after fumigation, tomato (var. HM 1823) seedlings were transplanted, and nutrients were applied one week after planting by fertigation. In Florida, the recommended rate of N fertilization for tomato production is 224 kg·ha^{-1} (Mylavarapu et al., 2022). Soil P (196 mg·kg^{-1}), Ca (431 mg·kg^{-1}), and pH (6.5) levels were sufficient and therefore application of P, Ca and lime was not done in any tomato season. After the termination of tomato, the area was tilled and cover crop grasses of winter rye and sorghum-sudan was drilled at 84–112 kg·ha^{-1} after fall tomato and spring tomato, respectively. Cover crop termination timing occurred in time for the bed preparation of the next tomato season. Therefore, cover crops were terminated after a growth period of approximately 90 days for winter rye and 21 days for sorghum-sudan.

A randomized complete block design (r = 3) was established, where soil samples were taken in between tomato plants. The sampling area was 7 m^2. Soil samples were taken at midseason and at harvest for the tomato seasons of Fall 2019, Fall 2020, and Spring 2021. The exception is Spring 2020, where only at harvest samples were taken due to the COVID-19 shutdown. During the cover crop seasons (Summer 2020, Winter 2020, and Summer 2021), soil samples were taken at harvest within the same sampling area, where tomatoes were previously grown. Soil samples in cores were analyzed for the soil physical indicators of saturated hydraulic conductivity (K_s) and bulk density (B_d). The K_s and B_d were calculated in the following equations:

$$Ks\left(\frac{cm}{min}\right) = \left(\frac{V_W \times l}{h \times A \times t}\right)$$

where V_w is the volume (mL) of water that passes through the soil core, l is the length (cm) of the soil core and constant headspace, h is the height (cm) of the soil in the core, A is the area (cm^2) of the soil core, and t is the time measuring water go through the core, which was 5 min; and

$$Bd\left(\frac{g}{cm^3}\right) = \frac{M_s}{V_t}$$

where M_s is the mass (g) of the oven dry soil (dried 24 h at 105 °C), and V_t is the volume of the soil core in cm^3 (Arias et al., 2005; Oliver et al., 2013).

Data were analyzed using statistical software (JMP 16.1; SAS Institute, Cary, NC). Background values of K_s and B_d were used as a control treatment in all analysis to show differences between managed and unmanaged soil. Since midseason and at harvest sampling occurred in the tomato seasons only, a two-way analysis of variance, with season and time as fixed factors and the block as a random effect, was applied to evaluate treatment effects on mean K_s and B_d in a randomized block design. To assess the effects of the tomato-cover crop rotations on mean K_s and B_d, only at harvest soil samples were considered and a one-way analysis of variance, with season as one fixed factor and the block as a random effect was done. The Tukey-Kramer honestly significant difference (HSD) post-hoc test at $\alpha = 0.05$ was used.

Results and Discussion

Depending on the production system, soil can undergo physical changes over a growing season as a result of tillage, compaction, wetting and drying cycles, and crusting (Moebius et al., 2007). Under typical soil management practices for tomatoes, soil K_s and B_d did not vary within a season or between seasons in the two-year study (Fig. 1 a, b). This could be due to the similarity in field preparation season to season, as machinery, such as the bed shaper, should result in similar B_d conditions within the plastic covered raised bed. Although it rains more during the spring season than fall, the soil surface protection from the totally impermeable plastic mulch seemed to have helped protect the soil structure. Additionally, the only substantial soil disturbance occurs during bed preparation (area is tilled and then beds are shaped, fumigated, drip tape is laid and covered in plastic simultaneously) three weeks before beds are planted so at the time of midseason and at harvest sampling, the lack of external disturbance is minimal.

Over the two-year tomato-cover crop rotation, K_s and B_d did not vary season to season when compared to background levels (Fig. 2 a, b). Due to the course texture of sandy soils, low B_d and high K_s is likely (USDA, 2019), thus, finding differences between background (unmanaged land) and managed soil K_s and B_d values was minimal under the short-term that cover crops were introduced to the tomato rotation. Spring 2020 tomato season was the only season, where B_d values were higher (more compacted) from the background, however, B_d did not vary in Spring 2021. Therefore, assessment of soil physical properties will require longer term observations for both overall and seasonal trends in physical properties changes. Despite the minimal findings of the physical aspect, it is expected that the microbiological, soil N and C may improve after additions of cover crops in the system.

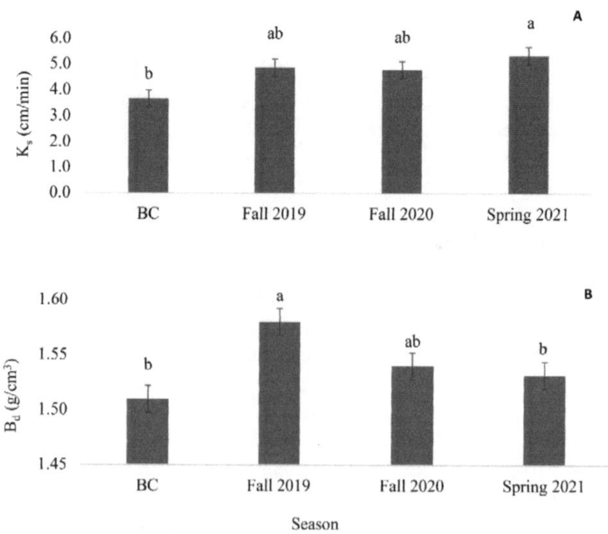

Fig. 1. (**A**) soil saturated hydraulic conductivity (K_s) and (**B**) bulk density (B_d) samples taken at midseason and at harvest in fall and spring tomato seasons. Only season as the main effect was significant, where timing and the interaction effect of season*timing were not significant according to Tukey-Kramer honestly significant difference (HSD) post-hoc test at $\alpha = 0.05$. BC is the background values of K_s and B_d taken before the start of the project, where the field had been left fallow previously.

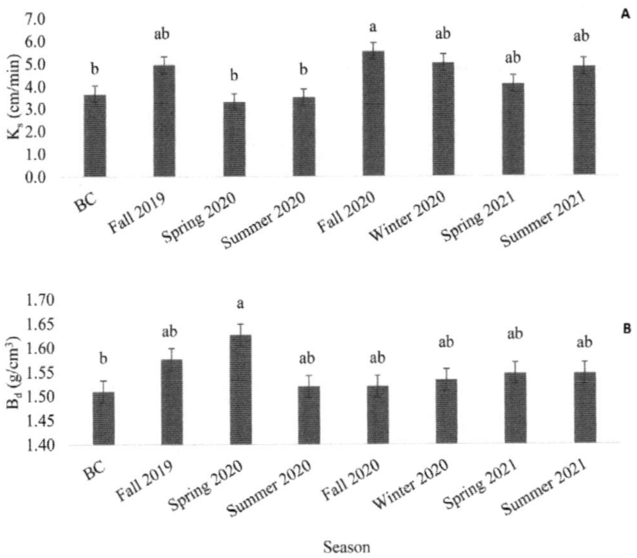

Fig. 2. (**A**) soil saturated hydraulic conductivity (K_s) and (**B**) bulk density (B_d) samples taken at harvest in a two-year rotation of fall tomato-winter rye-spring tomato-sorghum-sudan. BC is the background values of K_s and B_d taken before the start of the project, where the field had been left fallow previously.

Conclusions

When assessed individually, K_s and B_d indictors show the inclusion of cover crops into commercial vegetable production did not impact the physical component of soil, particularly sandy soils in the short-term. A long-term study may be needed to study the physical aspect of soil health under vegetable production systems.

Literature Cited

Arias M.E., González-Pérez J.A., González-Vila F.J., & Ball A.S. 2005. Soil health: A new challenge for microbiologists and chemists. *International Microbiology*. 8:13–21.

Bagnall D.K., McIntosh W.A., Morgan C.L.S., Woodward R.T., Cisneros M., Black M., Kiella E.M., & Ale S., 2020. Farmers' insights on soil health indicators and adoption. *Agrosystems, Geosciences and Environment*. 3(1):e20066. https://doi.org/10.1002/agg2.20066

USDA. 2019. Soil survey manual (U.S. Department of Agriculture Handbook No. 18). (n.p.): Lulu Press, Inc.

Jalpa L. & Mylavarapu R. 2021. Nitrogen management and recovery efficiency and response to nitrogen sources in drip irrigated tomato. *Proc. Fla. State Hort. Soc.* 134:99–102.

Jalpa L. & Mylavarapu R. 2023. Current tomato production practices and their effects on plant and soil carbon and nitrogen dynamics. *J. Plant Nutrition*. 46(16):3905–3917. https://doi.org/10.1080/01904167.2023.2220708

Jalpa L., Mylavarapu R.S., Hochmuth G.J., Wright A.L., & van Santen E., 2020. Apparent recovery and efficiency of nitrogen fertilization in tomato grown on sandy soils. *HortTechnology*. 30:204–211. https://doi.org/10.21273/HORTTECH04480-19

Jalpa L., Mylavarapu R.S., Hochmuth G.J., Wright A.L., & van Santen E. 2021. Recovery efficiency of applied and residual nitrogen fertilizer in tomatoes grown on sandy soils using the ^{15}N technique. *Scientia Horticulturae*. 278:109861. https://doi.org/10.1016/j.scienta.2020.109861.

Lehmann J., Bossio D.A., Kögel-Knabner I., & Rillig M.C., 2020. The concept and future prospects of soil health. *Nature Reviews Earth and Environment*. 1(10):544–553. https://doi.org/10.1038/s43017-020-0080-8

Moebius B.N., Van Es H.M., Schindelbeck R.R., Idowu O.J., Clune D.J., & Thies J.E., 2007. Evaluation of laboratory-measured soil properties as indicators of soil physical quality. Soil Science, 172(11):895–912. https://doi.org/10.1097/ss.0b013e318154b520

Mylavarapu R.S. 2003. Role of an extension soil testing program in the development of best management practices: A Florida case study. *Journal of Extension*. 45(4). Accessed 30 June 2023. https://archives.joe.org/joe/2003august/a7.php

Mylavarapu R., Hochmuth G., and Liu G. 2022. UF/IFAS Standardized nutrient recommendations for vegetable crop production in Florida. CIR1152. Soil and Water Science, IFAS Cooperative Extension Service, University of Florida, Gainesville, FL. https://doi.org/10.32473/edis-cv002-2021

Mylavarapu R. Mitchell C.C., & Savoy H.J. 2014. Soils of the southeastern US. In: Sikora F.J., & Moore, K.P. (eds.). Soil test methods from the southeastern United States. Southern Cooperative Series Bulletin. 419:54–58. Accessed 30 June 2023. https://aesl.ces.uga.edu/sera6/?PUB/MethodsManualFinalSERA6.pdf

Norris C.E., & Congreves K.A. 2018. Alternative management practices improve soil health indices in intensive vegetable cropping systems: A review. In Frontiers in Environmental Science (Vol. 6, Issue JUN). Frontiers Media S.A. https://doi.org/10.3389/fenvs.2018.00050

Oliver D.P., Bramley R.G.V., Riches D., Porter I., & Edwards J. 2013. Review: Soil physical and chemical properties as indicators of soil quality in Australian viticulture. *Australian Journal of Grape and Wine Research*. 19:129–139. https://doi.org/10.1111/ajgw.12016.

Tahat M.M., Alananbeh K.M., Othman Y.A., & Leskovar D.I. 2020. Soil health and sustainable agriculture. *Sustainability*. (Switzerland), 12(12). https://doi.org/10.3390/SU12124859

Zotarelli L., Scholberg J.M., Dukes M.D., Muñoz-Carpena R., & Icerman J. 2009. Tomato yield, biomass accumulation, root distribution and irrigation water use efficiency on a sandy soil, as affected by nitrogen rate and irrigation scheduling. *Agr. Water Manage*. 96:23–34.

Proc. Fla. State Hort. Soc. 136:116. 2023.

—Scientific Note—

LED Lights Influence the Volatile and Nutrition Quality of Amaranth Microgreens

MARLEE TRANDEL-HAYSE[1], JINHE BAI[1*], KRISTEN JEFFRIES[1], GAVIN POOLE[3], MICHAEL HENSLEY[1], WESLEY SCHONBORN[1], FRANCESCO DI GIOIA[2], AND ERIN ROSSKOPF[1]

[1]*USDA-ARS, U.S. Horticultural Research Laboratory, Fort Pierce, FL*

[2]*Department of Plant Science, Pennsylvania State University, University Park, PA*

[3]*Center for Integrated Pest Management, North Carolina State University, Raleigh, NC*

Keywords. amaranth, carotenoid, LED light, microgreen, metabolite, volatile profile

Amaranth microgreens are one of the most preferred due to the texture, flavor and eating quality (Maes et al., 2020). Microgreen production has shifted into controlled environment agriculture via light-emitting diodes (LEDs). Light spectra and intensity play a critical role in plant growth and phytochemical production. Previous assessment of red:blue (R:B) light with varying light intensity and photoperiods have been conducted (Alrifai et al., 2019; Chang et al., 2016; Maes et al., 2020; Kamal et al, 2020). Increased anthocyanin content was observed when growing amaranth under R80:B20 and R70:B30 (Maes et al., 2020). The addition of green (G) LED to R:B increased fresh weight and decreased hypocotyl length in brassica, but little is known about the influence of green light on amaranth. The objective of this work was to assess how LED light treatments of R:B, R:G:B, white and ambient light affect the volatile profile, flavonoid, carotenoid and vitamin content of amaranth microgreens.

Findings

Microgreens were grown in a growth chamber using Dyna-RX30 LED lights. Three LED light treatments of R:B 80:20, R:G:B 70:10:20 and white (LED control) at 300 $mmol \cdot m^{-2} \cdot s^{-1}$ and one ambient treatment (ambient greenhouse light) were used. Microgreens were subjected to a 13-hour photoperiod and trays were rotated 6.5 hours into the photoperiod. Before harvest, hypocotyl length (cm) and color (L*, a*, b*, hue angle and C*) were recorded. Immediately after harvest, microgreens were frozen at –80 °C and ground to a powder, then split into two subsamples for volatile analysis via GC-MS for flavonoid, carotenoid and vitamin quantification via LC-QQQ.

No significant differences were seen in amaranth yield (g) but hypocotyl length (cm) was lowest in amaranth grown under R:B and R:G:B. Leaf color differed and a* was highest and hue angle was lowest in R:B and R:G:B indicating enhanced red color. Thirty-one volatiles were identified in amaranth. The most abundant compounds were hexanal (22%), octyl 4-ethylbenzonate (17%), E-2-hexenal (11%) and Z-3-hexenal (9%). Principal component analysis (PCA) was performed and clearly separated R:B and R:G:B from the LED control and ambient light. Seven vitamins, six carotenoids and seventeen flavonoids were quantified. The most abundant compounds were Vitamin B6, Lutein and homeoriodictyol. PCA clearly separated the R:B and R:G:B treatments from the LED control and ambient light. Tangeretin, nobiletin, Vitamin E, prunin, zeaxanthin and lutein increased in amaranth grown under R:B and R:G:B. The addition of green light did not have a substantial impact on amaranth quality as plants absorb and utilize mainly red and blue portions of the light spectra and reflect most of the green (Kamal et al., 2016). R:B and R:G:B both decreased hypocotyl length and increased nutrient composition when compared to the LED control and ambient light.

*Corresponding author. Email: jinhe bai@usda.gov

Suggestions for Future Direction

The use of LED lights has the potential to enhance consumer acceptance, volatile profile and nutritional quality of microgreens. Future work focusing on sensory analysis is needed to address the changes in volatiles and flavonoids in microgreens grown under various LED light treatments. Additional work is needed to identify how the biochemical pathway differs under LED light treatments to upregulate genes and enhance microgreen flavor or nutrient composition.

Literature Cited

Alrifai O., Hao X., Marcone M., & Tsao R. 2019. Current review of the modulatory effects of LED lights on photosynthesis of secondary metabolites and future perspectives of microgreen vegetables. *J. Agric. Food Chem.* 67:6075–6090.

Chang S., Li C., Yao X., Chen S., Jiao X., Liu X., Xu Z., & Guan R. 2016. Morphological, photosynthetic and physiological responses of rapeseed leaf to different combinations of red an blue lights on the rosette stage. *Front. Plant Sci.* 7:1–12.

Kamal K., Khodaeiaminjan M., El-Tantawy A., Moneim D., Abdel Salam A., Ash-Shormilesy S., Attia A., Ali M., Herranz R., El-Esawi M., Nassrallah A., & Ramadan M. 2016. Evaluation of growth and nutritional value of Brassica microgreens grown under red, blue and green LEDs combination. *Physiol Planta.* 169:625–638.

Maes S., Luengwilai K., & Thongket T. 2020. Enhancing growth and phytochemicals of two amaranth microgreens by LED light irradiation. *Scientia Hortic.* 265:109204.

Proc. Fla. State Hort. Soc. 136:117. 2023.

—Scientific Note—

Exploring the Terroir of Florida Hops through Cone Quality Analysis and Beer Sensory Evaluation

Alvaro J. Bautista and Shinsuke Agehara*

Gulf Coast Research and Education Center, University of Florida, IFAS, Wimauma, FL

Keywords. alpha acid, alternative crop, craft beer, essential oil, *Humulus lupulus*

Hops (*Humulus lupulus*) are perennial, twining herbaceous plants that grow vertically on a trellis system. Global hop production is concentrated in temperate climatic regions between 45 °N and 50 °N. However, our previous studies developed production strategies that enable double-season production of hops under subtropical climatic conditions in Florida. The commercial value of hops is obtained from the female inflorescence, which is an essential ingredient that provides aroma and bitterness to beer.

The term terroir, well-studied in wine, refers to the environmental characteristics of a region that adds unique sensory qualities to the wine produced. Florida's climatic conditions, soils, and crop management practices may affect bitterness and aromatic profiles of hops, which, along with double-season production, could add a distinguished value to craft beer made with hops grown in Florida. In this study, we aimed to explore the terroir of Florida hops through cone quality analysis (α acids, β acids, and essential oils) and beer sensory evaluation. 'Cascade' hops were grown on a 5.5-m V-shaped trellis in west central Florida.

Data were collected during the spring growing seasons from 2020 (establishment year) to 2022 and the Fall 2021 growing season. Total α acid concentration was equivalent to or greater than the commercial average value in all three spring seasons, whereas total essential oil concentration reached the commercial average value only in the 2021 and 2022 Spring growing seasons.

In 2021, the spring growing season had higher α acid, β acid, and essential oil values than the fall growing season. Dried or fresh whole cones were used by local breweries to brew a variety of beer, including pale ale, lager, and Saison. American pale ale brewed using dried whole cones had a unique flavor profile with notes of cantaloupe and honeydew up front and ripe citrus in the background. This study provides a benchmark to characterize the terroir of Florida hops.

*Corresponding author. Email: sagehara@ufl.edu

Proc. Fla. State Hort. Soc. 136:118. 2023.

—Scientific Note—

Progress Toward Breeding Sweet Mini Peppers Improved for Bacterial Spot Resistance

Dominick Padilla[1]*, Lucas Manko[1], Jeff Jones[2], and Bala Rathinasabapathi[1]

[1]Horticultural Sciences Department, University of Florida, IFAS, Gainesville FL

[2]Plant Pathology Department, University of Florida, IFAS, Gainesville FL

Keywords. *Capsicum annuum*, *Xanthomonas euvesicatoria*, prebreeding

Pepper (*Capsicum annuum*) is a valuable vegetable crop, with Florida ranking as the second largest producer of bell peppers in the United States. However, the disease bacterial leaf spot (BLS) caused by *Xanthomonas euvesicatoria* can devastate plant growth of susceptible varieties. Breeding for resistance to *X. euvesicatoria* represents an environmentally sound nonchemical alternative to managing this pathogen. Resistance genes are available in the pepper germplasm but have not been deployed in modern bell pepper varieties. Mini peppers are novel specialty sweet peppers that are convenient to use. A breeding project was started to develop inbred sweet mini pepper cultivars with durable and broad-spectrum resistance to BLS.

Findings

A biparental F_2 breeding population derived from two sweet heirloom pepper varieties (HRLM0005 and HRLM0006) was grown under greenhouse conditions in Gainesville, FL during Spring 2021. Three F_2 plants, susceptible to BLSwere were selected based on performance and fruit quality for advancing through single-seed descent. In the following season of Fall 2022, all three plants were then used as the female parent to cross-pollinate with the bacterial spot resistant bell pepper variety *C. annuum* 'ECW50R' carrying the recessive resistance gene *bs5* (Jones et al., 2002). Among the progenies from three selected hybrids, one was selected based on fruit quality traits fitting the mini pepper type criteria to screen for BLS resistance. These three traits were: 1) fruit weight range of 10–20 g; 2) fruit length range of 5–10 centimeters;and 3) fruit width range of 2.5–5 centimeters. A screening population of 28 plants were grown under greenhouse conditions and inoculated with *X. euvesicatoria* race 6. The segregation of resistant, moderately susceptible, and susceptible plants fit an expected 1:2:1 ratio respectively. Among the 28 plants, 5 plants demonstrated high levels of resistance and were selected for advancing. One plant with moderate susceptibility was also selected. The six F_2-derived F_3 families each consisted of at least 10 plants, which were screened for fruit quality traits fitting the mini pepper type criteria.

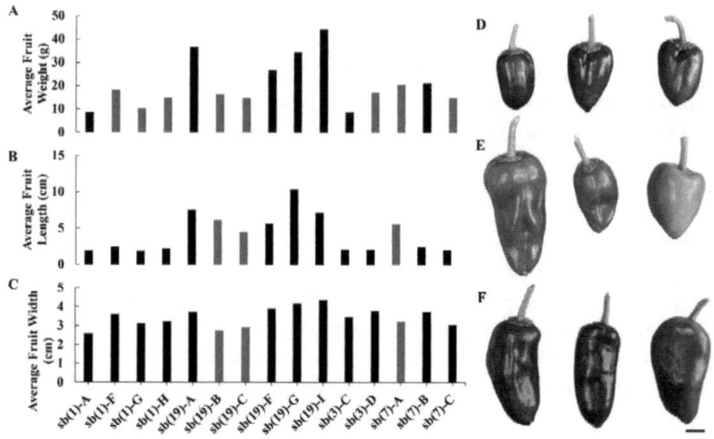

Fig. 1. Measured traits in selection index for sweet mini peppers. Columns highlighted in blue indicate values agreeable to selection index for mini pepper criteria. The F_3 plants selected are sb(19)-B, sb(19)-C, and sb(7)-A. (**A**) Average fruit weight (g) of evaluated F_3 plants. (**B**) Average fruit length (cm) of evaluated F_3 plants. (**C**) Average fruit width (cm) of evaluated F_3 plants. (**D**) Fruit of sb(19)-C. (**E**) Fruit of sb(7)-A. (**F**) Fruit of sb(19)-B. The scale bar is 2 cm.

All evaluated plants in the screening had high BLS resistance. Three lines, called sb(7)-A, sb(19)-B, and sb(19)-C, were selected for advancing (Fig. 1). They have 50% of the genome of 'Early California Wonder' (the donor of *bs5* gene) and 25% each of the genomes from two heirloom varieties.

Future Directions

The lines in the F_3 generation selected from this breeding cycle are expected to breed true for bacterial spot resistance (*bs5*/*bs5*) but need to undergo several rounds of recurrent selection to fix fruit quality traits and yield. Thus, new cultivars of specialty mini sweet pepper with durable and broad-spectrum resistance to BLS combined with desirable fruit quality and yield can be developed.

Literature Cited

Jones J.B., Minsavage G.V., Roberts P.D., Johnson R.R., Kousik C.S., Subramanian S., & Stall R.E. 2002. A non-hypersensitive resistance in pepper to the bacterial spot pathogen is associated with two recessive genes. *Phytopathology*. 92(3):273–277. https://doi.org/10.1094/phyto.2002.92.3.273

*Corresponding author. Email: dpadilla@ufl.edu

Proc. Fla. State Hort. Soc. 136:119. 2023.

—Scientific Note—

Comparing Corn Silk Fly Populations in Sweet Corn Fields and Surrounding Habitats

Larissa Pereira Lima[1], Julien Beuzelin[1]*, Dakshina Seal[2], and Sandra Allan[3]

[1]*Everglades Research and Education Center, University of Florida, IFAS, Belle Glade, FL*

[2]*Tropical Research and Education Center, University of Florida, IFAS, Homestead, FL*

[3]*Center for Medical, Agricultural and Veterinary Entomology, USDA-ARS, Gainesville, FL*

Keywords. *Euxesta stigmatias*, landscape ecology

Picture-winged flies (Diptera: Ulidiidae), referred to as corn silk flies in Florida, are major insect pests of sweet corn produced for the fresh market in the southern region of the state. Adults of three species, *Euxesta stigmatias*, *Euxesta eluta*, and *Chaetopsis massyla* lay eggs on sweet corn silks, resulting in maggots feeding on developing kernels. These adults are managed with insecticide applications to prevent oviposition. However, frequent applications are needed to meet market standards because corn silk fly adults quickly reinfest treated sweet corn fields. In Spring 2021 and 2022, eight sites on commercial farms in the Everglades Agricultural Area (EAA) were sampled to determine corn silk fly adult relative abundance in a sweet corn field, a sugarcane field, and a non-crop area.

For each habitat at each site, five Multilure traps baited with ammonium acetate and 1,4-dimethoxybenzene were deployed along a 245-m transect to collect adults weekly for three to four weeks. Additionally, sampling was conducted at the University of Florida, Institute of Food and Agricultural Sciences Everglades Research and Education Center, which is a research farm in the EAA. One hundred fifty-four sampling locations were selected throughout this research site following a grid pattern with inter-location distances of 90–115 m. These locations represented a mosaic of habitats similar to those sampled on commercial farms. However, unlike commercial fields, sweet corn research fields are not intensively treated with insecticides. A Multilure trap was deployed at each sampling location, and corn silk fly adults were collected weekly for four weeks.

Findings

In general, the number of adult corn silk flies on commercial farms was higher in non-crop habitats and sugarcane fields than in sweet corn fields, primarily due to the frequent applications of insecticides in sweet corn fields. The three species of corn silk flies were found in all habitats during the spring season; however, the proportion of *E. stigmatias* relative to *E. eluta* and *C. massyla* was greater in sweet corn fields than in the other two habitats. These observations might be partially explained by the lower susceptibility of *E. stigmatias* to pyrethroids, which are the main insecticides used for corn silk fly management (Schwan Resende et al., 2022).

On the research farm, *C. massyla* adults were clustered in tall sugarcane fields whereas *E. eluta* was found in a diversity of habitats, including sweet corn fields not treated with insecticides. Adults of *E. stigmatias* were clustered in sweet corn fields regardless of insecticide applications. Thus, observations for *E. stigmatias* on the research farm were consistent with observations on commercial farms, but aggregation in both treated and non-treated sweet corn fields suggests a preference for sweet corn habitats.

Suggestions for Future Direction

The three species of corn silk flies exhibit different behaviors and susceptibilities to insecticides. Thus, our results suggest that the effectiveness of management practices might change with corn silk fly species. Nevertheless, the development of practices preventing the movement of *E. stigmatias* populations into sweet corn fields might complement management with insecticides.

Literature Cited

Schwan Resende E., Beuzelin J.M., Dunkley V.E., Paula-Moraes S.V., Seal D.R., & Nuessly G.S. 2022. Pyrethroid susceptibility in field populations of picture-winged flies (Diptera: Ulidiidae) infesting fresh market sweet corn in Florida. *J. Econ. Entomol.* 115:1685–1692. https://doi.org/10.1093/jee/toac125

*Corresponding author. Email: jbeuzelin@ufl.edu

Proc. Fla. State Hort. Soc. 136:120. 2023.

—Scientific Note—

Effect of Cover Crop Species Proportion on Plant-parasitic Nematode Populations under Greenhouse Conditions in Florida

JEAN MAUDE LOUIZIAS[1]*, JOHAN DESAEGER[2], ROSALIE L. KOENIG[3], GABRIEL MALTAIS-LANDRY[4], AND CARLENE A. CHASE[1]

[1]*Horticultural Sciences Department, University of Florida, IFAS, Gainesville, FL*

[2]*Gulf Coast Research and Education Center, University of Florida, IFAS, Wimauma, FL*

[3]*Agromomy Department, University of Florida, IFAS, Gainesville, FL*

[4]*Department of Soil, Water, and Ecosystem Sciences, University of Florida, IFAS, Gainesville, FL*

Keywords. *Belonolaimus longicaudatus*, *Crotalaria juncea* L., *Meloidogyne hapla*, *Sorghum bicolor* Moench × *S. sudanense* (Piper) Stapf

Sunn hemp (SH; *Crotalaria juncea* L.) is a cover crop that is a poor host to plant-pathogenic nematodes that can adversely affect strawberry (*Fragaria ×ananassa* Duch.) production in Florida (Braz et al., 2016). Due to the high cost of sunn hemp seed, strawberry growers are interested in a biculture of sunn hemp with sorghum-sudangrass [SS; *Sorghum bicolor* Moench × *S. sudanense* (Piper) Stapf] to decrease the cost of SH use. Additionally, SS is valued as a catch crop, which prevents leaching of residual nutrients from a prior crop (Allar et al., 2023). However, SS cultivars are good hosts of sting nematode (*Belonolaimus longicaudatus* Rau) and vary in their susceptibility to root-knot nematodes (*Meloidogyne* spp.). Thus, we hypothesized that SH/SS bicultures would be less effective at suppressing plant-pathogenic nematodes than the SH monoculture. Therefore, our objective was to evaluate the effect of SS proportion in bicultures with SH on northern root-knot nematode (*M. hapla* Chitwood) and sting nematode populations.

Methods

In April 2022, two distinct replacement series experiments were initiated with six treatments: five SH:SS proportions (SH monoculture, SH-75:SS-25, SH-50:SS-50, SH-25:SS-75, and SS monoculture) and a susceptible control. Corn (*Zea mays* L.) and tomato (*Solanum lycopersicum* L.; cv. HM1823) were used as the control treatments for the *B. longicaudatus* and *M. hapla* experiments, respectively. The treatments were arranged in a randomized complete block design with five replications for the *B. longicaudatus* experiment and four replications for the *M. hapla* experiment. Naturally infested soil obtained from the Gulf Coast Research and Education Center in Wimauma, FL was used in greenhouse pot studies in Gainesville, FL. Plant density was four plants per pot. The SH and SS cultivars were Tropic Sun and Sweet Six BMR, respectively. At 10 weeks after planting, data were collected from each pot on shoot biomass per species. Sting nematodes and juvenile (J2) root-knot nematodes were extracted from 100 cm^3 soil samples and quantified.

Findings

In both trials, SH and SS monocultures produced a similar amount of shoot biomass, whereas the three bicultures all produced significantly higher amounts of biomass than the SH and SS monocultures. Shoot biomass did not differ among the biculture treatments. Sting nematode population was highest with corn, the susceptible control. Neither the SH monoculture nor the SH/SS bicultures resulted in sting nematode populations that were significantly higher than the SS monoculture. Interestingly, the SH-75:SS-25 treatment had a lower sting nematode population than the SS monoculture ($P < 0.0001$). Whereas the SH monoculture and the three bicultures had significantly lower root-knot nematode populations than the susceptible tomato control. The root-knot nematode population with the SS monoculture was not significantly different from the control. Therefore, in both trials no increase in sting and root-knot nematodes populations occurred as the SS proportion increased in bicultures.

Suggestions for Future Direction

These preliminary results suggest that the lower cost bicultures, which contain as much as 75% SS have the potential to provide the ecological services that are typical with the SH monoculture. The greenhouse studies will be repeated in 2023 to determine whether these results are consistent. A field study also will be conducted with the same SH:SS proportions but at three different total densities. The results will be used to develop recommendations that will allow farmers to make more informed decisions when using bicultures with the two cover crop species to attain specific goals.

Literature Cited

Allar J., Mallinger R., Liu C., Grabau Z., & Maltais-Landry G. 2023. Different cover crops have a limited impact on marketable yields and biogeochemical cycling but secondary effects on pollinators and plant-parasitic nematodes in Florida organic vegetable systems. *Front. Sustain. Food Syst*. 7:1148866. https://doi.org/10.3389/fsufs.2023.1148866.

Braz G.B.P., Oliveira Jr. R.S., Crow W.T., & Chase C.A. 2016. Susceptibility of different accessions of *Crotalaria juncea* to *Belonolaimus longicaudatus*. *Nematropica*. 46:31–37.

*Corresponding author. Email: jlouizias@ufl.edu

Proc. Fla. State Hort. Soc. 136:121. 2023.

—Scientific Note—

Effect of Insecticide Residue on the Survival Rate of *Catolaccus hunteri*, Biocontrol Agent of Pepper Weevil *Anthonomus eugenii* Cano

Nagamani Kanchupati[1]*, Dakshina R. Seal[1]*, Oscar E. Liburd[2], Julien Beuzelin[3], and Bruce Schaffer[1]

[1]*Tropical Research and Education Center, University of Florida, IFAS, Homestead, FL*

[2]*Department of Entomology and Nematology, University of Florida, IFAS, Gainesville, FL*

[3]*Everglades Research and Education Center, University of Florida, IFAS, Belle Glade, FL*

Keywords. biocontrol agent, insecticides, integrated pest management, pepper weevil

Pepper, *Capsicum* annum L., is one of the five most-cultivated species of the plant genus *Capsicum* in tropical and temperate regions. Pepper weevil, *Anthonomus eugenii* Cano is the most harmful insect pest of pepper. The use of broad-spectrum insecticides is the primary management tool where insecticide application is targeted to kill adults leaving all immature stages, as their development occurs inside the fruits, unharmed.

In the present study, efforts have been made to find out the effect of an alternative strategy, integrating insecticides with biological control agent, *Catolaccus hunteri*. In order to avoid harmful effect of insecticides, *C. hunteri* will be released at different intervals after spraying insecticides.

As a preliminary step, I conducted a laboratory bioassay using four different insecticides: azadirachtin (Neemix, botanical); Esfenvalerate (Asana, Pyrethroid); acetamiprid (Assail, Neonicotinoid); and cyantraniliprole (Exirel, Diamide). In this study, *C. hunteri* adults (48 hours-old) were exposed to insecticide-treated Petri dishes 24 hours, 3 days, 5 days, and 7 days post-insecticide application.

Based on the bioassay results, *C. hunteri* exposed to 7-day-old residues of each insecticide showed less mortality having normal biological activity. In the next step, *C. hunteri* will be released in the pepper field on the seventh day after spraying insecticides to control pepper weevil.

*Corresponding authors. Email: kanchupati.n@ufl.edu; dseal3@ufl.edu

Proc. Fla. State Hort. Soc. 136:122. 2023.

—Scientific Note—

Evaluation of Tropical Pumpkin for Yield and Horticultural Traits

PRERNA SABHARWAL, YUQING FU, SWATI SHRESTHA, AND GEOFFREY MERU*

Tropical Research and Education Center, University of Florida. IFAS, Homestead, FL

Keywords. brix index, calabaza, fruit parameters, yield

Tropical pumpkin (*Cucurbita moschata*), also known as calabaza, fulfills lucrative emerging markets for the crops in the United States. A recent consumer survey indicated fruit quality (flavor, texture, flesh color, and shape) and price point as the most important factors in buying decision, while growers emphasized yield, earliness, growth habit and tolerance to pests and diseases. The market for calabaza in Florida according to the most current ethnic market analysis was estimated at $5 million, while the total value of imports into the country was estimated at $30 million (Moreno et al., 2023). The objective of this study is to evaluate different calabaza cultivars for the abovementioned parameters and work towards finding cultivars that will perform best in Florida due to the surge in consumer demands for calabaza.

Findings

The findings of this project span the duration of three years. In the years 2020 and 2021, we evaluated the yield and horticultural performance of 6 breeding lines and 3 commercial tropical pumpkin cultivars (La Estrella, Soler, and Taina Dorada) under conventional practices at two locations in Florida (Homestead and Live Oak). We collected data on flowering time (days to male or female flowering), fruit size, fruit shape, disease/insect tolerance, brix index (the sweetness of the fruit), and yield. We analyzed the data using the GLM procedure in SAS. The average fruit yield per plant ranged from 1–3 fruit/plant, while the mean yield per acre across locations ranged from 6760 lb/acre to 41,760 lb/acre. Among the small-size pumpkins (< 6 lb), UFTP8 had the highest yield followed by UFTP22, while for medium-large size pumpkins (> 6 lb), UFTP38, UFTP42, and 'Taina Dorada' had superior yields. UFTP8 and UFTP22 showed the best resistance to silverleaf disorder, while 'Soler' and UFTP57 were the most susceptible. Brix values ranged from 3.7 to 7.7, highest in UFTP8 and UFTP24, but lowest in 'Soler'. In the year 2022, we evaluated 19 breeding lines and 10 commercial cultivars. We had 1–4 fruit/plant with the mean yield per acre ranging from 3466 lb/acre to 29,240 lb/acre. UFTP8 had the highest yield followed by cultivar 1990; 'Fortuna' had the lowest yield for small-sized pumpkins (< 5–6 lb). For medium-large size pumpkins (> 6 lb), UFTP38, UFTP10, and UFTP42 had superior yields; 'Soler' and UFTP26 had the lowest yields. Brix values ranged from 2.5 to 7.7 and were highest in JW6823, 'Waltham Butternut', and UFTP8, but lowest in 'Soler' and 'Dickinson'.

Suggestions for Future Directions

The objective of this study was to assess and compare the different traits of new calabaza hybrids with those of commercial cultivars that served as references. The results from this study indicate a wide variation in yield and horticultural performance among the germplasm evaluated. The extensive parameter evaluation being conducted in this study will provide an overall assessment of a cultivar for growers. These results can be used to release one or more potentially good calabaza cultivars into the market that grows well in Florida, have a good flesh quality, taste, and resistance to diseases and pests. These genotypes might increase the variety of calabaza cultivars that can be grown commercially in tropical and subtropical regions.

Literature Cited

Moreno S.R., Yazdanpanah M., Huang T., Sims C.A., Chase C.A., Meru G., Simonne A., & MacIntosh A.J. 2023. Comparative analysis of qualitative attributes for selection of calabaza genotypes in the southeast United States. *Horticulturae*. 9(3):409. https://doi.org/10.3390/horticulturae9030409

*Corresponding author. Email: gmeru@ufl.edu

Proc. Fla. State Hort. Soc. 136:123. 2023.

—Scientific Note—

Exploring the Host Range *Xanthomonas hortorum* pv. *vitians* the Causal Agent of Bacterial Leaf Spot of Lettuce

Byron Manzanero, Katia Xavier, Calvin Odero, and Germán Sandoya*

Everglades Research and Education Center, University of Florida, IFAS, Belle Glade, FL

Keywords. alternative hosts, asteraceae, bacterial leaf spot, cultivated crops, Everglades Agricultural Area, hypersensitivity response, sporadic, weed species

Bacterial leaf spot (BLS), caused by *Xanthomonas hortorum* pv. *vitians* (*Xhv*), is a severe and destructive disease that affects lettuce (*Lactuca sativa* L.). All lettuce commercial cultivars are susceptible to this disease. Its management is challenging due to the absence of effective chemical control methods. Bacterial leaf spot's sporadic nature needs constant and rigorous preventive measures. The Florida lettuce industry, concentrated in the Everglades Agricultural Area (EAA) in southern Florida, is particularly vulnerable to the unpredictability of BLS. The disease thrives in the conducive environmental conditions of the region, facilitating its development and fast spread when outbreaks occur. Elucidating the complete host range of *Xhv*, identifying alternative hosts, and understanding their role in the disease cycle are essential for effective BLS management, particularly because of high weed pressure in lettuce fields. Previous research indicates the presence of the pathogen in weeds neighbouring lettuce fields and the diverse leafy vegetables grown alongside lettuce during the season. To address this knowledge gap, the objective of this study was to identify potential alternative hosts of *Xhv* among weed species and leafy vegetables crops that are planted in conjunction with lettuce in southern Florida.

To achieve this, Florida bacterial *Xhv* isolate Sc8B was spray-inoculated onto seven common weed species and 17 crops from different plant families. These included weed species from Malvaceae (n = 1), Portulacaceae (n = 1), Amaranthaceae (n = 2), Asteraceae (n = 3), Brassicaceae (n = 1), and Poaceae (n = 1). Vegetable crops included Asteraceae (n = 6), Brassicaceae (n = 7), and Apiaceae (n = 5). The inoculated plants were initially incubated for two days in a growth chamber set at 100% relative humidity and were subsequently relocated to a glasshouse environment with 60% shade, featuring an automated misting system programmed to activate at 15-minute intervals, with each misting lasting 30 seconds. Disease severity was assessed 12 days post-inoculation (DPI) using a rating scale of 0–5 based on symptomatic foliage area, ranging from 0 (absent) to 5 (> 50% leaf surface with symptoms). During the incubation period, initial BLS symptoms were observed on the susceptible lettuce 'Okeechobee' at 2 DPI as water-soaked lesions. These symptoms progressed into severe infection with an average rating score of 4 at 12 DPI. Similar symptoms were observed in crop plants belonging to the Asteraceae and Apiaceae families. The median disease severity (DS) scores were as followed for infected plants: Endive (*Cichorium endivia* L.) = 3; escarole (*C. endivia* L. *var. latifolium*) = 3; and coriander (*Coriandrum sativum* L.) = 2. These infected plants exhibited features similar to bacterial leaf spot (BLS) symptoms in lettuce, including water soaking, chlorosis, and leaf necrosis. However, other crop species within the Brassicaceae and all tested weed species remained asymptomatic.

This suggests that *Xhv* persistence in the environment is due to its enhanced epiphytic capabilities and possible intrusion into plant leaves. To validate the presence or intrusion of *Xhv* in lettuce, leaf samples were collected from three plants of each accession per replication from both symptomatic and asymptomatic leaves. Leaves were surface sterilized and plated on glucose nutrient agar, followed by incubation at room temperature. The recovered colonies, selected across various plant species, were yellow resembling *Xhv*. Pathogenicity confirmation for the collected bacteria was attained through hypersensitivity and compatibility response tests. Lettuce leaves were infiltrated with the recovered bacteria, using resistant PI 358001-1 and susceptible 'Okeechobee' lettuce, respectively. The recovery rates of *Xhv* varied among plant accessions within the distinct crop families tested. For the Apiaceae, rates ranged from 33% to 56%, while the Brassicaceae exhibited rates from 33% to 67%, and Asteraceae displayed rates from 44% to 78%. Among weed species, the recovery rates for different families were as follows: Amaranthaceae (0% to 44%), Asteraceae (0% to 67%), Brassicaceae (0%), Poaceae (44%), and Portulacaceae (33%). The results indicate that certain plant species, particularly those belonging to Asteraceae and Apiaceae, may act as alternative hosts for *Xhv*. Moreover, both leafy vegetable crops and weed species have the potential to harbor this pathogen, indicating potential sources of inoculum.

In conclusion, *Xhv* may infect plants beyond the *L. sativa* species, including *C. endivia* and *C. sativum*. Furthermore, the pathogen demonstrates robust survival in an epiphytic state and within the internal structures of leaves as an endophyte.

*Corresponding author. Email: gsandoyamiranda@ufl.edu

Proc. Fla. State Hort. Soc. 136:124. 2023.

—Scientific Note—

Tackling Nitrogen Fertilization for Improving Organic Celery Production: Fertilizer Source, Preplant Fertilization, and In-season Fertigation

Zachary Ray[1], Lincoln Zotarelli[1], Gabriel Maltais-Landry[2], and Xin Zhao[1]*

[1]*Horticultural Sciences Department, University of Florida, Gainesville, FL*

[2]*Department of Soil, Water, and Ecosystem Sciences, University of Florida, Gainesville, FL*

Keywords. celery, mineralization, nitrogen management, organic

Timing is crucial in fertility management for organic vegetable production, as the release of mineral nitrogen (N) from nutrient sources rich in organic N must be synchronized with crop demand. Despite emerging trends in organic celery (*Apium graveolens* var. *dulce*) production in North and Central Florida, research-based recommendations for N application timing are scarce. Improved nutrient management has been identified as a research priority for organic crop production (Snyder et al., 2022), with N management in sandy soils posing particular challenges for organic producers throughout the Southeast. With a broad range of organic fertilizers and raw N sources available, it is challenging to predict mineralization rates and seasonal availability of N to develop organic N management programs. In addition to organic fertilizer properties, factors like soil temperature, moisture, and organic matter content can affect mineralization of organic N (Grzyb et al., 2020) and the retention of mineral N in the soil. The objective of this experiment was to assess celery growth, yield performance, and N uptake under different ratios of preplant N fertilization to in-season N fertigation using contrasting granular organic fertilizers. We hypothesize that integrating preplant fertilization with in-season fertigation could be a cost-effective way to help improve overall N use efficiency by enhancing synchrony of N supply with crop demand.

A field trial was conducted from November 2022 to March 2023 on certified organic land at the University of Florida Plant Science Research and Education Unit (Citra, FL) to compare preplant fertilization regimes and organic fertilizer sources using a split plot design with four replications. Whole plots included six treatments consisting of preplant fertilization at 0 (PF_0), 20 (PF_{20}), 40 (PF_{40}), 60 (PF_{60}), 80 (PF_{80}), and 100% (PF_{100}) out of a total N rate of 280 $kg \cdot ha^{-1}$ using Aqua Power 5N–0.4P–0.8K liquid fish fertilizer (JH Biotech Inc., Ventura, CA) for in-season fertigation as needed. Two preplant fertilizer sources including bone-, blood-, feather meal-based Nature Safe 10N–0.9P–6.6K (Darling Ingredients Inc., Irving, TX) and heat-processed poultry litter Everlizer 3N-1.3P-2.5K (Organic Growing Solutions, Mayo, FL) were randomized in subplots. Crop performance was evaluated throughout the production season by monitoring leaf relative chlorophyll content (SPAD-502 Plus; Konica Minolta, Inc., Osaka, Japan) and normalized difference vegetation index (NDVI) on a canopy-scale (Green Seeker; Trimble Inc., Westminster, CO). At 82 and 110 days after transplanting (DAT), crop yield was assessed and leaf blade and petiole tissues were separated to quantify partitioning of biomass and N. Data were analyzed following a linear mixed model (GLIMMIX) in SAS (Version 9.4, SAS Institute, Cary, NC), with multiple comparisons using Fisher's least significant difference test ($\alpha = 0.05$).

Leaf relative chlorophyll content was affected by preplant N regime, although no effect of source was observed. At 48 DAT, SPAD values at PF_{80}-PF_{100} were higher than PF_0-PF_{40} but at 68 DAT, SPAD was highest at PF_{40}, significantly higher than PF_0 and PF_{100}. At 110 DAT, PF_{20} resulted in SPAD values higher than PF_{40}, PF_{80}, and PF_{100}. Preplant N regime and source exhibited a significant two-way interaction effect on NDVI at a canopy level at 68 and 110 DAT. At 68 DAT, NDVI was highest at PF_{80} when using Everlizer, greater than all N regimes except PF_{60}. The NDVI values from PF_{20}-PF_{80} were similar when using Nature Safe and significantly higher than PF_0 and PF_{100}. At 110 DAT, NDVI values at PF_0-PF_{60} were similar under both preplant fertilizer sources, significantly higher than PF80 and PF100 with Everlizer and higher than PF_{100} with Nature Safe. Everlizer also improved NDVI relative to Nature Safe with PF_{60}-PF_{100} at 68 DAT and with PF_{60} and PF_{100} at 110 DAT. Crown diameter at PF_{40} was greater than PF_0, PF_{80}, and PF_{100}, while thicker crowns were observed in Everlizer vs. Nature Safe across all N regimes at 82 DAT. At the final assessment (110 DAT), only PF_{100} resulted in significant reductions in crown diameter. There was a significant two-way interaction between preplant N regime and source on celery yield at 82 DAT. Yields were higher at PF_{40}-PF_{80} than for other regimes with Everlizer and greater yields at PF_{20}-PF_{80} with Nature Safe. Meanwhile, Everlizer improved yields in contrast to Nature Safe at preplant N regimes greater than PF_{20}. At 110 DAT, PF_0-PF_{60} resulted in yields significantly higher than PF_{100}, although no significant effect of fertilizer source was observed. Total aboveground N accumulation data from 110 DAT supported trends in yield and biometric data, where PF_0-PF_{60} accumulated more N in leaf blade and petiole tissues as compared with PF_{80}-PF_{100}.

Nitrogen application timing can play a critical role in enhancing celery productivity and N uptake efficiency in organic systems. Future work is needed to evaluate different organic sources of N, with special attention to N rates and application timing to improve organic crop performance.

Literature Cited

Grzyb A., Wolna-Maruwka A., & Niewiadomska A. 2020. Environmental factors affecting the mineralization of crop residues. *Agronomy*. 10(12):1–18. https://doi.org/10.3390/agronomy10121951

Snyder L., Schonbeck M., Velez, T. & Tencer, B. 2022. 2022 National organic research agenda. Organic Farming Research Foundation, Santa Cruz, CA.

This work is supported by the Organic Agriculture research and Extension Initiative grant no. 2019-51300-30243 from the USDA National Institute of Food and Agriculture.

*Corresponding author. Email: zxin@ufl.edu

Proc. Fla. State Hort. Soc. 136:125. 2023.

—Scientific Note—

Comparing Aluminum Tolerance of Two Primary Potato Cultivars Grown in Florida

Marina Curtis Gluck, Qiansheng Li, and Guodong Liu*

Horticultural Sciences Department, University of Florida, IFAS, Gainesville, FL

Keywords. aluminum toxicity, potato production, hydroponic potatoes

Florida's fields grow both chipping and table-stock potato cultivars. These potatoes require intensive phosphorous (P) fertilization regardless of legacy soil P levels. This is in part due to high levels of aluminum (Al) from the soil's parent material. Much of the legacy P content is alkaline-extractable P bound to Al (Liao et al., 2015). Aluminum is toxic to potatoes, but there is a biodiverse range of reactions to its presence in soil. We aimed to identify the responses of two different cultivars to Al toxicity.

Findings

Two experiments were run in parallel with the chipping cultivar Atlantic and the table-stock cultivar Red LaSoda both in pots as per Lee et al. (1971) and in hydroponics as per Tabaldi et al. (2007). Treatments were 0 ppm, 50 ppm, 100 ppm, or 200 ppm Al as aqueous $AlCl_3{\cdot}6H_2O$. Ten pots per cultivar and treatment were each fertilized in split applications according to University of Florida/IFAS recommendations. Al treatments were delivered via daily irrigation with 300 ± 50 mL water balanced to 3.5 pH. The seed potatoes germinated in the pots and treatment began on March 14. Plants across both cultivars and all treatments began to show edema symptoms in early April, therefore treatments were slowed to every other day starting April 25. The potted potatoes were harvested on May 12. Five hydroponic containers per cultivar and treatment were filled with 1 L Hoagland solution containing all essential nutrients, as well as the Al treatments. The seed potatoes were germinated in sand and transplanted on March 16. Every week, the water level was refilled to 1 L, the pH was balanced to 4.0 with HCl and NaOH, and 500 μL H_2O_2 were added to oxygenate the solution. However, each week the pH would revert to as high as 6.7 for the control and as low as 3.8 for the highest aluminum treatment. The containers were harvested on May 1.

The external factors in each experiment caused too much statistical noise to draw conclusions from most of the collected data. None of the harvest statistics for the pots were significant by treatment. More potted 'Atlantic' plants germinated and survived to harvest than 'Red LaSoda'. Roots and tubers of both cultivars tended to decay in the hydroponic solution. 'Red LaSoda' plants were generally smaller than 'Atlantic' in hydroponic containers. Controlling for solution pH, higher ppm Al significantly decreased root mass and tuber maturation in both cultivars grown hydroponically. This is in line with our findings from the previous year's potted potato experiment, in which potato growth and development varied significantly by treatment in both cultivars. We cannot currently conclude that either cultivar performs better, and further experiments are required.

Suggestions for Further Direction

In future experiments, plants must be harvested before May to prevent heat stress. The potted plants should receive less water overall. Water requirements varied greatly mid-season due to different canopy and root sizes. Soil moisture monitoring would help quantify any statistical differences and prevent edema. A low pH is necessary to suspend Al in the hydroponic solution. A different pH buffer may decrease root damage. Larger containers may encourage tuber growth and elucidate the differences between the treatments and cultivars.

Previous field experiments varied P fertilization levels in high Al soil for 'Atlantic' potatoes (Qiu et al., 2022). Therefore, field experiments in plots with varying Al levels would be useful. We welcome collaboration with any farmers concerned about Al toxicity in their fields.

Literature Cited

Lee C. 1971. Influence of aluminum on plant growth and tuber yield of potatoes. *Agron. J.* 63:363–364. https://doi.org/10.2134/agronj1971.00021962006300030004x

Liao X., Liu G., Hogue B., & Li Y. 2015. Phosphorus availability and environmental risks in potato fields in North Florida. *Soil Use Manag.* 31(2):308–312. https://doi.org/10.1111/sum.12186

Qiu Y., Fall T., Su Z., Bortolozo F., Mussoline W., England G., Dinkins D., Morgan K., Clark M., & Liu G. 2022. Effect of phosphorus fertilization on yield of chipping potato grown on high legacy phosphorus soil. *Agronomy* 12(4):812. https://doi.org/10.3390/agronomy12040812

Tabaldi L., Nicoloso F., Castro G., Cargnelutti D., Gonçalves F., Rauber R., Skrebsky E., Schetinger M., Morsch V., & Bisognin D. 2007. Physiological and oxidative stress responses of four potato clones to aluminum in nutrient solution. *Braz. J. Plant Physiol.* 19(3):211–222. https://doi.org/10.1590/S1677-04202007000300005

*Corresponding author. Email: guodong@ufl.edu

Proc. Fla. State Hort. Soc. 136:126. 2023.

—Scientific Note—

Effects of Oxygen and Phosphorus Fertilization on Tuber Yield of Potato Grown in a Preliminary Study in Northeast Florida

Marina Gluck, Qiansheng Li, and Guodong Liu*

Horticultural Sciences Department, University of Florida, IFAS, Gainesville, FL

Keywords. oxygen fertilization, phosphorus application, potato tuber yield

Potato (*Solanum tuberosum*, L.) is a prominent spring crop in Florida. This crop is sensitive to low-oxygen stress in soil, particularly when deep-well water is used for irrigation or when seawater intrusion raises the irrigation water's salinity. The crop requires high phosphorous (P) levels due to the limited uptake capability of its shallow, low-density, and minimally branching root structure. In October 2022, the University of Florida, Institute of Food and Agricultural Sciences (UF/IFAS) announced its provisional P management recommendation for application rates independent of preplant P soil test levels in commercial potato production. The objective of this preliminary study was to evaluate the effects of applying preplant solid oxygen fertilizer and phosphorus fertilizer on potato tuber yield of 'Atlantic' chipping potato grown in Northeast Florida.

Oxygen fertilizer is any substance that can release bioavailable oxygen to the growth medium. Calcium peroxide is one of the solid oxygen fertilizers we identified. This peroxide is insoluble and can keep releasing bioavailable oxygen for as long as 180 days when it is moisturized. We used a randomized complete block design (RCBD) with four replications at the UF/IFAS Hastings Agricultural Extension Center (HAEC). The plot size was 533.3 ft^2 including four 40-foot rows with 40-inch spacing. Plant spacing was eight inches. There were two solid O_2 fertilizer rates of 0 and 60 lb/acre as calcium peroxide (CaO_2) and five phosphorus rates of 0, 80, 120, 160, and 200 lb/acre of phosphorus pentoxide (P_2O_5) as triple superphosphate (TSP).

The chipping potato was planted on January 13, 2023, and harvested on April 21, 2023. The central 20 feet of the middle two rows were harvested for data collection. The results of tuber yield from this preliminary study are shown in Fig. 1.

Potato tuber yield increased with phosphorus application significantly as compared to the control without any phosphorus fertilization. There were significant differences in the treatments with 80 through 200 lb/acre phosphorus pentoxide even though the 200 lb/acre phosphorus pentoxide had the greatest tuber yield this year. None of the oxygen fertilization treatments produced significant difference in tuber yield as compared to the corresponding phosphorus treatments. The data from this preliminary study shows that oxygen fertilization did not significantly contribute to potato tuber yield. We also measured the phosphorus content in both shoots and tubers. The control without P application had the lowest P content of 0.15% in shoots but had a similar P content of 0.20% in tubers as compared with that (0.21%) of all the tubers with P applications. Tubers (0.21% ± 0.013%) had a greater P content than shoots (0.19% ± 0.016%) for all treatments excluding the control. The average P content of all shoots with oxygen application was (0.20% ± 0.006%), which was greater than the corresponding P content of those (0.18% ± 0.010%) without oxygen fertilization. Tubers had a stable P content of (0.21% ± 0.013%) for those with oxygen fertilization and (0.21% ± 0.016%) for those without oxygen fertilization. Because this oxygen fertilization is a brand-new concept for minimizing hypoxic stress in potato fields with seepage irrigation using deep well water as irrigation water source, we may need to further investigate the application method of solid oxygen fertilizer to keep the released bioavailable oxygen in the root zone.

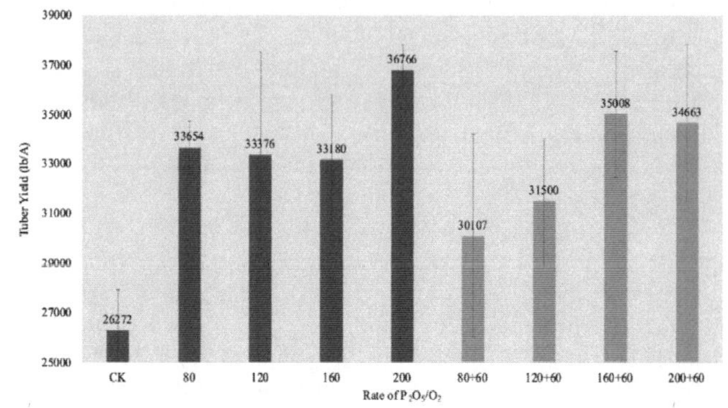

Fig. 1. Chipping potato ('Atlantic') tuber yield vs. phosphorus and oxygen fertilization. All phosphorus applications had significantly greater tuber yields than the control without phosphorus application. Oxygen fertilization had approximately 2000-lb greater potato tuber yield only for the 160 lb/acre P_2O_5 treatment. All the other treatments with oxygen fertilization had lower potato tuber yields than the corresponding treatments without oxygen fertilization.

*Corresponding author. Email: guodong@ufl.edu

Proc. Fla. State Hort. Soc. 136:127–130. 2023.

Boosting Farm Incomes with Maize Production in Zambia

Tiberious Etyang[1], Godfrey Ejimakor[2]*, and Leonard Sonnenschein[1]

[1]*Salvation Farming Solutions LLC, Boca Raton, FL*

[2]*Department of Agribusiness, Applied Economics and Agriscience Education, North Carolina A&T State University, Greensboro, NC*

Keywords. food security, maize, productivity, soil carbon, soil nitrogen

Food insecurity in Zambia is related to low incomes and farm productivity in the country. This study sought ways to increase the production of maize which is a stable food item in Zambia. Salvation Farming Solutions methods, which is a nature-based system, were tested on experimental plots in Kabwe, Zambia. The methods resulted in substantial increases in maize (*Zea mays* L.) yield, soil organic carbon and soil nitrogen. Productivity on maize farms in Zambia could be increased by the adoption of Salvation Farming Solutions methods. In addition to increasing farm productivity, adoption of the method also improves soil health.

Zambia is a lower middle-income country with a per capita gross domestic product (GDP) of $1,100 in 2020. The country had a poverty rate of 58% in 2015 (World Bank, 2021). About 50% of the 19 million Zambians are employed in agriculture, which accounted for only 2.73% of the country's GDP in 2020 (USAID, 2021). Zambia has a low food security index score of 36.6%, which is very low when compared with a score of 80% or more for many developed countries. The rural poor, who make up about 54% of the population, face the brunt of food insecurity. Developing the agricultural sector is key to increasing per capita GDP, increasing rural employment, increasing food security, the resilience of rural households, and better managing the country's natural resources.

This study sought to contribute to the enhancement of farm incomes in Zambia by increasing the productivity of maize (Zea mays L.) farms in the country through the introduction and application of Salvation Farming Solutions (SFS) methods that use a nature-based approach. The SFS methods involve the application of one or more treatments. The treatments include treating seeds prior to planting, treating the soil prior to planting, and applying organic fertilizer. All the products used for the treatments are derived from vitamins and are nature-based. The products were developed and provided by Salvation Farming Solutions, LLC, which is a farm service company.

Nature-Based Agriculture

In agriculture, nature-based solutions (NbS) are considered more environmentally sound in the long-run, cost efficient and effective at addressing sustainable land resource management and climate change issues (Miralles-Wilhelm, 2021; Isemani & Miralles-Wilhelm 2021). Such practices improve water availability, restore ecosystems and soil health in a way that improves food security by increasing land ecosystem production and productivity (Abdelmagied, 2020). The benefits associated with the use of NbS include climate change adaptation and disaster risk reduction, climate change mitigation potential, provision of non–carbon ecosystem services, resource use efficiency without the use of mineral fertilizers and other external inputs, food security and income generation, and social benefits (GCA, 2019). NbS can help farming communities to establish a resilient food production system by improving soil health and water retention, and enhancing food and nutrition security. They can reduce the use of chemical additives, which reduces production costs and creates safe foods (GCA, 2019). Nature-based solutions can play a critical role in mitigating climate change by reducing carbon emission from the agricultural sector by enhancing carbon sequestration in ways that enhance the carbon retained in plants and soils (Pinner et al., 2020). Studies by the Food and Agriculture Organizations of the United Nations (FAO) show NbS can provide 37% of climate change mitigation through 2030 and 20% between then and 2050 (FAO, 2014). Also, NbS can further enhance the ecosystem and species by improving habitat diversity while enhancing the quality and reliability of water resources.

According to Seddon et al. (2020), NbS lead to a healthy ecosystem that is more resilient. It further increases soil health, nutrient availability, water retention, soil biodiversity and reduces soil degradation which increases crop yields with less chemical additives resulting in a safer and more nutritious food (Sonnenschein, 2017). According to Pearson (2007), an increase in diversity above ground improves below ground biodiversity which contributes to nitrogen fixation, nutrient cycling with ability to regulate the dynamics of greenhouse gas emission. Sonnenschein et al. (2019) reported that the agricultural sector is primarily responsible for excess nitrogen in the form of ammonia, nitrite and nitrate, phosphorus, pesticides, and pollution of water bodies in agricultural zones. Nitrogen and phosphorous cause eutrophication in water bodies and affecting aquatic life (Juma et al. 2014). Nature-based solutions can be a valuable option for intercepting and treating various streams of nutrient pollution from agriculture (Rizzo et al., 2023) and improving crop yields.

Methods

Study Site. The study was carried out at Kabwe Research Station (KRS) situated in the Kabwe District of Zambia (Fig. 1). The KRS is approximately 142 km from Zambia's Capital City,

*Corresponding author. Email: ejimakor@ncat.edu

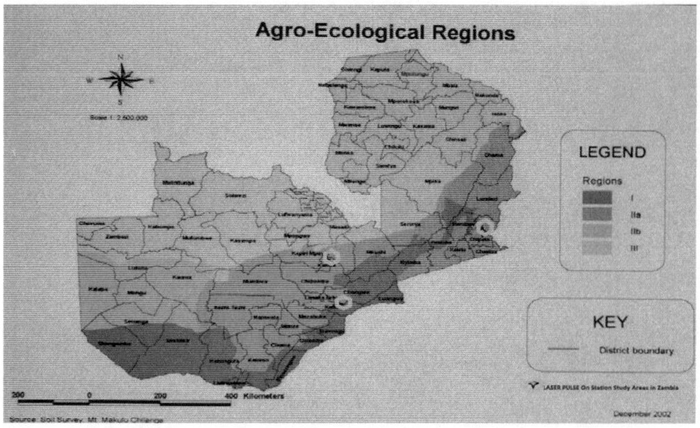

Fig. 1. Map of Zambia's agro-ecological zones showing study location.

Lusaka, along the Great North Road. The station lies at latitude –14.3954 and longitude 28.4934, with a gross area of over 500 ha. The area receives rainfall of between 800–1000 mm per annum with temperatures ranging between 20 to 34°C.

Experimental design. An on-station demonstration field was established at the Zambia Agricultural Research Institute (ZARI) in Kabwe, Zambia in February 2023. The field was used to demonstrate to stakeholders how to use SFS products as well as evaluate the efficacy of its different applications on maize, growth and yields. The SFS method consists of the application of soil and fertilizer treatment, seed powder treatment, and organic pesticide.

A type 1 randomized complete block design (RCBD) experiment with a split-split plot treatment structure with three replications was used in the experiment. The test crop was maize (*Zea mays* L*)*. The selected treatments comprised of three main plots with no chemical fertilizer (no fertilizer), 20% of the full rate fertilizer, and full rate fertilizer application (200 kg·ha^{-1} Basal and 200 kg·ha^{-1} for top dressing, for 100% of the fertilizer use recommendation for maize in Kabwe) that were designated as main plots. The main plot size was 12.25 m × 15 m. Eight sub-plot factors were used to gather data on SFS nature-based methods and included two subplots with no SFS treatments, two with SFS seed treatments, two with SFS soil and fertilizer treatments, and two with SFS seed treatment with soil and fertilizer treatment. The sub-plot size was 5.9 m × 3 m. The treatments were replicated three times on the experimental field. Inorganic fertilizer was applied in designated sub plots (0%, 20%, and 100% rates), followed by seeds and then sprayed with soil and fertilizer treatment.

SFS field protocol. Each SFS Trial Kit provided enough sample products for maize growth evaluation for 1+ acre of land. The SFS Trial Kit included the seed, soil, and fertilizer treatments.

Seed treatment powder preparation. Maize seeds to be planted were put in a 20-liter bucket. All the contents in the seed treatment powder were sprinkled into the bucket with the seeds. A cup of treated water was added to the bucket, and the contents were mixed with wooden spoons. The bucket contents were stirred slowly until all the seeds were uniformly coated with the seed treatment. The seeds were planted one hour after mixing.

Soil and fertilizer treatment concentrate. A backpack sprayer was filled with water, then a full bottle of the soil and fertilizer treatment concentrate was added to the backpack. An even moist application was added to the soil.

Field soil sampling and analysis. A zigzag sampling framework was used to sample composited soils at the plot level (0–20cm depth) before planting, 9 weeks after planting, and after harvesting for soil nutrient analysis. This followed the methods outlined in Okalebo, et al. (2002). To determine total organic carbon, the soil sample was oxidized using acidified dichromate at 150 °C for 30 minutes followed by calorimetric determination. Nitrogen (N) was analyzed using the Kjeldahl method (Ryan et al., 2001).

Table 1. Yield ((kg·ha^{-1}) comparison with % change with and without SFS treatments.

Fertilizer application rate (%)	No SFS treatment (kg·ha^{-1})	SFS treatment (kg·ha^{-1})	Change (%)
0	400	1823	356
20	1010	5892	583
100	1580	6310	399

Results

Maize yields. The average maize yields from the project's control and treatment plots are presented in Table 1. The no treatment yield without the application of fertilizer was 400 kg·ha^{-1}. Applying SFS methods increased the yield per hectare to 1823 kg with no fertilizer application. This represents a 356% increase over the no SFS treatment yield and is within the range of yield 1000–2000 kg·ha^{-1} outcomes that has been reported for Zambia by Nicholas et al. (2023). The yield was 1010 kg·ha^{-1} when 20% of the required fertilizer is applied without SFS treatment. Combining the SFS treatment with the 20% fertilizer application increased the yield to 5892 kg·ha^{-1}, which represents a 583% increase over the no SFS treatment yield. The highest yield of 6310 kg·ha^{-1} was recorded for the SFS treatment with 100% fertilizer application, which is a 399% increase over the no treatment (control) with a 100% fertilizer application.

The maize yield without SFS products was within the country's current average of 1–2 t·ha^{-1} while the inclusion of SFS products increased the maize yield significantly though it was still below the country's maize yield potential of 9.8 t·ha^{-1} (GYGA, 2018). The addition of SFS products at least tripled the yield of maize which represents an economic benefit to the farming communities in Zambia. Reduction in fertilizer used by farmers by up to 80% also reduced production costs and soil toxicity making farming affordable to many farmers while significantly increasing the farmers' economic position and making the soils healthier for subsequent cropping seasons. The yield numbers indicate that the opportunity exists for farmers to increase revenue from maize production while simultaneously eliminating or reducing expenditures on fertilizer. For example, a farmer who uses 20% of the required fertilizer application with the SFS treatment will reduce fertilizer expenses by 80% while increasing the revenue from maize by 583%. Such revenue increases and cost reductions translate to increased net income, which has the potential to make farmers, their families and communities more food secure and resilient.

Soil organic carbon. Table 2 shows the comparison in percent change in soil organic carbon (SOC) after planting and after harvesting.

Significant soil organic carbon sequestration with SFS treatments was observed as shown in Fig. 2 which suggests a more resilient food system for generational impact. Plots with SFS soil and fertilizer treatments with seed treatment led to higher changes in soil organic carbon (SOC) while those without SFS treatments

Table 2. Percent change in soil organic carbon after planting and after harvest.

Field description	After planting	After harvesting	Change after planting (%)	Change after harvesting (%)
No SFS treatments	3	6	100	100
Seed treatment	9	24	300	400
Soil and fertilizer treatment	12	30	400	500
Seed, soil, and fertilizer treatment	18	43	600	717

Table 3. Percent change in soil nitrogen after planting and after harvesting.

Field description	After planting	After harvesting	Change after planting (%)	Change after harvesting (%)
No SFS treatments	5	6	100	100
Seed treatment	13	18	260	300
Soil and fertilizer treatment	16	23	320	383
Seed, soil, and fertilizer treatment	17	38	340	633

had the smallest changes in SOC. The soil carbon levels represent a more resilient future food system.

Soil Nitrogen. Table 3 shows the percentage change in soil nitrogen after planting and after harvesting under different field treatments.

The percentage changes in soil nitrogen showed significant increase in natural nitrogen production due to SFS treatments as shown in Fig. 3. Soil and fertilizer treatments with seed treatment contributed to the highest percentage increase in soil N, while the control (no SFS treatment) contributed to the least increase in soil N during the crop growth phase. Use of SFS treatments significantly improved soil health and nitrogen sequestration.

Discussion

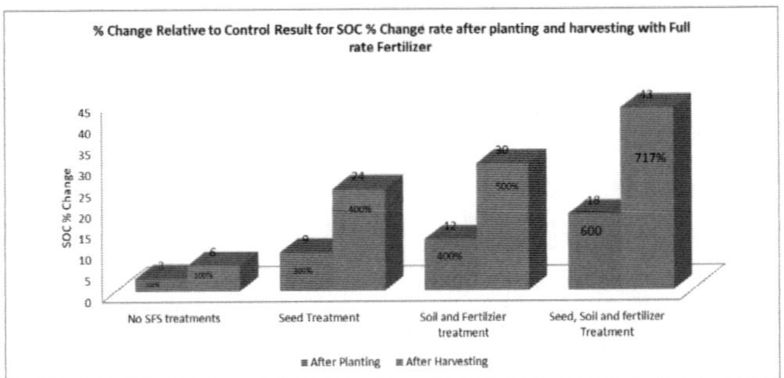

Fig. 2. Percent change relative to control result for soil organic carbon (SOC). Percent change rate after planting and harvesting with full rate fertilizer.

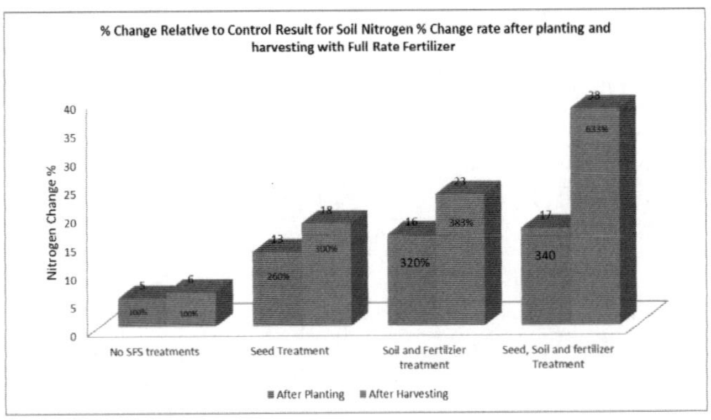

Fig. 3. Percent change relative to control for soil nitrogen. Percent change rate after planting and harvesting with full rate fertilizer.

The SFS treatments can at least double the yield of maize on farms without fertilizer application. The reduction in fertilizer used by farmers by up to 80% translates to reduced production costs thereby making farming more affordable to many farmers while significantly increasing the farmers' economic position and making the soils healthier for the subsequent cropping seasons. The NbS SFS treatments improved soil health and regenerative potential of the soil. The resulting soil carbon sequestration represents a potential for establishing a more resilient food system. The SFS treatments substantially increased maize yields by as much as 200%. SFS treatments also improve soil health and nitrogen sequestration which demonstrate their ability to make farming to be more climate resilient.

Conclusion

Over 63.4% of smallholder farmers in Zambia use fertilizer. However, maize growers in the country have average yields of 1000–2000 kg·ha^{-1}, which indicates that a yield gap exists. Better management, including the use of SFS methods, can close this gap. The results from this study indicate that SFS methods could increase the income of maize farmers, by both increasing yield, and reducing or eliminating the need for fertilizer and the associated cost. Adopting the practice could improve farm incomes and food availability for small farmers and reduce the level of food insecurity in Zambia. Our results have demonstrated the potential of SFS treatments for carbon sequestration and also increasing soil nitrogen especially in a developing world setting where farming is not intensive. Use of NbS such as SFS methods are beneficial in establishing sustainable farming practices. Results from the studies in Zambia demonstrate the potential in building a resilient farming system in Sub-Saharan Africa with the potential to cope with challenges associated with climate change.

Literature Cited

Abdelmagied M, Mpheshea M, 2020, Ecosystem-based adaptation in the agriculture sector A nature-based solution (NbS) for building the resilience of the food and agriculture sector to climate change. Rome, FAO.

Burke WJ, Jayne TS, Black JR. 2017. Factors explaining the low and variable profitability of fertilizer application to maize in Zambia. *Agricultural Economics*. 48(1):115–126.

FAO. 2014. Building a common vision for sustainable food and agriculture: Principles and approaches. 56 p. Food and Agriculture Organizations of the United Nations, Rome, Italy ISBN 978-92-5-108471-7

Global Commission on Adaptation (GCA). 2019. Adapt Now: A Global Call for Leadership on Climate Resilience. World Resources Institute and Global Commission on Adaptation. https://files.wri.org/s3fs-public/uploads/GlobalCommission_Report_FINAL.pdf

Iseman T. & Miralles-Wilhelm F. 2021. Nature-based solutions in agriculture—The case and pathway for adoption. Virginia. FAO and The Nature Conservancy. 53 p.

Juma D.W., Wang H, & Li F. 2014. Impacts of population growth and economic development on water quality of a lake: case study of Lake Victoria Kenya water. Environmental Science and Pollution Research 21: 5737-5746.

Miralles-Wilhelm F. 2021. Nature-based solutions in agriculture: sustainable management and conservation of land, water, and biodiversity. FAO and The Nature Conservancy.

Okalebo J.R., Gathua K.W., & Woomer P.L. 2002. Laboratory methods for soil and plant analysis: A working manual. 2nd ed. The Sustainable Agriculture Centre for Research Extension and Development in Africa, Bungoma and Nairobi, Kenya.

Pearson C.J. 2007. Regenerative, semi closed systems: A priority for twenty-first-century agriculture. *BioScience*. 57:409–418.

Pinner D., Rogers M., & Samandari H. 2020. Addressing climate change in a post-pandemic world. McKinsey & Company Quarterly. Accessed 15 December 2023. https://www.mckinsey.com/capabilities/sustainability/our-insights/addressing-climate-change-in-a-post-pandemic-world

Rizzo A., Sarti C., Nardini A., Conte G., Masi F., & Pistocchi A. 2023 Nature-based solutions for nutrient pollution control in European agricultural regions: A literature review. *Ecological Engineering*. 186:106772. Accessed 15 December 2023. https://doi.org/10.1016/j.ecoleng.2022.106772

Ryan J.G., Estefan J., & Rashid A. 2001. Soil and plant analysis laboratory manual. International Center for Agricultural Research in the Dry Areas (ICARDA), Islamabad, Pakistan.

Seddon N., Chausson A., Berry P., Girardin C.A.J., Smith A., & Turner B. 2020. Understanding the value and limits of nature based solutions to climate change and other global challenges. *Phil. Trans. R. Soc.* B37520190120. Acceessed 15 December 2023. http://doi.org/10.1098/rstb.2019.0120

Sonnenschein L. 2017. If we are what we eat, how have decreases in nutritional densities of food affected health. *Biomed J Sci & Tech Res*. 1(2):2017. BJSTR. MS.ID.000179. Accessed 15 December 2023. https://doi.org/10.26717/BJSTR.2017.01.000179

Sonnenschein L. & Etyang T.B. 2019. Dealing with the sins of the past and work towards a nature-based future: New innovative technologies for sustainable agriculture. *World J Agri & Soil Sci*. 2(1): 2019. WJASS. MS.ID.000526. http://dx.doi.org/10.33552/WJASS.2019.02.000526

United States Agency for International Development (USAID). 2019. Zambia Country Development and Cooperation Strategy. July 16, 2019–July 15, 2024. https://www.usaid.gov/zambia/cdcs

World Bank. 2021. Zambia—Poverty reduction strategy paper (PRSP) Annual progress report and joint staff assessment. World Bank Group. Washington, D.C. http://documents.worldbank.org/curated/en/713811468167971790/Zambia-Poverty-Reduction-Strategy-Paper-PRSP-Annual-Progress-Report-and-Joint-Staff-Assessment

Proc. Fla. State Hort. Soc. 136:131. 2023.

—Scientific Note—

Exploring Computer Vision Approaches toward Water Vapor Condensation Detection on Fresh Produce Surfaces

Spencer Serrano* and Ziynet Boz

University of Florida Agricultural and Biological Engineering, University of Florida, IFAS, Gainesville, FL

Keywords. circularity, cold chain, computer vision, convolutional neural network, deep learning, object detection, produce spoilage, postharvest, water vapor condensation

The motivation for this work aligns with a broader push toward circularity in food systems. Here we investigate approaches to detect water vapor condensation droplets on the surface of fresh produce using opensource machine learning and computer vision tools. The three primary objectives in this study are, to establish methods for collecting images of produce with condensation droplets, to build image datasets of condensation afflicted produce, and to use this data to identify the best model for the task.

Findings

Object detection. First an initial dataset of 164 images of apples with spray applied water droplets was collected with a USB webcam. This set was labeled with bounding boxes and used to retrain a pretrained object detection model built with TensorFlow Lite. The model was then installed on a Raspberry Pi computer with a webcam to detect simulated droplets on apple surfaces from video feed in real time. The model qualitatively identified droplet regions from live video feed.

Binary classification. With the goal of photographing real condensation on fruit surfaces, a DSLR camera outfitted with a 120 mm zoom lens was used to build a new set image of apples with and without condensation droplets. This dataset was then used to train a convolutional neural network built in tensor flow for a binary classification task. Five separate training and validation experiments were run to optimize the hyperparameters of the model. Early on, overfitting was identified, and it was determined that increasing the dimensions of the input layer significantly alleviated this issue. Model accuracy was found to be as high as 97% during these trials.

Suggestions for Future Direction

The convolutional neural network showed the most promise for being able to correctly predict condensation states form images. Expanding the dataset is paramount to increase prediction accuracy and avoid overfitting the model to extraneous image features. To generalize this approach, it will be necessary to add images of a variety produce items in addition to apples. Once the model has been sufficiently generalized, incorporation of the predictions into a cold chain digital twin or an internet of things platform could provide insights into spoilage of fresh produce from water vapor condensation.

*Corresponding author. Email: serranos@ufl.edu

Proc. Fla. State Hort. Soc. 136:132–133. 2023.

—Scientific Note—

Understanding & Bridging the Gap Between Consumers, Growers, Researchers, and Stakeholders

Alejandro Bolques[1]*, Zhifeng Gao[2], Derek Farnsworth[2], and Md Azhar Uddin[2]

[1]*Research and Extension Center, Florida A&M University, Quincy, FL*

[2]*Food and Resource Economics Department, University of Florida, Gainesville, FL*

Keywords. agricultural BMPs, small and medium sized farm, willingness to pay

New marketing opportunities for small and medium-sized farms are being identified by understanding consumer preferences for produce grown using conservation and best management practices (BMPs). Agricultural BMPs are broadly defined as sustainable practices aiming for optimum crop production with minimal adverse environmental impact. They are tools that help farmers to balance water quality improvements and agricultural productivity. This study focuses on BMPs for specialty fruit and vegetables that have been broadly categorized by the Florida Department of Agriculture and Consumer Services (FDACS) as: Nutrient Management, Irrigation Management, Sediment and Erosion Control, Water Resource Protection, Integrated Pest Management, and Recycling and Waste Management BMPs (FDACS, 2023). In Florida, as in other US States, BMPs are continually updated as new scientific data becomes available to improve water quality and environmental sustainability.

Product labeling

Food labels are important because they provide consumers with information about product attributes. Consumers depend on labels to make an informed product purchase decision. The United State Department of Agriculture (USDA) organic logo and labeling is a well-known food label in the marketplace. The logo indicates that the organic food item has undergonen strict production standards and certification. Other product labels certified by USDA include Good Agricultural Practices (GAP) and Good Handling Practices (GHP).

Ag BMPs label

Here we introduce the idea that agricultural BMPs are sustainable production practices that help protect the environment, particularly water quality and conserve natural resource. so that BMP-produced food could generate price premiums similar to those of organic food labels. A certified BMP production label would also help incentive small and medium-sized farm producers to adopt their state's BMPs if consumers were willing to pay premiums for products with BMP labels (Fig. 1).

Fig. 1. Sample labels that could be used to certify the use of BMPs on farms compared with the USDA Organic label.

Willingness to Pay

In Florida and other states, Ag BMP adoption is supported with cost-sharing incentives. It has been shown that Floridian are willing to pay for water quality improvements that directly impact them locally. Since much of the produce grown by small and medium-sized farms is sold in local markets, new marketing

*Corresponding author. Email: alejandro.bolques@famu.edu

methods and BMP labeling may provide a vehicle for converting consumers' willingness to pay (WTP) for water quality improvement or water conservation into an actual price premium for the producers' protecting and improving local water quality and quantity. Survey was developed to estimate consumer WTP for BMP labels and other popular USDA labels in the market, such as USDA Organic, and the GAP/GHP labels using strawberries and tomatoes as the fresh produce of interest. A grower survey was also developed to estimate specialty crop growers' willingness to pay to join a BMP labeling program that allows them to use the BMP labels on their products. Consumer and grower surveys can help identify demand for BMP-labeled fresh produce. Information obtained could help policymakers evaluate the feasibility of introducing a BMP labeling program and its potential cost of benefit.

Conclusion

Agricultural BMPs are essential approaches to reducing the adverse environmental impact of agricultural production while aiming for optimum crop production. Most states provide cost-share programs to reduce the financial burn on growers who implement BMPs because of the cost of equipment and applying BMP practices. Because BMPs are environmentally friendly practices and many consumers are willing to pay a premium for products that are produced in an environmentally friendly way, it is possible that BMP-labeled products could receive price premiums. These would help develop a market-oriented approach to promote BMPs among growers, thus reducing the cost of promoting BMPs using the cost-sharing programs. Two major pieces of information are critical for the stakeholders to move forward to introduce the BMP labels into the market: whether consumers are willing to pay a premium for BMP labels and whether growers are willing to join the BMP labeling programs to add such labels to their products. This research aims to answer the two crucial questions, providing essential information for policymakers and stakeholders to decide whether to introduce the BMP label into the market and, if so, how to effectively promote this label among consumers and encourage growers to join the labeling programs.

Literature Cited

Florida Department of Agriculture and Consumer Services (FDACS). 2023. Agricultural best management practices. Accessed 25 May 2023. https://www.fdacs.gov/Agriculture-Industry/Water/Agricultural-Best-Management-Practices

Proc. Fla. State Hort. Soc. 136:134–137. 2023.

Chemical Delivery is Critical to Successful Control of Root-knot Nematodes in Squash: A Case Study of Fluazaindolizine

QINGCHUN LIU[1] AND SHOUAN ZHANG*

Tropical Research and Education Center (TREC), Department of Plant Pathology, University of Florida, IFAS, Homestead, FL

Keywords. drench, fluazaindolizine, *Meloidogyne incognita*, root-knot nematode, squash

Root-knot nematode (RKN) (*Meloidogyne incognita*) is a common pest problem on vegetable crops including tomatoes and squash in south Florida that can lead to significant yield loss when not managed properly. Chemical control is often needed in the management of RKN due to lack of effective soil fumigants. The object of this study was to evaluate the efficacy of fluazaindolizine (Recklemel) in control of RKN on squash. Squash (cv. Enterprise) was seeded in beds covered with plastic, in which the soil was inoculated with RKN 1 week before seeding. Treatments were applied twice by drip injection, one immediately after seeding for all treatments and the second at 2 weeks after the first treatment. In addition, the first four plants at the south end of each plot of the treatments with Recklemel were drenched with 50 mL of a solution of the same chemical to the base of each seedling right after emergence. The initial nematode population was high as some seedlings in several plots wilted soon after emergence or grew poorly. Although Recklemel at 23.0 fl oz/A applied at seeding had the best overall performance, there were no significant differences among the treatments when the chemical was applied through the dripline. However, the plants drenched after emergence grew significantly better and showed much lower levels of root infection by RKN compared to those treated with the chemical only through dripline. The results indicated that Reklemel is effective against RKN in squash but the delivery method is key, which can greatly affect its efficacy. Further research will be needed to develop the best delivery systems for this nematicide for improved RKN control in sustainable vegetable production.

Squash is an important vegetable crop in Florida. In 2019, a total of 7700 acres of squashes were harvested in Florida with a value of $33.7 million (NASS, 2021). The majority of squash production is in south Florida, particularly in Miami-Dade County, where approximately 90% of squashes are planted (Seal et al, 2022). There are many factors limiting squash production in Florida, including root-knot nematode (RKN) caused by *Meloidogyne incognita* (Seal et al., 2022).

M. incognita, also called southern root-knot nematode, mainly causes problems in tropical and sub-tropical areas (Mitkowski and Abawi, 2011). *M. incognita* can cause serious problems in vegetable crops, including tomato and squash (Mitkowski and Abawi, 2011). Because RKN has a wide range of hosts, it is difficult to manage it using crop rotation (Mitkowski and Abawi, 2011; Seal et al., 2022). Though RKN obtains nutrients from host roots through forming galls and rarely kills hosts, it seems seldomly significantly affect early plant growth and fruit yield (Liu and Zhang, personal observations; Mitkowski and Abawi, 2011). However, young seedlings could be killed when the density of RKN in the soil is high without proper control, which leads to a significant loss of seedlings in the field (Liu and Zhang, personal observations, Mitkowski and Abawi, 2011).

The second-stage juvenile (J2) is the only infectious stage for RKN. After J2s enter inside the roots and establish their feeding site, they stay at the location permanently where a group of giant cells develops into gall that we see on the infected roots and is the location RKN competes for nutrients with the host. In addition, the galls are also the location where adult female RKN produce eggs. Because RKN moves slowly inside the soil, the proximity of RKN to the host roots would be critical to affecting plants by the nematodes and the root-knot development.

Chemical control is a major approach for RKN management through fumigation, injection, drenching, and foliar spray (Hajihassani, 2018). When applied to the soil, non-fumigant chemicals, such as fluazaindolizine, are dispersed by movement in soil water. However, the efficacy of non-fumigant compounds is affected greatly by quick leaching of the active ingredients, particularly in sandy soils, which may reduce their effectiveness (Hajihassani, 2018). This could be a major concern for RKN control in squash with soil treatment in Miami-Dade County because most of the soils are calcareous soil that drains rapidly (Li, 2001). A lack of dripline (Liu and Zhang, personal observations). In a preliminary test in the greenhouse, okra plants sprayed with fluazaindolizine onto the soil surface after emergence grew significantly better than those treated at seeding by spraying to soil after covering (Liu and Zhang, unpublished). In this field study, our objectives were: 1) to evaluate programs consisting of fluazaindolizine for their control of RKN in squash when applied through dripline; 2) to determine the effect of fluazaindolizine in control of RKN in squash when applied through drench.

Materials and Methods

Field trial setup. The field trial was conducted on the farm at University of Florida, Institute of Food and Agricultural Sciences Tropical Research and Education Center (UF/IFAS TREC) in

[1]Current address: UF/IFAS Extension Miami-Dade, Department of Parks, Recreation & Open Spaces, Homestead, FL33030

*Corresponding author. Email: szhang0007@ufl.edu

Spring 2022. The beds were prepared for vegetable cultivation that was 30 inches wide, centered 6 feet apart, and covered with white-on-black plastic mulch. There were six treatments including a commercial standard (Velum Prime) and an untreated control (Table 1). Each plot consisted of a 25-foot section of a single bed with a 3-foot buffer zone between two adjacent plots. The trial was designed as a randomized complete block (RCB) with five replications for each treatment. Thirteen squash plants were planted in each plot, and a total of 65 plants were grown for each treatment. Squash plants were irrigated through two driplines, one at each side of a plant, and sprayed for other common pests and diseases on squash in the area. Fertilizer was applied according to the Vegetable Production Handbook in Florida (Dimar et al., 2021).

Soil inoculation of M. incognita. On April 26, 2022, the soil was inoculated with *M. incognita* by mixing small pieces of infected okra roots, an average of 2 roots per hole, into the topsoil to a depth of 4 inches in each planting hole. The okra roots used for inoculation were collected from a local field in Homestead, FL. where okra plants were heavily infected with *M. incognita* and had conspicuous galls on the roots.

Treatment application. On May 2, 2022, yellow squash (cv. Enterprise) was seeded directly into the inoculated soil in each hole, 2-feet apart within rows. The first soil treatment through dripline injection was conducted on May 2, right after seeding squash. The amount of chemicals used for treating a plot was calculated based on the application rate of the treatments and the area of each plot. The chemicals were mixed in 1 gallon of water and was injected through one dripline for each plot. The chemicals were then flushed out of the driplines with irrigation water to the soils after all treatments were injected. The second treatment through dripline injection was accomplished on May 16, i.e. 2 weeks after the first treatment. When squash seedlings emerged, the first 4 plants at the south end of each plot in treatments 1, 2, 3, and 4 were drenched by hand on May 9 with 50 mL of a chemical solution prepared the same way for dripline injection to the base of each seedling.

Data Collection

Plant vigor. Plant vigor was visually assessed on June 8 (mid-season) and June 24, based on a 1–10 scale where 1= worst in plant growth and 10 = best in plant growth.

Root infection. In the mid-season evaluation on June 8, four squash plants were dug out in each plot from south to north, i.e. plants #1, 7, 12, and 13. The severity of squash root infection caused by *M. incognita* was assigned with a 0–10 scale according to the diagrammatic root-knot scoring chart described by Bridge & Page (1980), in which 0 = no infection and 10 = more than 90% of the roots have galls. The trial was terminated on Jun 27 because many plants started dying due to excessive rainfall and the final root infection was assessed for the remaining plants in each plot.

Fruit yield. Marketable fruit was harvested every 2–3 days from June 5 to June 24.

Nematode populations. The initial RKN population in the inoculated soil was sampled by randomly collecting soil samples from each block across all the treatments on May 1, one sample per block, right before planting. The final RKN population in the soil was sampled for each treatment by collecting soil samples

Table 1. Effect of Reklemel and delivery method on root-knot nematodes, plant vigor, and yield on squash.

					Crop response[y]			
Code	Treatment	Application	Rate (fl oz per acre)	Application time[z]	Plant vigor (final rating)[w,u]	Root infection (final gall rating)[x,u]	#Juvenile (per 100 cc soil)[t,u]	Yield per plot (kg)[v,u]
1	Reklemel	Drip line	15.4	01	65 ab	91 a	1275 a	4.42 a
		Drench	15.4		48 bc	40 b	2010	
2	Reklemel	Drip line	23.0	01	55 abc	93 a	1340 a	4.99 a
	Reklemel	Drench	23.0		48 bc	16 c	1800	
3	Reklemel; Reklemel	Drip line	15.4 7.7	01; 02	70 a	92 a	1815 a	4.11 a
	Reklemel	Drench	15.4		50 bc	15 c	810	
4	Reklemel; Vydate	Drip line	7.7 64	01; 02	65 ab	93 a	2217 a	4.90 a
	Reklemel	Drench	7.7	–	44 c	37 b	1200	
5	Velum Prime	Drip line	6.5	01	67 ab	93 a	1955 a	2.91 a
6	Untreated Control	–	–	–	63 abc	93 a	2107 a	4.34 a

[z]01 = a single application at seeding through dripline on May 2, 2022; 02 = a sequential drip application 14 days post-planting on May 16, 2022.

[y]Crop response (CR) to treatments was calculated for plant vigor and root infection based on their ratings, in which 0 = no injury, and 100 = dead plants.

[x]Root-knot gall rating was conducted according to the diagrammatic root-knot scoring chart described by Bridge & Page (1980). Gall rating was conducted on Jun 8 and 27, 2022. The formula to calculate CR for root infection: $CR\_gall = 10 \times$ gall rating.

[w]Plant vigor was determined according to a 1–10 scale, where 1 = worst in plant growth, and 10 = best in plant growth. Plant vigor was rated on Jun 8 and 24, 2022. The formula to calculate CR for plant vigor: $CR\_vigor = 10 \times (10 -$ vigor rating$)$.

[v]Squash fruit were harvested once every 2–3 days from Jun 5-24, 2022.

[u]Means within a column followed by the same letter(s) are not significantly different according to LSD ($P = 0.05$).

[t]Soil samples were collected for plants received drench treatments and mixed across five replications to obtain one soil sample for each of treatments 1, 2, 3, and 4 after the final gall rating.

from root zones of 3 randomly selected plants in each plot right after the last gall evaluation on June 27. A total of 300 cc soil collected from 3 plants in each plot that were treated through dripline injection were mixed in a plastic bag. For plants that received drench treatment, one soil sample of 300 cc was collected per treatment from one plant per plot. All soil samples were processed to determine the number of *M. incognita* juveniles at the LLH Ag & Research Service in Tifton, GA.

Data analysis. Crop response (CR) to treatments was calculated for plant vigor and root infection based on their ratings, in which 0 = no injury, and 100 = dead plants. The formula to calculate CR for plant vigor is as follows: CR_vigor = 10 × (10 – vigor rating); the formula to calculate CR for root infection is: CR_gall = 10 × gall rating (according to collaborator's suggestion). Data of CR_vigor, CR_gall, squash yield per plot, and the number of *M. incognita* juveniles per 100 cc soil were subjected to analysis of variance (ANOVA). Mean separation tests [least significant difference test (LSD)] were performed using the Statistical Analysis System (SAS) package version 9.4 (SAS Institute, Inc., Cary, NC).

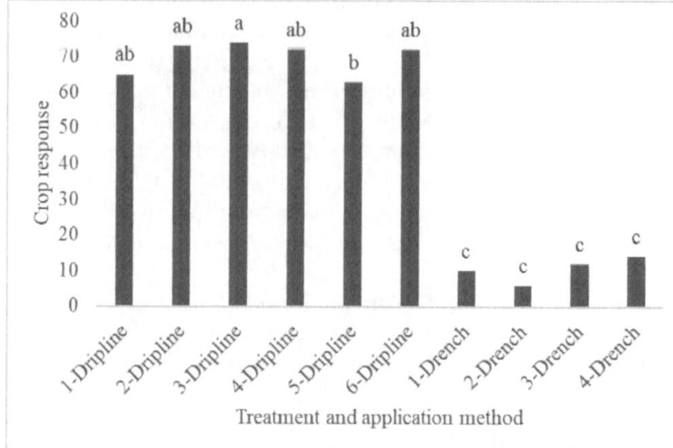

Fig. 2. Mid-season squash root infection responded to treatments applied through dripline injection and drench. Root infection with root-knot nematode was assessed at mid-season on June 8, 2022. Crop response for root infection was calculated as 10 × gall rating. The list of treatments was shown in Table 1. Bars labeled with the same letter were not significantly different ($P = 0.05$).

Results and Discussion

Crop response (CR) as plant vigor to treatments. At the mid-season assessment conducted on June 8, 2022, the treatment 2, i.e. Recklemel at 23.0 fl oz/A only applied at seeding was the only treatment applied through dripline that significantly lowered CR as plant vigor compared to the untreated control (Fig. 1). CR was significantly decreased for the treatments applied by drench compared to by dripline overall, except for treatment 2 applied through dripline. The average CR in treatments applied through drench was lowered by 66.5% compared to that through dripline. At the end of the trial, though overall CR was lower for treatments applied through drench, there was no such dramatic difference in CR over treatments (Table 1).

Crop response (CR) as root infection to treatments. At the mid-season assessment conducted on Jun 8, 2022, there was no significant difference in CR as root infection of all treatments applied through dripline compared to the untreated control (Fig. 2). However, CR was dramatically lower in treatments applied through drench than through dripline. The average CR for the treatments applied through drench was lowered by 85.3% compared to that through dripline. As the season continued, root infection continued to increase and the same for CR values of all treatments. However, such an increase of CR values for the treatments applied through drench was slower than those applied through dripline (Table 1). At the end of the field trial, the average CR for treatments applied through drench was lowered by 70.8% compared to that through dripline. Soil drench with the treatment 2 had lower CR values at both assessments for root infection.

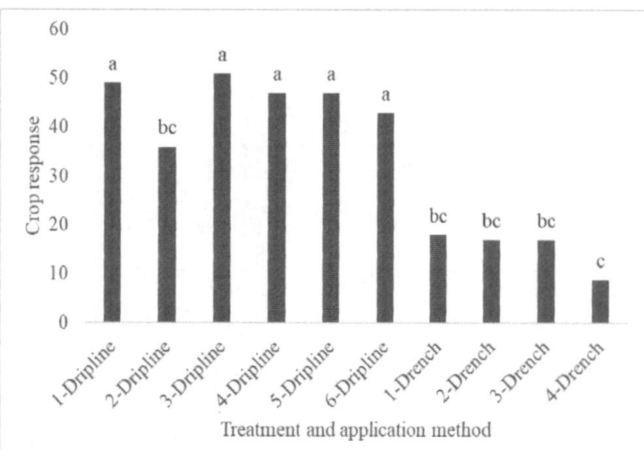

Fig. 1. Mid-season squash plant vigor responded to treatments applied through dripline injection and drench. Plant vigor was assessed at mid-season on June 8, 2022. Crop response for plant vigor was calculated as 10 × (10 – vigor rating). The list of treatments was shown in Table 1. Bars labeled with the same letter were not significantly different ($P = 0.05$).

Nematode population. When seeding the squash, the average juvenile RKN population over 5 blocks was 3244 per 100 cc soil. The initial RKN population seemed very high as some seedlings in several plots wilted soon after emergence or grew poorly. At the end of the field trial, there was no significant difference in the RKN population among the treatments applied through dripline (Table 1). The populations of RKN varied greatly for the treatments applied through drench. The overall RKN population in the treatments applied through drench was relatively lower than that applied through dripline, but the trend was inconsistent.

Fruit yield. The fruit were harvested from all plants treated through either dripline or drench in each plot. No significant differences were found in fruit yield among the treatments (Table 1). Overall, the fruit yield of treatments 2 and 4 were the top two among six treatments and were 15.0% and 12.9% higher than that of the untreated control.

In this field trial, treatments applied through dripline did not significantly affect the crop response in plant vigor and root infection, lower the final nematode population, or increase fruit yield compared to the untreated control. However, treatments applied through drench significantly reduced the crop response in plant vigor and root infection, particularly at the mid-season assessment. When treatments were applied through dripline, treatment 2 (23.0 fl oz/A) performed relatively better than other treatments in reducing crop response as plant vigor and increasing fruit yield. Interestingly, treatment (7.7 fl oz/A) applied as drench performed similarly as other two higher rates in reducing crop response as plant vigor and root infection. When treatments were applied through dripline, there was a distance from the dripline to the plants. Certain parts of the chemicals might leach through the calcareous soil before they reached the plant root zones and affected RKN. When treatments were applied through drench, much less chemicals leached off during the application, so more chemicals remained in the root zones and affected RKN. Therefore, the chemical could be applied at further lower rate which may provide satisfactory protection to the plant roots when they are applied through a soil drench. On the other hand, the distribution of RKN in the root zone may not be uniform, which may affect the accuracy in sampling RKN populations at the end of the trial. Thus soil RKN populations in this field trial soils may not accurately reflect the efficacy of the treatments against RKN. Mixing the soils around the root zone before sampling may help reduce inconsistency among the samples for better estimation of the RKN population. It is clear in this field trial that the treatment delivery method greatly affected the efficacy of Reklemel against root infection by RKN. This warrants further investigation into delivery methodology and application timing to maximize the effects of Reklemel on managing RKN in vegetable production.

Literature Cited

Bridge J. & Page S.L.J. 1980. Estimation of root knot nematode infestation levels in roots using a rating chart. *Tropical Pest Management*. 26:296–298. https://doi.org/10.1080/09670878009414416

Dimar P.J., Agehara S., & Dufault N.S. (eds). 2021. Vegetable production handbook of Florida. https://edis.ifas.ufl.edu/publication/CV29. https://doi.org/10.32473/edis-cv100-2022

Hajihassani H. 2018. Chemical nematicides for control of plant-parasitic nematodes in Georgia vegetable crops. https://secure.caes.uga.edu/extension/publications B 1502_1

Li, Y. 2001. Calcareous soils in Miami-Dade County. University of Florida/IFAS Extension #SL 183. http://edis.ifas.ufl.edu

Mitkowski N.A. and G.S. Abawi. 2003. Root-knot nematodes. *The Plant Health Instructor*. https://doi.org/10.1094/PHI-I-2003-0917-01 (Revised 2011)

NASS, 2021. www.nass.usda.gov

Seal D., Zhang, S., Dittmar P., Li Y., Klassen W., Wang Q., and Olczyk T. 2022. Summer squash production in Miami-Dade County, Florida. University of Florida/IFAS Extension, #HS-861 (revised July 2022). https://doi.org/10.32473/edis-tr012-2002

Proc. Fla. State Hort. Soc. 136:138. 2023.

—Scientific Note—

Chemical Control of a New Weevil Pest of Celery in Florida

JULIEN BEUZELIN[1]* AND ANNA MÉSZÁROS[2]

[1]*Everglades Research and Education Center, University of Florida, IFAS, Belle Glade, FL*

[2]*Palm Beach County Extension, University of Florida, IFAS, West Palm Beach, FL*

Keywords: foliar application, insecticide, *Listronotus sparsus*, transplant tray drench

Listronotus sparsus (Say) (Coleoptera: Curculionidae) has emerged as a new weevil pest of celery in Florida. Female adults insert eggs under the surface of petiole and crown tissues, where larvae subsequently feed and reduce the quality of the crop. Foliar applications of oxamyl have been used to control *L. sparsus* in Florida celery. However, oxamyl is a highly hazardous insecticide which is required by the label to be incorporated into the soil using water or mechanical means. Thus, alternatives to oxamyl are needed to manage *L. sparsus* in celery. Field experiments were conducted in Belle Glade, FL, to determine the efficacy of potential alternatives to oxamyl.

Findings

In a first experiment evaluating two foliar applications made two weeks apart in January–February 2022, oxamyl was associated with a strong trend ($P < 0.1$) suggesting *L. sparsus* control across four post-treatment sampling dates with a decrease of 52% in the proportion of plants with crown injury and of 82% in the number of larvae infesting plants. In addition, cyantraniliprole and novaluron were associated with trends for a decrease in *L. sparsus* injury and infestations two weeks after the second insecticide application (last sampling date). This experiment was repeated in December 2022–January 2023. Although the patterns of *L. sparsus* injury and infestation levels were consistent with those observed in the first experiment, *L. sparsus* pest pressure was low, and differences between insecticide-treated and non-treated plots were not detected ($P > 0.05$). Another experiment conducted in Fall 2022 evaluated the efficacy of a transplant tray drench using cyantraniliprole or clothianidin. However, extremely low *L. sparsus* pest pressure prevented the observation of conclusive results three and five weeks after transplanting.

Suggestions for Future Direction

Although additional research is needed to develop effective alternatives to oxamyl, the present study suggests that cyantraniliprole and novaluron might play a role in *L. sparsus* management in celery in Florida. Because novaluron is not registered for use on celery, results of the present study will support requests for registration.

*Corresponding author. Email: jbeuzelin@ufl.edu

Proc. Fla. State Hort. Soc. 136:139. 2023.

—Scientific Note—

Management of *Listronotus sparsus* in Parsley with Foliar and Drench Insecticide Applications

Anna Mészáros[1]*, Julien Beuzelin[2], and Craig Frey[3]

[1]*Palm Beach County Extension, University of Florida, IFAS, West Palm Beach, FL*

[2]*Everglades Research and Education Center, University of Florida, IFAS, Belle Glade, FL*

[3]*Hendry County Extension, University of Florida, IFAS, LaBelle, FL*

Keywords. insect pest management, insecticides, parsley, weevil

Listronotus sparsus (Say) is a geographically widespread weevil species that has recently become damaging in celery and parsley in Florida. Like the carrot weevils *Listronotus oregonensis* and *Listronotus texanus* (Edelson, 1985; Justus & Long, 2019), *L. sparsus* females deposit eggs on petioles and the base of plants. As larvae develop, they injure petioles, the crown, and roots, and eventually pupate in the soil. In celery, insecticide trials showed that oxamyl reduced *L. sparsus* injury (Mészáros & Beuzelin, 2023). However, oxamyl is not registered for use on parsley. Two trials were conducted on a commercial farm in Hendry County, Florida to evaluate insecticides and application methods for *L. sparsus* control on parsley.

Findings

The first trial compared the efficacy of foliar applications of conventional and biological insecticides. Novaluron (Diamond or Rimon, 12 fl oz/acre), a premix of novaluron and acetamiprid (Cormoran 12 fl oz/acre), thiamethoxam (Actara 3.5 oz/acre), cyantraniliprole (Exirel 20.5 fl oz/acre), a tank mix of thiamethoxam + cyantraniliprole (Actara 3.5 oz/acre + Exirel 20.5 fl oz/acre), and beta-cyfluthrin (Baythroid XL 3.2 fl oz/acre) were applied twice, two weeks apart, starting three weeks after transplanting. *Burkholderia rinojensis* strain A396 (MBI-306 20 fl oz/acre) was applied four times on a weekly schedule. Treatments were applied to plots 26 ft long and 5 ft wide, each containing 4 rows of parsley, and were replicated four times.

The second trial evaluated the efficacy of a single drench application of conventional insecticides One week after transplanting, each plant hole was treated with 2 fl oz of treatment solution. Cyantraniliprole (Verimark 13.5 fl oz/acre), thiamethoxam (Platinum 3.67 oz/acre), a tank mix of cyantraniliprole + thiamethoxam (Verimark 13.5 fl oz/acre + Platinum 3.67 oz/acre), and a premix of chlorantraniliprole + thiamethoxam (Durivo 13.5 floz/acre). For both trials, eight plants were collected per plot, one, three, and five weeks after the initial application to assess insecticide efficacy by recording the percentage plants with crown and root injury, as well as the number of larvae per plant.

For the first trial, there was a difference among treatments for the percentage of plants with injury to crowns ($P = 0.003$) and roots ($P = 0.006$), and for the number of larvae per plant ($P = 0.016$). Across sampling dates, Rimon, Cormoran, Exirel, and Actara+Exirel decreased crown and root injury levels by > 61% and > 87%, respectively, and larval infestation levels by > 82%. The tank-mix of Actara + Exirel provided the greatest level of control. Beta-cyfluthrin and *B. rinojensis* suppressed weevil infestations.

For the second trial, there was also a difference among treatments for the percentage of plants with crown and root injury, and for larval infestation ($P < 0.001$). All treatments were effective, with Durivo providing the greatest level of control. Durivo decreased crown injury levels by 75%, root injury levels by 97%, and larval infestation levels by 89% across sampling dates.

Suggestions for Future Direction

These field trials were the first to support management recommendations for control of *L. sparsus* in Florida parsley. Cyantraniliprole (Exirel for foliar applications, Verimark for drench applications) is registered for use on parsley and is effective against *L. sparsus*. Novaluron (Diamond, Rimon) is currently not registered for use on parsley, but results suggest that foliar application of this insecticide control *L. sparsus*. Thus, results will support requests for the registration of this reduced-risk insecticide. More studies determining *L. sparsus* seasonal population dynamics and crop host preference are planned. In addition, the threat to carrot production in Florida needs to be assessed.

Literature Cited

Edelson J.V., 1985. Biology of a carrot weevil, *Listronotus texanus* (Coleopter.: Curculionidae): Range and seasonality of infestations. *J. Econ. Entomol.* 78:895–897.

Justus E.J. & Long E.Y., 2019. Biology and management of the carrot weevil (Coleoptera: Curculionidae) in North America. *J. Integr. Pest Manag.* 10:8. https://doi.org/10.1093/jipm/pmz007.

Mészáros A. & Beuzelin J., 2022. *Listronotus* sp., an emerging weevil pest of celery and parsley in southern Florida. *Proc. Fla. State Hort. Soc.* 135:130.

*Corresponding author. Email: ameszaros@ufl.edu

Proc. Fla. State Hort. Soc. 136:140. 2023.

—Scientific Note—

Water Use Efficiency and Soil Carbon and Nitrogen Dynamics for Tomatoes Produced in Conventional and Compact Bed Geometries

Shahd Alrabia and Rao Mylavarapu*

Soil, Water, and Ecosystem Sciences Department, University of Florida, IFAS, Gainesville, FL

Keywords. carbon and nitrogen dynamics, conventional-compact beds, water use efficiency

Tomato (*Solanum lycopersicum L.*) makes a significant contribution to the agricultural economy in the state of Florida. Consequently, optimization of water and nutrient management efficiencies is crucial. A field experiment was conducted to evaluate the suitability of conventional and compact bed geometries in north central Florida to enhance water use efficiency (WUE) and the dynamics of total carbon (TC) and total nitrogen (TN) in the soil. Seasonal differences in WUE, TC and TN were observed during the study period.

Findings

Over the Spring and Fall 2021 seasons, water productivity and soil TC and TN were tested under two raised plastic mulch bed geometries. Conventional beds were raised to a height of 0.15 m and a length of 0.76 m, while compact beds were raised to a height of 0.25 m and a length of 0.61 m. Tomato seedlings were transplanted on March 25 in spring and September 16 in fall. Irrigation was run three times a day at one-hour duration at a flow rate of 31.5 cm^3/s per 30.5 m of drip tape and a pressure of 69 kPa. Applied irrigation amounts were estimated by measuring water flow at multiple emitters at random points and times in the field for a duration of 15 minutes. WUE was calculated as the yield obtained per unit of water applied per bed type. Additionally, soil samples were collected at 15 cm intervals up to a depth of 45 cm for TC and TN determination.

The Florida Automated Weather Network station located in the University of Florida/IFAS Plant Science Research and Education Unit, Citra, was accessed for the Spring (March 25–June 28) and Fall (September 16–January 18) periods. During the Spring and Fall seasons, the average temperatures were 22.20 °C and 20.33 °C, respectively, with soil temperatures of 26.03 °C and 23.53 °C, and average precipitation of 151.00 mm and 80.50 mm, respectively (FAWN, 2023).

Statistical analysis was performed using JMP PRO 15.0 to assess the effects of bed geometries on crop response variables. A Student's *t*-test was utilized to compare WUE between both bed types. The results indicated that in the spring, WUE was higher in compact beds compared to conventional beds. However, during the fall, the trend was reversed, with WUE being lower in compact beds compared to conventional beds (Table 1). Similarly, TC and TN were observed to follow the same pattern, where TC and TN were higher during the Spring and lower during the Fall (Fig. 1).

Table 1. Water Use efficiency (WUE) for conventional and compact bed geometries for Spring and Fall 2021.

	WUE ($kg \cdot ha^{-1} \cdot L^{-1}$)	
Bed type	Spring	Fall
Conventional	150.90 (b)	177.46 (a)
Compact	208.83 (a)	132.24 (b)

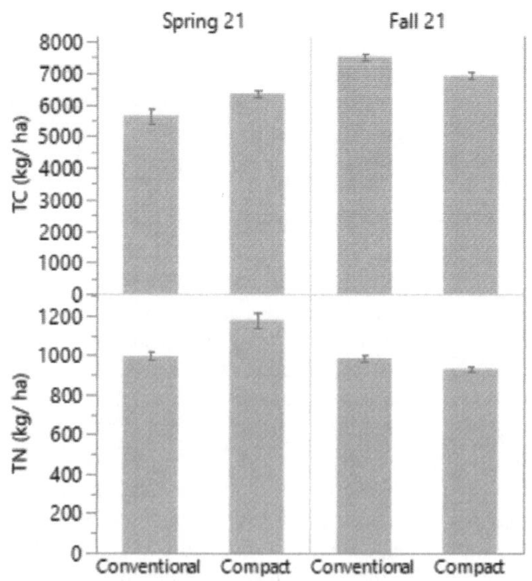

Fig. 1. Effects of conventional and compact bed geometries on total carbon (TC, top) and total nitrogen (TN, bottom) during Spring and Fall 2021.

Suggestion for Future Direction

Based on the findings of this study, future experiments should focus on evaluating the trends observed in WUE, TC, and TN and ultimately, tomato yields, as a result of the differences in bed geometries.

Literature Cited

Florida Automated Weather Network (FAWN), 2023. Report Generator. UF IFAS Extension Data Access Online. Accessed 21 May. 2023. https://fawn.ifas.ufl.edu/data/reports/.

*Corresponding author. Email: raom@ufl.edu

Proc. Fla. State Hort. Soc. 136:141. 2023.

—Scientific Note—

Introducing Horticulture Skills to Youth through the 4-H Blueberry Project

Julio Perez*

The University of Florida, IFAS, Extension Putnam County, East Palatka, FL

Keywords. blueberries, horticulture skills, youth

Blueberry production in Florida has increased to encompass around 5100 acres, with a value of utilized production estimated 62 million dollars (USDA–NASS, 2020). The 4-H Berry Project provides participants with insight into the horticultural needs of blueberries and experiences in budgeting, marketing, and selling a product. Since 2019, 51 youths have presented a pair of blueberries of different cultivars to the fair for judging. University of Florida, Institute of Food and Agricultural Sciences (UF/IFAS) extension agents and specialists, knowledgeable about blueberries, judged each entry on their own merits. Judges considered the health of the bushes and the horticultural skills demonstrated (proper pruning, repotting, and fertilizing). Blue, red, and white ribbons were awarded based on the judges' recommendations. Nine Cloverbuds, youth ages 5–7, were not judged and were only presented with participation ribbons. Forty participants were awarded blue and red ribbons demonstrating the horticultural skills needed to produce healthy blueberry bushes. Cloverbuds and entries that were awarded blue and red ribbons were allowed to sell their entries in a silent action to recoup the money they had invested in the project. Auctioned entries received between $30–$125.

Literature Cited

United States Department of Agriculture, National Agricultural Statistics Service (USDA–NASS). 2020. 2019 Blueberry statistics. Accessed June 6, 2021. https://www.nass.usda.gov/Statistics_by_State/New_Jersey/Publications/Blueberry_Statistics/NJ%202019%20Blueberry%20Summary.pdf

*Corresponding author. Email: pjulio@ufl.edu

Proc. Fla. State Hort. Soc. 136:142–143. 2023.

—Scientific Note—

Yield Productivity between Two Nitrogen Sources in Drip-irrigated Tomatoes Under Conventional and Compact Bed Geometrics

Shraddha Sharma and Rao Mylavarapu*

Soil, Water, and Ecosystem Sciences, University of Florida, IFAS, Gainesville, FL

Keywords. compact bed, controlled release urea, tomatoes, yield

This research aims to address the challenges of low nitrogen (N) use efficiency and environmental concerns by using controlled-release fertilizers (CRF), a preferred formulation over soluble N sources, which retain nutrients in the soil for a longer period and release them according to the plant's needs. A replicated field study comparing polymer-coated controlled-release urea and soluble urea was conducted with tomatoes (var. HM 1823) as a test crop, grown on sandy soils under two plastic-mulched bed geometry systems. The objective of the study is to compare the effects of N sources and yield components and yields between conventional and compact beds.

Findings

The study was located in the Plant Science Research and Education Unit in Citra, FL, where the soil is characterized as Candler sand (99% sand, 0.5% silt, and 0.5% clay). While the control treatment did not receive any N (0 kg·ha^{-1} N), the soluble urea (U) and the controlled release urea (CRU) plots received 140, 168, and 224 kg·ha^{-1}, respectively. Formulations with three release periods (CRU-60 days, 75 days, and 90 days) were used for all of the three applied rates. The experiment was designed as a randomized complete block design with 13 treatments and 3 replications. Specifics on field management was the same as described in Jalpa & Mylavarapu (2021). Tomatoes (var. HM 1823) were planted on March 25, 2021 (spring), September 16, 2021 (fall), May 6, 2022 (spring), and October 4, 2022 (fall). Mean yield estimation was done by based on two or three harvests each season. Due to varying planting and harvest dates in each season, the effects of the bed and N treatments were analyzed separately. Early results indicated that the use of CRF as a viable alternative N source for tomato cultivation, with CRU and U showing similar yields in Spring 2021, Fall 2021, and Fall 2022 (Fig. 1). Plants experienced higher temperatures during Spring 2022 due to late planting, resulting in shortened production season (Fig. 2). White mulch was used due to the high heat in that season. Whereas, in Fall 2022, tomato yields did not reach typical Florida yields (approx. 30,000 kg·ha^{-1}) due to late planting, which exposed plants to frost and decreased the number of harvests. The study demonstrated that there were seasonal differences in yields between the two bed geometries. Compact beds produced higher yields than conventional beds in Spring 2021, whereas conventional beds produced higher yields than compact beds in Fall 2021 (Fig. 3).

*Corresponding author. Email: raom@ufl.edu

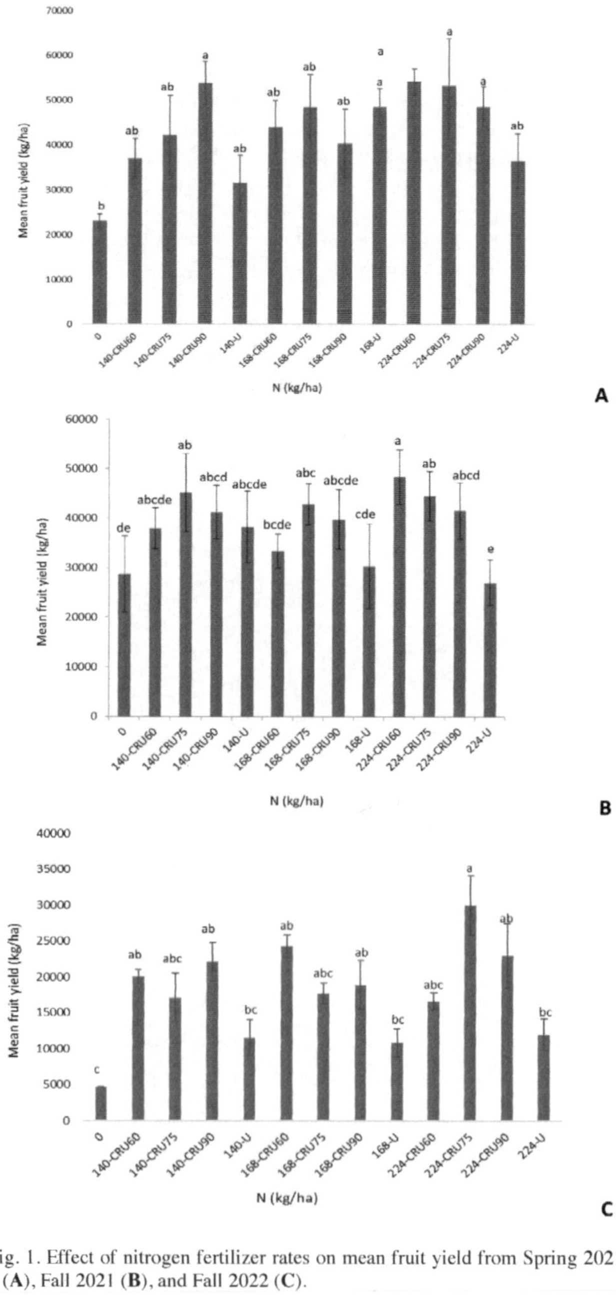

Fig. 1. Effect of nitrogen fertilizer rates on mean fruit yield from Spring 2021 (**A**), Fall 2021 (**B**), and Fall 2022 (**C**).

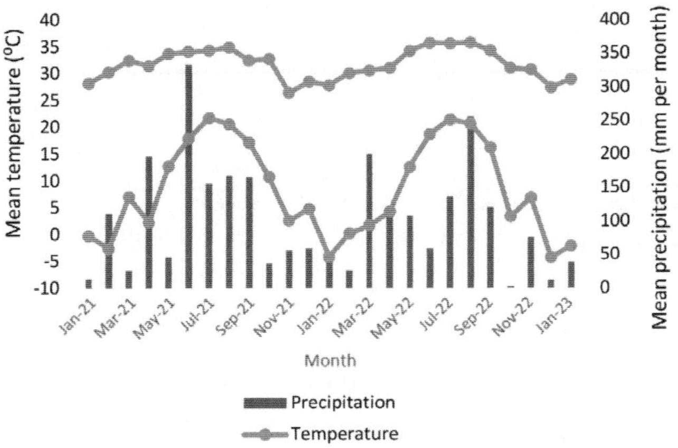

Fig. 2. Mean monthly precipitation (mm), maximum and minimum temperature (°C) during the experimental period in northern Florida. Source: Florida Automated Weather Network.

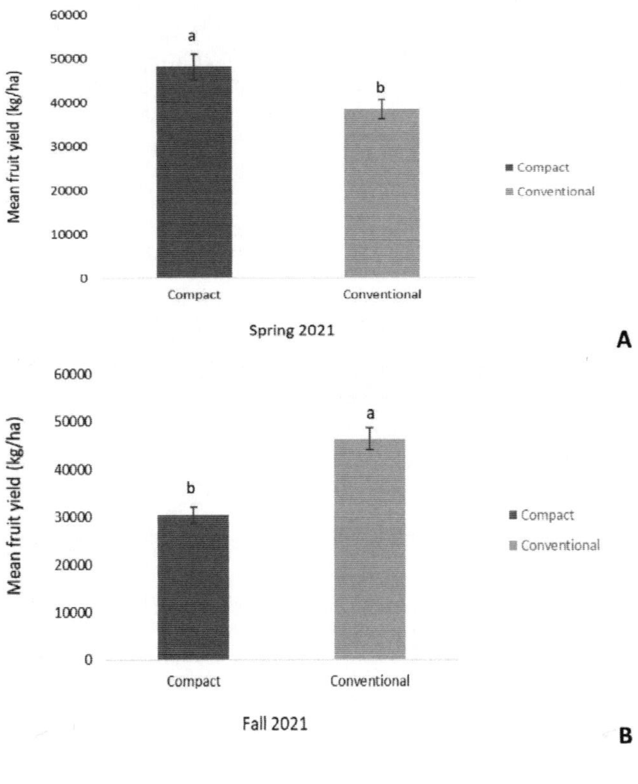

Fig. 3. Yield comparisons of compact and conventional in Spring 2021 (**A**) and Fall 2021 (**B**).

Suggestions for Future Direction

The preliminary result suggests that there is an economic benefit in terms of saving time, energy, and labor when using CRF, which is applied once preplant compared to multiple fertigation applications of soluble N sources. Future studies are needed for more seasonal data before making recommendations for using compact beds in tomato production in north-central Florida.

Literature Cited

Jalpa L. & Mylavarapu R., 2021. Nitrogen management and recovery efficiency and response to nitrogen sources in drip irrigated tomato. *Proc. Fla. State Hort. Soc.* 134:99–102.

Vegetable Section

Proc. Fla. State Hort. Soc. 136:144. 2023.

—Scientific Note—

Genome Editing in Cucurbits: Current Status and Future Prospects

Shallu Thakur and Geoffrey Meru*

University of Florida, IFAS, Tropical Research and Education Center, Homestead, FL

Keywords: Cas9, CRISPR, cucurbits, genome editing, PDS

Globally, the *Cucurbitaceae* family includes the most genetically diverse vegetables and fruits contributing greatly to human diet and nutritional security. Cultivation of cucurbit crops is affected by various abiotic and biotic stresses, amongst which powdery mildew (*Podosphaera xanthii*) (PM) and downy mildew (*Pseudoperonospora cubensis*) (DM) are devastating. The resistant sources against the PM and DM are not known to date in commercial cultivars except from *wild* species which have been difficult to transfer due to the lengthy breeding processes and linkage drag. Disruption of susceptibility genes (*S*-genes) is an attractive alternative for resistance breeding in cucurbits. The genome editing techniques especially the CRISPR-Cas9 system has potential to knock-down susceptible gene(s) and successfully employ this technique in crops. The objective of this research is to optimize the platform for genome editing employing the CRISPR-Cas9 system.

Findings

The sgRNAs specific to PDS gene targeting different regions were designed. Designed sgRNAs were synthesized and cloned in pHSE401 vector. Cotyledonary node explant cv. Black Beauty was used as the explant. Direct transformation was performed using gold particles 1 μm coated with plasmid pHSE401 carrying sgRNA specific to PDS gene (1100 psi). The presence of the transgene was confirmed by PCR in the transformants having bleached and albino phenotypes. Amplicon containing the sgRNAs regions were amplified from the screened positive transformants and sent for sequencing to confirm and check the edits. The 1-bp, 2bp and 3-bp deletion was observed in the transformants having bleached/albino phenotypes. It can be concluded from the study that the CpPDS knockout system generatee in the lab creates mutations in 6–8 weeks, which resulted in albino/bleached phenotype.

Suggestions for Future Research

A platform for genome editing employing CRISPR-Cas9 has been optimized. The optimized platform can be used to target susceptibility genes (*S*-genes). The long-term goal of our group is to create durable and broad-spectrum resistance to powdery and downy mildew by creating loss-of-function *S*-genes that act as recognition sites for the two pathogens in the host. *S*-genes were already identified in four species *viz. Cucurbita maxima*, *C. pepo*, *C. moschata* and *Citrullus lanatus* in earlier studies. Oligos were designed for all the identified *S*-genes. Expression studies are in progress in susceptible and resistant genotypes (challenged and control plants). This study intends to identify candidate *S*-genes for editing to achieve broad spectrum resistance against PM and DM.

*Corresponding author. Email: gmeru@ufl.edu

Proc. Fla. State Hort. Soc. 136:145. 2023.

—Scientific Note—

Dual-purpose Seed Pumpkin: Advances and Opportunities

YUQING FU, SWATI SHRESTHA, VINCENT N. MICHAEL, MARIE DORVAL, RIPHINE MAINVIEL, PRERNA SABHARWAL, SHALLU THAKUR, AND GEOFFREY MERU*

Tropical Research and Education Center, University of Florida, IFAS, Homestead, FL

Keywords. hull-less seeds, plant breeding, seed-oil

Pumpkin seeds/oil and flesh are of highly nutritive and have medical values due to heathy oil, fatty acids, protein, vitamins, minerals, and antioxidants which have benefits for prostate, heart and liver health. Dual-purpose seed pumpkin can be used for both seed/oil and flesh consumption. And hull-less (naked) seed pumpkins are advantageous for seed/oil production without dehulling costs and direct use in the food industry. Worldwide, the hull-less seed pumpkin varieties are the derivatives of a natural mutation of the Austrian Styrian pumpkin (*Cucurbita pepo* L.) discovered in late nineteenth century, and *C. moschata* hull-less seed pumpkin germplasm which is only available in China. Originally hull-less seed pumpkins were only used for seeds due to their bland flesh, nowadays dual-purpose varieties with hull-less seeds and tasty flesh are desirable so breeding for ecotype cultivars adapted to different growing environment is necessary. We developed the robust Kompetitive Allele Specific PCR (KASP) marker Ch12_3412046 associated with the recessive hull-less seed gene n, accelerating hull-less pumpkin breeding with marker-assisted selection (MAS) (Meru et al., 2022). This study was aimed to breed dual-purpose *C. pepo* varieties with MAS breeding, creating a niche market for growers (Meru & Fu, 2018) and developing a hull-less seed *C. moschata* through interspecific hybridization.

Materials and Methods

The hull-less seed 'Kakai' (*C. pepo*) was crossed with 5 diverse hulled seed *C. pepo* accessions of tasty flesh, different vine types, fruit shapes and colors including 'Table Gold Acorn', 'Winter Luxury', 'Dostal Cucumber Squash', 'Bush Delicata', and 'Cream of the Crop'. Breeding schemes of selfing, backcrossing, and MAS using the KASP marker Ch12_3412046 were employed. To introduce the hull-less seed trait from *C. pepo* into *C. moschata*, a three-way cross between the UF-created SS333 bridge line (32% *C. pepo* and 68% *C. moschata*) and the *C. pepo* F_1 progeny of 'Cream of the Crop' and 'Kakai' was created. Leaf DNA was extracted at the seedling stage and genotyped with marker Ch12_3412046. The seedlings of homozygous hull-less seed genotype (nn) and heterozygous hulled seed genotype (Nn) were chosen for further generation, while the homozygous hulled seed genotype (NN) was discarded.

Result and Discussion

The F_2 populations of the intraspecific crosses 'Kakai' × 'Dostal Cucumber Squash' (n = 54), 'Kakai × 'Winter Luxury' (n = 117), and 'Kakai' × 'Table Gold Acorn' (n = 143) were screened using marker Ch12_3412046, exhibiting the genotype ratio of 1NN:2Nn:1nn. The F_2:3 family of 'Kakai' × 'Table Gold Acorn' continued to show the hull-less seed trait and improved flesh color/quality compared to the 'Kakai' parent. The BC_2F_1 population of 'Kakai' × 'Dostal Cucumber Squash' was developed through backcrossing and MAS. In this study, the KASP marker Ch12_3412046 was further validated as a universal/powerful marker with different *C. pepo* background and can be extensively employed for *C. pepo* hull-less seed breeding.

We successfully introduced the hull-less seed gene from *C. pepo* into *C. moschata* with the aid of the UF-created bridge line SS333. The KASP marker Ch12_3412046 detected the *C. pepo* hull-less seed gene n in the progenies of the interspecific cross SS333-21 × ['Cream of the Crop' × 'Kakai']F_1 and their later generations (67% and 84% *C. moschata* background, respectively). Seed coat variations were observed showing different degrees of lignification such as semi-thick-coated and semi-thin-coated. To better understand the mechanism of hull-less seed, sequencing the NAC domain-containing gene (Cp4.1LG12g04350) which is associated with seed coat lignin biosynthesis could tell us whether the hull-less seed gene is in integrity after introgression.

With the facts of narrow genetic background of *C. pepo* hull-less pumpkins and *C. moschata*'s harboring multiple disease resistances for squash crown rot, zucchini yellow mosaic virus, and papaya ringspot virus, all of which commonly affect *C. pepo* pumpkins, the introgression brings more opportunities to broaden genetic background. In addition, the introgression of hull-less seed pumpkin (*C. pepo*) into a heat-tolerant tropical *C. moschata* species, calabaza, will highlight dual-purpose seed pumpkin breeding in south Florida.

Literature Cited

Meru G. & Fu Y. 2018. Yield and horticultural performance of naked-seed pumpkin in south Florida. UF/IFAS Extension, Gainesville. http:// edis.ifas.ufl.edu/HS1323

Meru G., Fu Y., Shrestha S., Michael V.N., Dorval M., & Mainviel R. 2022. Genomic position, markers and candidate genes associated with the hull-less seed trait in pumpkin. *Plants.* 11(9):1238. https://doi.org/10.3390/plants11091238

*Corresponding author. Email: gmeru@ufl.edu

Vegetable Section

Proc. Fla. State Hort. Soc. 136:146. 2023.

Key Quality Component and Bioactive Compound Analysis of Novel Calabaza Germplasm Lines

Megan Kinsman[1]*, Amarat Simonne[2], Steven Sargent[3], Geoffrey Meru[4], Carlene A. Chase[3], and Andrew MacIntosh[1]

[1]*Food Science and Human Nutrition Department, University of Florida, Gainesville, FL*

[2]*Family, Youth and Community Sciences Department, University of Florida, Gainesville, FL*

[3]*Horticultural Sciences Department, University of Florida, Gainesville, FL*

[4]*Tropical Research and Education Center, University of Florida, Homestead, FL*

—abstract—

Calabaza (*Cucurbita moschata* Duch.) is a nutritious specialty crop reported to possess health-promoting bioactive compounds, specifically β-carotene, ascorbic acid, and phenolics. Novel calabaza germplasm lines are currently being developed by the University of Florida that are under evaluation for conventional and organic cropping systems in the southeastern United States. Calabaza is considered an emerging crop and new cultivars adapted to U.S. cropping systems will reduce imports while satisfying consumer demand for calabaza from Latin American and Caribbean countries. Standard methods of analysis were used to determine moisture content (MC), total soluble solids (TSS), carotenoid content, ascorbic acid, and total phenolic content (TPC) in raw calabaza mesocarp from seven novel germplasm lines, two commercial calabaza cultivars, and Waltham butternut squash, a well-studied cultivar, used as a control. Average MC and TSS varied significantly among calabaza lines, with UFTP24 and UFTP46 ('La Estrella') having the lowest MC (84.1% and 76.3%, respectively) and the highest TSS (10.2 °Brix and 13.2 °Brix, respectively). Raw samples from each line were found to have high concentrations of provitamin A β-carotene, each providing ≥ 20% of the dietary reference intake (DRI) for retinol activity equivalents. Calabaza from line UFTP24 and 'La Estrella' could be considered "excellent" sources of ascorbic acid (≥ 20% DRI). UFTP45 ('Soler') had lower concentrations of TPC (193.6 µg GAE/g), while 'La Estrella' had the highest concentration of TPC among all lines analyzed (526.6 µg GAE/g). These results will be used to inform prospective breeding efforts aimed at including traits for high phytonutrient content in new calabaza cultivars.

*Corresponding author. Email: m.kinsman@ufl.edu

Proc. Fla. State Hort. Soc. 136:147. 2023.

Screening Chili and Sweet Pepper Lines for Drought Tolerance and Hybrid Vigor for Leaf, Stem, and Fruit Quality Traits

ALEX BUTLER, HAILEY MAURER, EMILY SANZ, AND BALA RATHINASABAPATHI*

Horticultural Sciences Department, University of Florida, IFAS, Gainesville, FL

—abstract—

Pepper (*Capsicum annuum*) is an important vegetable commodity grown in Florida. We are focused on developing improved pepper cultivars adapted to field production. For this purpose, we needed quick and efficient methods to screen large number of breeding lines for plant and fruit quality traits. Loss of water from leaves through transpiration is one aspect of whole plant tolerance to drought. To test variation in this trait, we screened 8 inbred lines grown in a greenhouse for the rate of water loss from excised leaves. In a 30-minute incubation period under light, excised leaves from the lines lost 13% ('Georgia Flame') to 19% ('Corno di Toro') of the initial water suggesting that this screen could be an efficient way to identify germplasm to test for whole plant tolerance to drought. To understand the degree of hybrid vigor of F_1 peppers for leaf size, stem branching and fruit quality traits, controlled pollinations between pairs of plants selected from three sweet pepper inbred lines and four chili pepper inbred lines were made. Distinctly different traits in these lines for leaf size, stem branching and fruit size and shape characteristics were recorded. We expect that the methods developed in this research will improve the efficiency in pepper breeding.

The authors acknowledge funding support by FDAC via a Specialty Crops Block Grant (AWD13579).

*Corresponding author. Email: brath@ufl.edu

Proc. Fla. State Hort. Soc. 136:148. 2023.

—Scientific Note—

Postharvest Performance of UF 'Marathon' Mandarin

Cuifeng Hu[1], Fred G. Gmitter[2], Jude Grosser[2], and Mark A. Ritenour[1]*

[1]*Indian River Research and Education Center, University of Florida, IFAS, Fort Pierce, FL*

[2]*Citrus Research and Education Center, University of Florida, IFAS, Lake Alfred, FL*

Keywords. *Lasiodiplodia* spp., postharvest decay, postharvest disease management

'Marathon' is an early season, easy peeling, seedless mandarin released by the University of Florida (UF) in 2019. Reported here are postharvest fruit quality, degreening performance, and shelf life of fruit from eight harvests of 'Marathon' spanning five seasons, starting Fall 2014 through 2023. Not all tests were conducted on each harvest. Between 20 to 50 fruit were evaluated at each harvest during the 2014–15 and 2015–16 seasons because the trees were young, and between 60 to 2600 fruit during subsequent seasons. After harvest, fruit external color, weight, total soluble solids (TSS), titratable acidity (TA), and storage quality were evaluated. Degreening treatments consisted of exposure to 3.5 to 5.5 ppm ethylene at 85 or 70 °F, with 95% relative humidity (RH). Control fruit were stored under the same temperature and RH but without ethylene. Fruit color was measured before and after degreening and the fruit stored at 40 °F.

Peel color was measured using a Minolta colorimeter (model CR-3000; Minolta Co. Ltd., Osaka, Japan) using 20 to 30 fruit, each fruit was measured three times around the equatorial zone. Color was recorded as L*, a*, and b* values, but converted to a*/b* ratio. As fruit peel changes from green to yellow/orange, a*/b* increases from negative to positive values. TSS were measured using a temperature-compensated refractometer (Atago, 3810, Tokyo, Japan). TA was measured by titrating juice samples to pH 8.3 with 0.3125 N NaOH using an automatic potentiometric titrator (model HI 901, Hanna Instruments Inc., Woonsocket, RI). Fruit harvested 21 Oct 2021 were run through an Invision 9000 system (TOMRA Food, Compac Sorting Equipment Inc., Visalia, CA) to measure fruit weight, diameter, and volume, with fruit size distribution being calculated for the population. Storage quality was evaluated bi-weekly by determining incidence of fruit decay and peel breakdown.

Findings

Spanning all seasons, fruit were harvested as early as 6 Sept and as late as 11 Dec with external peel color generally improving as the season progressed. Fruit harvested in September were usually green (a*/b* as low as –0.43). However, spot picking for color resulted in much better initial peel color (a*/b* = –0.08). Both sets of fruit responded to degreening treatments using ethylene, with the greener fruit needing longer treatment to achieve the same color. However, while degreening for 48 hours or less generally did not accelerate decay significantly, longer degreening time resulted in more decay. Fruit harvested in November or December had well-developed color at harvest (0.28 and 0.50, respectively).

Fruit harvested 6 Sept 2022 already had 10.4% TSS, 0.71% TA, and 14.5 TSS:TA ratio. Fruit from the latest harvest on 11 Dec 2014 had 14.1% TSS, 0.61% TA, and a 23.1 ratio. However, there were year-to-year variations with the lowest internal quality at harvest occurring on 26 Oct 2020 with 8.5% TSS, 0.72% TA, and 11.8 ratio.

Fruit harvested 30 Sept 2021 included over 2500 fruit from three trees showing variation in their fruit size distribution, with 26% of Tree 1 fruit peaking at size 120 (fruit per carton), while 21% of Tree 2 fruit peaked at size 150, and 31% of Tree 3 fruit peaked at size 246 or smaller. Average fruit weight ranged from 74.7 to 89.7 g depending on the tree.

Fruit of 'Marathon' held up well during storage, with fruit generally maintaining 92% or more healthy fruit after up to 8 weeks at 40 °F. However, there was some variation with fruit harvested 18 Oct 2022 having only 82% healthy fruit after 4 weeks in storage.

Suggestions for Future Direction

Results from these studies show that 'Marathon' fruit can be harvested with excellent quality attributes as early as September and into December. Early season fruit can also be spot-picked for better color which reduces the need for longer degreening durations, thus maximizing shelf life.

*Corresponding author. Email: ritenour@ufl.edu

Handling & Processing Section

Proc. Fla. State Hort. Soc. 136:149. 2023.

—Scientific Note—

Postharvest Evaluation of Elite Lettuce Germplasm in Non-traditional Areas of Florida

Adrian Berry[1], Steven Sargent[1], Jesse Murray[2], Christian Christensen[3], Robert Hochmuth[4], and Germán Sandoya[2]*

[1]*Horticultural Sciences Department, University of Florida, IFAS, Gainesville, FL*

[2]*Everglades Research & Education Center, University of Florida, IFAS, Belle Glade, FL*

[3]*Hastings Agricultural Extension Center, University of Florida, IFAS, Hastings, FL*

[4]*North Florida Research & Education Center-Suwannee Valley, University of Florida, IFAS, Live Oak, FL*

Keywords. breeding, hydroponics, sandy soils, shelf-life

Lettuce is a leading vegetable crop cultivated in Florida, typically grown on muck soils during the winter season (October to March). Iceberg (head) and romaine are the two main morphological types planted in Florida (USDA–NASS, 2017). However, growing less planted lettuce types in Florida including Boston, Latin, and leaf lettuce in sandy soils (SS), and in controlled environments (CE), could help diversify the Florida lettuce market. The University of Florida, Institute of Food and Agricultural Sciences (UF/IFAS) lettuce breeding program has identified germplasm that performs acceptably in SS and CE under warmer temperatures; however, their quality during postharvest storage is unknown. Therefore, postharvest quality and estimated shelf-life were evaluated among these UF/IFAS germplasms grown in SS and CE production systems.

In Winter 2022–23, ten lettuce germplasms (Envoy, Rex, 43007, Revolution, North Star, C1146, 60182, Red Hot Cos, Monte Carlo, 20c4426) were planted in sandy soil at the UF/IFAS Hastings Agricultural Extension Center (HAEC) and harvested in early March (H-1). A second group of eight germplasms (Green Forest, Taiwan, Floricos 83, 20c4426, North Star, Pavane, Darkland Cos, Monte Carlo) were planted in a hydroponic system at UF/IFAS North Florida Research & Education Center-Suwannee Valley (NFREC-SV) and harvested in mid-March (H-2). Head weight (g), marketable heads (%), bolting (%), and tipburn (%) were rated at harvest on a per-plot basis. Harvested lettuce heads were immediately transported to the UF/IFAS Horticultural Sciences Department Postharvest Horticulture Laboratory, in Gainesville, FL. The harvested heads were hydrocooled and stored on open trays for 13 d (H-1) or 12 d (H-2) at 50 °F (10 °C) and 90% relative humidity (RH). These storage conditions were based on the Accelerated Shelf-Life Testing protocol that simulates long-term storage at 34 °F (1.8 °C) (Belisle et al., 2021). Lettuce shelf-life was evaluated on a 9 to 0 scale, where 9 = excellent, essentially free from defects, 7 = good, minor defects, not objectionable, 5 = fair, slightly to moderately objectionable defects, lower limit of salability and 0 = extremely poor (Belisle et al., 2021).

*Corresponding author. Email: gsandoyamiranda@ufl.edu

Findings

Significant differences were identified among all lettuce accessions in H-1 for marketability ($P = 0.0067$) and for head weight within leaf ($P = 0.0001$) and Romaine ($P = 0.0006$); however, no differences were found for Boston ($P = 0.3452$) or Latin (non-calculated) lettuces. In H-2, significant differences ($P < 0.0001$) were found for both head weight and marketability. Similarly, significant differences ($P < 0.05$) were found for percentage of bolted plants and tipburn. 'North Star' and 'Monte Carlo' had the longest shelf-life for both production systems based on acceptable appearance, green color, and firmness. Those performing well were for H-1: 'Red Hot Cos', and for H-2: 'Darkland Cos', 'Green Forest', 'Taiwan', and 20c4426. The other germplasm for H-1 became unmarketable after 6 to 11 d.

Breeding lines and cultivars with longer shelf-life identified in both experiments are useful for either commercial purposes or as breeding material for nontraditional settings for lettuce production. These findings offer plant breeders germplasm to improve the shelf-life of lettuce grown in SS and CE, and helps growers comprehend the duration for which lettuce can be effectively marketed.

Suggestions for Future Direction

Additional experiments throughout the lettuce season should be conducted to understand the interaction with different environments for shelf-life in both SS and CE. Likewise, exploring additional lettuce germplasm can help to identify newer sources of breeding material with extended shelf-life.

Literature Cited

Belisle CE, Sargent SA, Brecht JK, Sandoya GV, & Sims CA. 2021. Accelerated shelf-life testing to predict quality loss in romaine-type lettuce. *HortTechnology*. 31:490–499. https://doi.org/10.21273/HORTTECH04812-21

US Department of Agriculture National Agricultural Statistics Service Information (USDA–NASS). 2017. Census of Agriculture, United States Summary and State Data. Accessed July 16, 2023. https://www.nass.usda.gov/Publications/AgCensus/2017/Full_Report/Volume_1,_Chapter_1_US/usv1.pdf

Handling & Processing Section

Proc. Fla. State Hort. Soc. 136:150. 2023.

Total Phenolic Content of Peel and Flesh of Muscadine Grape Genotypes: A Comparative Study

Fariborz Habibi[1]*, Ali Sarkhosh[1], Patrick J. Conner[2], and Jeffrey K. Brecht[1]

[1]*Horticultural Sciences Department, University of Florida, Gainesville FL*

[2]*Department of Horticulture, University of Georgia, Tifton GA*

Keywords. bioactive compounds, muscadine, nutritional value, quality

—abstract—

Muscadines (*Muscadinia rotundifolia* Michx) have a thick peel that is often discarded by consumers. To better understand the potential nutritional loss due to this practice, total anthocyanin concentration (TAC), total phenolic content (TPC), total antioxidant activity (TAA), and chemical attributes in peel and flesh of eight purple muscadine genotypes were evaluated to determine the nutritional value of each part. Muscadine genotypes that were used in this study included two commercial cultivars ('Supreme' and 'Paulk') and six selections ('Ga.10-1-222', 'Ga.10-1-294', 'Ga.10-1-329', 'Ga.12-5-46', 'Ga.12-3-22', 'Ga.13-4-79'). Bioactive compounds, antioxidant activity, and titratable acidity (TA) were higher in the peel of all muscadine genotypes, while the flesh had more sucrose, glucose, and fructose. In the peel, TAC (20.68–37.53 mg·L^{-1}), TPC (382.84-577.31 mg·L^{-1} GAE), ellagic acid (5.44–15.5 μg·mL^{-1}), TAA (436.07-1258.26 μmol·L^{-1} TE), total soluble solids (TSS, 11.45–16.62%), TA (0.64–1.17%), TSS/TA ratio (10.92–24.46), pH (3.42–3.82), sucrose (13.73–69.49 mg·mL^{-1}), glucose (58.42–77.69 mg·mL^{-1}), and fructose (19.26–26.84 mg·mL^{-1}) varied significantly among different cultivars. Similarly, in the flesh, the ranges of TAC (0.38–2.33 mg·L^{-1}), TPC (11.48–26.52 mg·L^{-1}), ellagic acid (0.02–0.3 μg·mL^{-1} GAE), TAA (14.65–21.62 μmol·L^{-1} TE), TSS (12.25–17.3%), TA (0.43–0.73%), TSS/TA ratio (19.44–35.34), pH (3.44–3.73), sucrose (39.33-124.02 mg·mL^{-1}), glucose (71.75–92.86 mg·L^{-1}), and fructose (24.25–32.25 mg·L^{-1}) were significantly different among different cultivars. The 'Supreme' peel exhibited the highest TAC, TPC, and TAA, while the peel of 'Ga.12-3-22' had the lowest values. In 'Supreme' fruit, the TAC values ranged from 0.382 mg·L^{-1} in the flesh to 37.53 mg·L^{-1} in the peel. The TPC values showed variability between 17 and 577 mg·L^{-1} GAE in both flesh and peel. Furthermore, TAA values ranged from 1258 to 16 μmol·L^{-1} TE in both peel and flesh. The peel contained over 90% of the ellagic acid in the fruit, with 'Ga.12-5-46' peel having the highest ellagic acid overall. The ellagic acid content in the peel of 'Ga.12-5-46' was 15.5 μg·mL^{-1}, which was 144 times greater than that in its flesh. 'Ga.13-4-79' had the highest TA in both peel and flesh. The highest sucrose content in both peel and flesh occurred in 'Ga.10-1-329' and 'Ga.12-3-22', while 'Ga.12-5-46' and 'Ga.13-4-79' had comparatively lower sucrose in flesh. Given the substantial phytochemical content present in muscadine peel, discarding the peel constitutes a missed opportunity of considerable magnitude to obtain the full nutritional benefits of muscadines.

The full paper can be found at: https://doi.org/10.1007/s11694-024-02404-1

*Corresponding author. Email: fhabibi@ufl.edu

Proc. Fla. State Hort. Soc. 136:151–152. 2023.

—Scientific Note—

A Comparison of the Effects of 1-MCP Treatment on Climacteric and Nonclimacteric Guava Fruits

Morgan M. Madison[1], Jeffrey K. Brecht[1]*, Steven A. Sargent[1], Moshe Doron[1], and Jonathan H. Crane[2]

[1]*Horticultural Sciences Department, University of Florida, IFAS, Gainesville, FL*

[2]*Tropical Research and Education Center, University of Florida, IFAS, Homestead, FL*

Keywords. decay, quality, *Psidium guajava*, ripening, shelf life, storage life

We evaluated the effectiveness of 1-MCP controlled release technology from Hazel Technologies for extending the shelf life of two guava (*Psidium guajava*) cultivars in terms of ripening inhibition and quality retention. Guava is a climacteric (Brown and Wills, 1983) round or oval shaped fruit, with several cultivars that are considered to be nonclimateric or possibly exhibiting suppressed climacteric ripening behavior (Azzolini et al., 2005; Biale and Barcus, 1970; Porat et al., 2009). The two guava cultivars included a climacteric cultivar called 'Pera' (pink-flesh) guava and a nonclimacteric or suppressed climacteric cultivar called green 'Crispy' guava. The fruit can be eaten fresh at two stages: Mature green, when the sensory character resembles unsweet apples with white flesh and some crispiness, and fully ripe, when the fruit have a light-yellow skin, white to bright-red flesh, very sweet flavor, and a strong, unmistakable fragrance.

Materials and Methods

Guava fruit were harvested at commercial maturity, representing a range of ripeness stages, at a commercial farm in Homestead, FL. Fruit were inserted into foam sleeves at the packing facility and then wrapped in bubble wrap and placed inside nylon-lined, waxed fiberboard cartons. Each carton was put inside a plastic cooler and shipped overnight by mail. Upon arrival in Gainesville, the fruit were evaluated visually and sorted according to external color, then placed into unlined, waxed fiberboard cartons. Two Hazel 1-MCP sachets (Hazel100™ 0.25 g, batch# 495A14) were placed inside each 1-MCP treatment carton after transfer into storage on shelving racks with plastic covers in a controlled temperature room at 8 °C and 95% relative humidity (RH), to prevent 1-MCP cross-contamination with the control treatment, which was placed in a separate room at the same temperature and RH. After 7 or 14 storage days, the fruit were transferred to 20 °C and 95% RH for 7 days ripening. Fruit quality and shelf life evaluations were performed upon fruit arrival, at fruit transfer, and after 7 days ripening (i.e., shelf life). For non-destructive evaluations, a subset of 15 guava fruit per treatment representing the external color range of the whole population was identified and maintained for repeated evaluations throughout the trial. For destructive evaluations, 15 fruit per treatment were sampled initially, at each transfer, and after shelf life, and used for fruit quality analyses.

The non-destructive evaluations included weight loss, external color development, and external decay incidence and severity. For color development, CIE L*a*b* measurements were taken on the same peel area using a chromameter (Minolta CR-400 Chroma Meter). For decay incidence and severity, each fruit was visually inspected and rated based on a 1 to 9 scale of decay severity. Individual fruit weight was measured using a digital scale (Mettler Toledo, model XA1502S). For the destructive evaluations, fruit were first photographed, then external and subepidermal colorimetric measurements were taken at the widest area of each fruit equator. Internal decay incidence and severity were counted and evaluated on fruit halves that were cut at the equator and photographed. For other quality analyses, three composite samples (reps) of five fruit halves were taken per treatment on each evaluation day, placed into Ziploc® bags, and frozen at –30 °C (–22 °F) until processed. Upon thawing, the fruit halves were blended and then centrifuged at 4 °C (39.2 °F) for 20 minutes at 22,217 $\times g_n$. After centrifuging, samples were filtered, and the juice used for analysis. A digital refractometer (model AR200, Reichert Analytical Instruments) was used to determine soluble solids content (SSC, °Brix); pH and total titratable acidity (TTA, % citric acid basis) were measured using an automatic potentiometer (model 719 S Titrino, Metrohm). The TTA was determined by diluting 1.0 mL of juice with 50 mL deionized water, then titrating with 0.1 N sodium hydroxide (NaOH) to an endpoint of pH 8.2 and expressing it as percent citric acid.

Results and Discussion

The 1-MCP sachets were effective in extending the storage of the climacteric 'Pera' guavas by reducing the rates of fruit softening and external and subepidermal color development after 7 days at 8 °C. Postharvest external decay was unaffected by 1-MCP treatment, preventing assessment of its effect on ripening at 20 °C. In future studies, effective humidity and decay control practices should be used to allow the potential benefits of 1-MCP treatment in extending the quality and shelf life of these fruit to be evaluated. The SSC, TTA, and SSC/TTA did not show significant differences between the control and 1-MCP treatments, but more testing could be done to evaluate 1-MCP effectiveness with other factors such as temperature, storage conditions, and pre- and

*Corresponding author. Email: jkbrecht@ufl.edu

postharvest treatments included to reduce decay incidence that impedes expected ripening reactions.

The 1-MCP sachets did not have any effect on green ('Crispy') guava external color ratings. Even though there was significantly higher weight loss in the 1-MCP treatment compared to the control treatment, fruit at the end of the storage period (after 22 days) were significantly firmer in the 1-MCP treatment compared to the control treatment, suggesting the 1-MCP delayed ripening/senescence. While the first external decay incidence appeared in the 1-MCP but not the control after 14 days at 8 °C, there may have been a minor positive effect of 1-MCP on external decay incidence during ripening at 20 °C with more external decay incidence in the control compared to the 1-MCP at the end of the study. Unlike external decay incidence, internal decay severity progressed in both treatments at a similar rate with higher severity in the 1-MCP treatment by the end of the study. Fruit quality results for both treatments on crispy guavas showed no significant differences between treatments for SSC and SSC/TTA, but TTA was significantly higher in the 1-MCP treatment compared to the control on the last day of the study. Internal decay incidence and severity shared a similar trend, appearing first in the 1-MCP treatment on day 15 and having almost the same degree of internal decay severity in both treatments by the end of the study. The results indicating 1-MCP inhibition of ripening suggest that green guava may be actually a "suppressed climacteric" rather than nonclimacteric fruit type.

Literature Cited

Azzolini M., Jacomino A.P., Bron I.U., Kluge R.A. & Schiavinato M.A. 2005. Ripening of "Pedro Sato" guava: Study on its climacteric or non-climacteric nature. *Braz. J. Plant Physiol. 17(3):299-306*.

Biale J.B. & Barcus D.E. 1970. Respiratory patterns in tropical fruits of the Amazon basin. *Trop. Science*. 12:93–104.

Brown B.I. & Wills R.B.H. 1983. Post-harvest changes in guava fruit of different maturity. *Sci. Hortic*. 19:23–243.

Porat R., Weiss B., Zipori I., & Dag A. 2009. Postharvest longevity and responsiveness of guava varieties with distinctive climacteric behaviors to 1-methylcyclopropene. *HortTechnology*. 19:580–585. https://doi.org/10.21273/HORTSCI.19.3.580

Proc. Fla. State Hort. Soc. 136:153. 2023.

—Scientific Note—

Functional Characteristics of Aldehyde Dehydrogenase and Its Involvement in Aroma Volatile Biosynthesis in Banana Fruit

Yoshinori Ueda[1], Wei Zhao[2], Hideshi Ihara[3], Yoshihiro Imahori[4], Eleni Tsantili[5], Sumithra K. Wendakoon[6], Alan Chambers[7], and Jinhe Bai[2]*

[1]*Center for Research & Development of Bioresources, Osaka Prefectural University, Osaka, Japan*

[2]*USDA, ARS, Horticultural Research Laboratory, Ft Pierce, FL*

[3]*Department of Biological Science, Osaka Prefecture University, Osaka, Japan*

[4]*Graduate School of Life & Environmental Science, Osaka Prefecture University, Osaka, Japan*

[5]*Department of Crop Science, Agricultural University of Athens, Athens, Greece*

[6]*Department of Bioresource Science, Ryukoku University, Shiga, Japan*

[7]*Tropical Research and Education Center, University of Florida, IFAS, Homestead, FL*

Keywords. ALDH, ester, enzyme characteristics, *Musa* AAA

Many enzymes regarding the conversions between volatile alcohols, aldehydes/ketones, acids, and esters have been studied extensively in fruits and plants, and/or adapted from microorganism studies. In banana fruit ripening, gene expression and enzyme activity of alcohol dehydrogenase (ADH, short and medium chains) which facilitates the interconversion between alcohols to aldehydes (Yang et al., 2011), acyl-CoA synthetase (ACS) which activates carboxylic acids to acyl-CoAs (Wendakoon et al., 2005) thus can be used to biosynthesize esters, and alcohol acetyl transferase (AAT) which catalyzes ester biosynthesis, have been intensively studied (Beekwilder et al., 2004; Ueda et al., 1992). Aldehyde dehydrogenase (ALDH) dehydrogenizes aldehydes to carboxylic acids in different plant tissues (Brocker et al., 2013). Our group first reported the function of aromatic volatile biosynthesis in bananas (Ueda et al., 2022).

Findings

Straight and branched chain C2-C6 aldehyde vapor feeding to ripe banana pulp produced their corresponding carboxylic acids. Hexanal had the highest conversion rate, followed by C3-C5 aldehydes, and ethanal had the lowest conversion rate. Generally, straight chain aldehydes had higher conversion rates than branched chain ones when the total carbon number was the same in the molecule. The conversion is an oxygen-independent metabolic pathway and is enzyme-catalyzed with NAD+ as the cofactor. Crude ALDH was extracted from ripe banana pulp, and the interference from ADH was removed by two procedures: 1) washing off elutable proteins that contain 95% of ADH, but only about 40% of ALDH activity, with the remaining ALDH extracted from the pellet residues at the crude ALDH extraction stage; 2) adding an ADH inhibitor in the reaction mixture. The optimum pH of the ALDH was 8.8, and the optimum phosphate buffer concentration was higher than 100 mM.

*Corresponding author. Email: jinhe.bai@usda.gov

Suggestions for Future Directions

Further research is needed to confirm if ALDH is an enzyme in the routine pathway for volatile production associated with fruit ripening or just a consequence of aldehyde scavenging. Plants encompass a total of fourteen ALDH protein families. Nevertheless, the existing data on published superfamily genes and enzymes do not indicate any functions related to the dehydrogenization of carbon 2-6 straight and branched-chain aldehydes or their subsequent synthesis into esters in bananas and other fruits.

Literature Cited

Beekwilder J., Alvarez-Huerta M., Neef E., Verstappen F.W.A., Bouwmeester H.J., & Aharoni A. 2004. Functional characterization of enzymes forming volatile esters from strawberry and banana. *Plant Physiol.* 135:1865–1878. https://doi.org/10.1104/pp.104.042580

Brocker C., Vasiliou M., Carpenter S., Carpenter C., Zhang, Y., Wang X., Kotchoni S.O., Wood A.J., Kirch H.-H., & Kopečn D. 2013. Aldehyde dehydrogenase (ALDH) superfamily in plants: gene nomenclature and comparative genomics. *Planta*. 237:189–210. https://doi.org/10.1007/s00425-012-1749-0

Ueda Y., Tsuda A., Bai J., Fujishita N., & Chachin K. 1992. Characteristic pattern of aroma ester formation from banana, melon and strawberry with reference to the substrate specificity of ester synthetase and alcohol contents in pulp. *J. Japan. Food Sci. Technol.* 39:183–187.

Ueda Y., Zhao W., Ihara H., Imahori Y., Tsantili E., Wendakoon S.K., Chambers A., & Bai J. 2022. Functional characteristics of aldehyde dehydrogenase and its involvement in aromatic volatile biosynthesis in postharvest banana ripening. *Foods* .11:347. https://doi.org/10.3390/foods11030347

Wendakoon S.K., Ueda Y., Imahori Y., & Tomimasu Y. 2005. Presence of acetyl-CoA synthetase in ripened bananas. *Food Preservation Sci.* 31:75–80.

Yang X., Song J., Fillmore S., Pang X., & Zhang Z. 2011. Effect of high teperature on color, chlorophyll fluorescence and volatile biosynthesis in green-ripe banana fruit. *Postharvest Biol. Technol.* 62:246–257. https://doi.org/10.1016/j.postharvbio.2011.06.011

Handling & Processing Section

Proc. Fla. State Hort. Soc. 136:154. 2023.

—Scientific Note—

Effect of 1-methylcyclopropene and Cooling Method on Broccoli Quality during Storage

Adrian Berry*, Steven Sargent, and Merce Santana

Horticultural Sciences Department, University of Florida, IFAS, Gainesville, FL

Keywords. MCP, postharvest, shelf life, SmartFresh™

Demand for broccoli has increased in recent years to become the sixth most consumed vegetable in the U.S. Broccoli has a high nutritional value containing significant levels of fiber, ascorbic acid, and antioxidants. However, since it is harvested as immature flowers (florets) with a high respiration rate, broccoli is extremely perishable. One of the main issues that reduces quality and marketability during postharvest handling is yellowing due to chlorophyll degradation. Improper handling can further reduce visual quality and nutritional content. Previous research has shown that exposure to ambient temperatures or to exogenous ethylene exhibits accelerates broccoli senescence; however this can be mediated by the use of 1-methylcyclopropene (1-MCP) (Able et al., 2002; Fan & Mattheis, 2000; Yuan et al., 2010). This study aimed to evaluate the potential of 1-MCP to extend broccoli shelf life and maintain quality during simulated retail storage.

Findings

In February 2020, freshly harvested broccoli (*Brassica oleracea var. italica* cv. Bay Meadows) was field packed into waxed 1 1/9 bushel cartons and subjected to the following commercial cooling treatments: forced-air cooling (FAC) for 2–3 hours at 1 °C, forced-air cooling with shrink wrap (FAC-W) and slush-ice cooling (SIC). Broccoli was then transported by refrigerated trailer to the Postharvest Horticulture Laboratory in Gainesville, where 1-MCP (AgroFresh Solutions Inc, SmartTabs™, Philadelphia, PA) was applied overnight at 1 °C to half of the FAC broccoli (FAC-MCP). The remaining FAC broccoli was stored untreated in a separate room also at 1 °C along with the FAC-W and SIC broccoli that served as controls. A commercial handling scenario was simulated, with relative humidity at 90 to 95%, as follows; Inventory storage and shipping: 14 d at 1 °C + Retail backroom: 7 d at 4 °C + Retail display: 7 d at 21 °C (Fig. 1). Throughout storage, various quality indices were evaluated using subjective ratings as well as objective color measurements and analysis of ascorbic acid content.

Quality deterioration was minimal up to 14 d at 1 °C then increased rapidly after transfer to 21 °C due to yellowing and softening. Floret hue* angle remained around 124° for 21 d until transfer to 21 °C then decreased to 108 to 117° for FAC-W, FAC and SIC, respectively. However, the FAC-MCP broccoli retained a hue* angle similar to the initial value (124°) for 2 d at 21 °C. After 14 d at 1 °C FAC-W broccoli was firmer, however after transfer to 21 °C for 7 d exhibited more yellowing and off-odor compared to the other treatments. Decay was more pronounced after transfer to 21 °C but remained below 5%. After 2 d at 21 °C, the FAC-MCP broccoli maintained slightly higher ascorbic acid content (70 mg/100 g) compared to the other treatments (62 mg/100 g).

Suggestions for Future Direction

The use of 1-MCP has potential to reduce yellowing in broccoli, especially if exposed to warmer temperatures during handling/transport. In addition, application of 1-MCP in conjunction with FAC could be more energy-efficient and sustainable compared to the traditional SIC method. Other formulations and concentrations of 1-MCP could be evaluated in the future.

Literature Cited

Able A.J., Wong L.S., Prasad A., & O'Hare T.J. 2002. 1-MCP is more effective on a floral brassica (*Brassica oleracea* var. Italica L.) than a leafy brassica (*Brassica rapa* var. Chinensis) *Postharvest Biol. Technol.* 26(2):147–155. https://doi.org/10.1016/S0925-5214(02)00011-X

Fan, X., & Mattheis, J.P. 2000. Yellowing of broccoli in storage is reduced by 1-methylcyclopropene. *HortScience*. 35(5):885–887. https://doi.org/10.21273/HORTSCI.35.5.885

Yuan G., Sun B., Jing Yuan J. & Wang Q. 2010. Effect of 1-methylcyclopropene on shelf life, visual quality, antioxidant enzymes and health-promoting compounds in broccoli florets. *Food Chem.* 118(3):774–781. http://dx.doi.org/10.1016/j.foodchem.2009.05.062

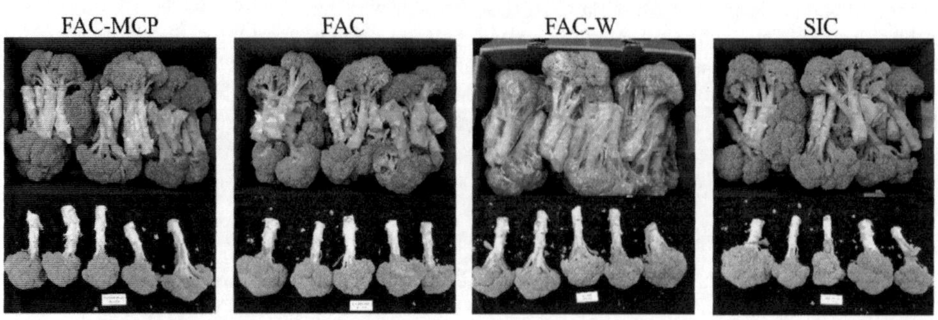

Fig. 1. Images of bunched and separated broccoli heads from each treatment after simulated commercial handling (14 d at 1 °C + 7 d at 4 °C + 2 d at 21 °C).

This research was supported by AgroFresh Solutions Inc. and the Horticultural Sciences Department, University of Florida/IFAS.

*Corresponding author. Email: adberry@ufl.edu

Handling & Processing Section

Proc. Fla. State Hort. Soc. 136:155–156. 2023.

—Scientific Note—

Evaluation of Akorn Semipermeable Fruit Coatings to Extend Mango Shelf Life by Creating an Internal Modified Atmosphere

Faisal Shahzad, Jeffrey K. Brecht*, Steven A. Sargent, and Moshe Doron

Horticultural Sciences Department, University of Florida, IFAS, Gainesville, FL

Keywords. chilling injury, *Mangifera indica*, optimum atmosphere, quality, shelf life, tree ripe

Mango export potential from production regions to the international markets is hindered because of the fruit's perishable nature, pest and disease infestations, and susceptibility to chilling injury when stored at lower than recommended temperatures (<10-12 °C depending on fruit developmental stage) (Brecht & Yahia, 2009). Various approaches (such as controlled atmosphere, modified atmosphere, and use of anti-ethylene compounds) are employed to overcome limitations encountered during transit and storage in order to maintain acceptable fruit quality (Brecht & Yahia, 2017). Edible coatings, being environment friendly, are gaining popularity since these form a transparent film on the fruit surface that acts as a barrier to gases (O_2 and CO_2) and water, imparts gloss, retains attractive color and minimizes microbial spoilage, ultimately attaining desirable fruit quality with extended storage life (Blancas-Benitez et al., 2022). In this study, mango fruit were coated with formulations from Akorn Technology Inc. (Berkeley, CA) plus coating with coconut oil in comparison with uncoated control fruit for evaluation of internal atmospheres and their effect on fruit ripening, quality, and shelf life. In the initial test, Florida-grown 'Keitt' mangos at maturity stage 3.5 on a 1–5 flesh color development scale where 1 = no yellow color to 5 = 100% yellow or orange color, were left uncoated or coated with coconut oil or one of three Akorn formulations and stored at 7 °C and 95% relative humidity (RH) for 10 days with daily measurements of internal oxygen (O_2), carbon dioxide (CO_2), and ethylene. In a second test, Ecuadoran 'Tommy Atkins' mango fruit (stage 3.5) were either left uncoated or coated with one of two Akorn formulations (T_1 and T_2) containing coconut oil and stored at 7 °C and 95% RH for 34 days, followed by 1 week of shelf life at 20 °C and 95% RH.

Findings

In the initial tests, all Akorn coatings resulted in lower O_2 and higher CO_2 inside the fruit compared to the uncoated control fruit, with coconut oil resulting in more extreme gas modification than the Akorn coatings. No significant differences were found in internal ethylene concentration across all the treatments. In the long-term storage test at 7 °C, results from physical, biochemical, and sensory quality attributes are shown in Table 1. Control (uncoated) mango fruit had higher O_2 and ethylene, and lower CO_2 compared to the Akorn coatings. During the shelf-life period at 20 °C, no differences were found in O_2 concentrations among all treatments, but the coated fruit had lower CO_2 concentrations compared to the control, suggesting ripening inhibition by the coatings. In the long-term storage test, it was found that coated fruit developed less chilling injury, lenticel discoloration, and decay compared to the uncoated control. Coatings also slowed ripening, with coated fruit retaining higher hue and L* values (peel, subepidermal, and near the stone) and higher firmness (compression and puncture resistance forces) with higher acid content when compared to the control fruit on D34 of storage and during shelf life. Fruit sensory quality results also suggested that the coatings inhibited ripening, with higher scores for firmness (texture) and acidity, along with lower scores for juiciness and ripeness, suggesting a promising role of fruit coatings in prolonging shelf life of mango fruit by slowing ripening.

Suggestions for Future Directions

Taken together, it can be concluded that Akorn coatings create a beneficial modified atmosphere within the fruit and seem promising to extend the storage life of mangos with desired quality traits during shipment to distant high-end markets.

Literature Cited

Blancas-Benitez F.J., Montaño-Leyva B., Aguirre-Güitrón L., Moreno-Hernández C.L., Fonseca-Cantabrana Á., del Carmen Romero-Islas L., & González-Estrada R.R. 2022. Impact of edible coatings on quality of fruits: A review. *Food Control*. 139:109063. https://doi.org/10.1016/j.foodcont.2022.109063.

Brecht J.K. & Yahia E.M. 2009. Postharvest physiology, p. 484–528. In: Litz R.E. (ed.). The Mango, Botany, production and uses, 2nd ed., CABI, Wallingford, U.K.

Brecht J.K. & Yahia E.M. 2017. Harvesting and postharvest technology of mango, p. 105–129. In: Siddiq M., Brecht J.K., and Sidhu J.S. (eds.). Handbook of Mango fruit: Production, postharvest science, processing technology and nutrition. Wiley-Blackwell Publishing Co., Oxford, UK

*Corresponding author. Email: jkbrecht@ufl.edu

Table 1. Summary of coatings long-term storage test (34 days at 7 °C followed by 1 week storage at 20 °C) for tree ripe 'Keitt' mangos.[z]

	Day 34 of 7 °C storage			Day 4 of 20 °C shelf life		
Factor	Control	T_1	T_2	Control	T_1	T_2
Respiration	↑					
Ethylene production	↑					
Chilling injury	↑					
Decay	↑			↑		
Lenticel spotting	↑			↑		
Firmness (compression)		↑	↑		↑	↑
Firmness (penetration)		↑	↑	↑	↑	
Peel hue						
Peel chroma						
Peel L*						
Subepidermal flesh hue						
Subepidermal flesh chroma						
Subepidermal flesh L*		↑	↑		↑	↑
At stone hue						
At stone chroma						
At stone L*		↑				↑
TSS						↓
TA						↑
TSS/TA ratio					↑	
pH					↑	
Mango aroma						↑
Texture					↑	
Juiciness					↑	
Sweetness						
Acidity						
Overall eating quality						↑
Ripeness					↑	
Internal O_2		↓	↓			
Internal CO_2		↑	↑			
Internal ethylene	↑	↑		↑		

[z]The treatments were: Control = uncoated fruit; T_1 = Akorn formulation 1; T_2 = Akorn formulation 2. Upward-pointing arrows indicate increases and downward-pointing arrows indicate decreases of a given factor relative to coating treatments for the control and relative to the control for the coating treatments.

Proc. Fla. State Hort. Soc. 136:157. 2023.

—Scientific Note—

Assessing the Effects of Gaseous Ozone Application During Cold Storage on Fresh Peach Quality

Orestis Giannopoulos[1]*, Ramsey Corn[1], Camille Esmel McAvoy[1], Dario Chavez[2], and Angelos Deltsidis[1]*

[1]*Department of Horticulture, University of Georgia, Tifton, GA*
[2]*Department of Horticulture, University of Georgia, Griffin, GA*

Keywords. decay incidence, gaseous ozone, peaches

Ozone has demonstrated strong sanitizing properties that are potentially useful in enhancing shelf-life and maintaining the quality of fresh produce. Gaseous ozone in low concentrations (0.5–3 ppm) has properties which suppress microorganisms and protect fruits and vegetables from brown rot, gray mold, and other fungal pathogens, resulting in lower decay incidence.

Findings

Our team evaluated the potential benefits of gaseous ozone application on fresh peach quality. Four peach cultivars; Scarlet Prince, July Prince, O'Henry, and August Lady, were stored at a temperature of 0.5 °C at 90 to 95% relative humidity (RH) plus varying levels (0.5–5 ppm) of gaseous ozone. After 28 days of storage, both 3 and 5 ppm gaseous ozone treatments had similar effects on 'Scarlet Prince' and 'July Prince' that manifested as high decay incidence. More specifically, the Scarlet Prince cultivar treated with 5 ppm ozone had an increased decay incidence by 13, 20, and 40% after 14, 21, and 28 days of storage, respectively, compared to the control. The lower ozone concentration of 3 ppm followed a similar increasing pattern with the highest decay incidence of 20% after 28 days of storage. 'July Prince' and 'Scarlet Prince' had a continuously increasing pattern of decay incidence for both ozone applications of 3 and 5 ppm compared to control.

On the other hand, lower ozone concentrations (0.5 and 1 ppm) in the second experiment were less detrimental to peach quality, resulting in lower levels of decay incidence. 'O'Henry' peaches treated with 0.5 ppm ozone resulted in a 0% decay incidence during the first 21 days of storage. On the last evaluation day (28), the decay incidence of peaches treated with 0.5 ppm gaseous ozone increased slightly to 3.5% compared to 0% for cold storage and 7% for 1 ppm application. On the 28th day of storage, 'August Lady' peaches treated with 0.5 ppm ozone had a 7% decay incidence, whereas 1 ppm treated fruit resulted in 1% decay incidence. Both 1 and 5 ppm concentrations resulted in increased pitting incidence in all four varieties tested, which was visible after the 14th day of storage.

The addition of 3 and 5 ppm of gaseous ozone concentrations resulted in softer peaches. On the 28th day of storage, a similar pattern was observed for both 'O'Henry' and 'August Lady' where peaches stored in cold storage had statistically significant differences with increased firmness values compared to all ozone treatments. This most likely occurred due to the phytotoxicity of ozone that is detrimental to the skin tissue of peaches. Ozone-treated peaches (3 and 5 ppm) shrank, dried, and wrinkled externally with symptoms becoming visible on the 21st day of storage and extending to the 28th day of storage. The firmness of peaches stored under 0.5 ppm ozone was higher compared to all other treatments, including the control, after 21 and 28 days of storage.

Overall, the lowest ozone concentration (0.5 ppm) exhibited the most promising visual results as a possible add-on to traditional cold storage. The titratable acidity analysis showed that the acidity of the 0.5 ppm ozone-treated fruit remained the same as the control. Decay incidence remained low (3.5 to 7%) and firmness values were higher than control for 'O'Henry' peaches on the 21st and 28th day of their storage life. Most importantly, peaches treated with 0.5 ppm ozone retained their visual quality with no visible defects as is common with ozone treatments. The results of this work provide valuable information on the efficacy of ozone and the ozone tolerance levels of fresh-market peaches. This study resulted important insights on the effectiveness of ozone and the tolerance levels of fresh-market peaches. However, additional research is required to explore the possibilities of using ozone in conjunction with current storage methods and to fully investigate the potential of ozone for postharvest disease suppression.

*Corresponding authors. Email: orestis.g@uf.ledu; adeltsidis@ugu.edu

Proc. Fla. State Hort. Soc. 136:158. 2023.

—Scientific Note—

Salmonella Cross-contamination Risks Between Tomatoes and Harvest Bins during Harvesting

Mari Schroeder* and Michelle Danyluk*

Citrus Research and Education Center, University of Florida, IFAS, Lake Alfred, FL

Keywords. cross contamination, *Salmonella*, transfer

Fresh tomatoes have been frequently associated with salmonellosis in the United States. Contaminated food contact surfaces are a concern as they can be a source of cross-contamination. Harvest bins are specific food contact surfaces that have the potential to lead to cross-contamination, but information on the transfer between tomatoes and harvest bins is limited. Understanding how *Salmonella* moves between contact surfaces and tomatoes is important information, especially for growers and other stakeholders when deciding what type of harvest bin to use. This study aimed to evaluate *Salmonella*'s transfer potential between three common harvest bin materials: high-density polyethylene (HDPE), eucalyptus wood and cardboard, and tomatoes.

Tomatoes or an HDPE, wood, or cardboard coupon (5 × 5 cm) were spot inoculated with a Rifampicin-resistant *Salmonella* cocktail (105) and dried for 1h under a bio-safety cabinet. Uninoculated tomatoes or coupons were placed in contact with inoculated items, and a 1-lb weight was placed on top of each tomato to mimic the pressure of a 25-lb bin. The tomatoes and coupons with the weight were left for 10 minutes, 3, 6, and 24 hours at ambient temperature (70 °C). At each sample time, the coupons and tomatoes were placed into separate WhirlPak bags with 0.1% peptone. A shake, rub, shake (30 seconds each) was followed by dilutions, plating onto non-selective media with Rifampicin, and incubation (35 °C, 24 hours). The experiment was replicated 3 times with 3 samples (n = 9). Transfer coefficients (TCs) were calculated by dividing the CFU/mL *Salmonella* on the uninoculated surface by the CFU/mL *Salmonella* on the initial inoculated surface and reported as log %TCs. The more negative the log %TCs, the lower the transfer of *Salmonella* from the inoculated surface to the uninoculated surface. In contrast, the higher the log%TCs, the greater the transfer of *Salmonella* from the inoculated surface to the uninoculated surface

Salmonella transfer between tomatoes and HDPE was significantly higher ($P \leq 0.05$) than between tomatoes and wood or cardboard; tomato to HDPE: 1.4 to 4.2, from 10 minutes to 24 hours, and HDPE to tomato: 0.2 to 2, from 10 minutes to 24 hours. Transfer between tomatoes and cardboard ranged from –1.3 to 2, from 10 minutes to 6 hours (tomato to cardboard) and –0.34 to 1.5, from 10 minutes to 6 hours (cardboard to tomato); no transfer was seen after 6 hours in either case. Transfer from tomatoes to wood ranged from –1.9 to 1.5, from 10 minutes to 3 hours; transfer did not increase after 3 hours. Transfers from wood to tomato ranged from –0.6 to 1.3 at 10 minutes; transfer did not increase after 10 minutes.

The TCs between the tomatoes and HDPE were higher than transfers from cardboard and wood, indicating that using HDPE may result in higher cross-contamination. Low TCs between cardboard and wood indicate a lower risk for cross-contamination, but the survival of *Salmonella* may be greater on these surfaces compared to HDPE. This can be attributed to the porous nature of the wood and cardboard that can aid in protecting *Salmonella* during activities such as cleaning and sanitizing. High-density polyethylene remains a viable choice for a harvest bin material as it is considered a more cleanable and sanitizable surface than cardboard or wood.

*Corresponding authors. Email: mari.schroeder@ufl.edu; mddanyluk@ufl.edu

Proc. Fla. State Hort. Soc. 136:159. 2023.

—Scientific Note—

Use of Modified Carton Boxes Improves Airflow during Forced-air Cooling Operations

RAMSEY CORN*, ORESTIS GIANNOPOULOS, CAMILLE ESMEL MCAVOY, AND ANGELOS DELTSIDIS*

CAES Department of Horticulture, University of Georgia, Tifton, GA

Keywords. air circulation, bell pepper, field heat, squash

Timely field heat removal is important to preserving the postharvest quality of produce by slowing down metabolic processes. Forced air cooling (FAC) is a common industry technique for rapid field heat removal by placing pallets of produce into a cold room and forcing cold air through the boxes. In local operations, the standard duration for this system is set to 45–60 minutes regardless of the commodity, packaging, or target temperature. This can cause insufficient field heat removal which ultimately affects the final quality of the produce. The cooling efficiency is often reduced due to inadequate airflow which is caused by a low vent area. The objective of this study was to develop a standard guide for FAC systems for various commodities. The specific ones used in the trials were squash and bell peppers. The method of pallet stacking used often disrupts the airway channels when vent openings are blocked or misaligned as it impedes cool air circulation through the containers. Modified containers to increase the vent area by 6% were compared alongside a factory recommended pallet stacking design, returnable plastic containers (RPCs), and a control. Modified containers used were the normal packing boxes that had been marked and hand-cut to increase the vent area to 12% open area. The pallet stack orientation was also compared alongside the packing containers. The stacks consisted of the control (packing house stacking), a pallet with boxes aligned to the open area vents, and a pallet with the box manufacturer's recommendation orientation. A commercial FAC system was used to cool the fruit for up to two hours, the system operated on 45-minute intervals. A 7/8ths temperature reduction was targeted when cooling the fruit. The 7/8ths target temperature is calculated using the initial temperature and suggested storage temperatures of the product. Thermocouples attached to data loggers were inserted into squash and bell peppers located at different elevation levels on the pallet. All fruit measured were on the inside of the FAC tunnel (warmest part). The trials showed that the current standard practices were insufficient in removing the field heat. Modified containers saw significant decreases in pulp temperature and reduced the FAC time required to achieve the target temperature.

Findings

Standard practice at the packinghouse involves placing stacked pallets of produce into a commercial sized FAC system for 45 minutes before shipping. The recommended storage temperature for squash and bell pepper is 45 °F. Using the storage temperature and the initial pulp temperature of the fruit we were able to calculate the 7/8ths target temperature for cooling. The calculated target temperature for both squash and bell pepper was 49 °F. The target temperature was never met during the cooling trial. Two hours was an insufficient amount of time for all carton options. The greatest decrease in temperature measured for squash and bell pepper was 20 °F and 10 °F, respectively, within the two-hour period. It was determined that the control was the most ineffective in cooling the produce. Temperature achieved after 45 minutes of cooling being a 2 °F decrease for bell peppers from 64 °F to 62 °F in the control cartons. Squash control cartons showed a 10 °F decrease over the 45-minute period from 76 °F to 66 °F. The modified carton boxes were more efficient in removing the field heat from the produce. Pulp temperature of fruit in modified boxes had decreased by 5 °F from 63 °F to 58 °F in bell pepper and by 10 °F from 75 °F to 65 °F in squash. Forced-air cooling is a powerful tool. Using it too little or too much can have adverse effects on product quality. Further research will be done on other commodities such as zucchini to compile a guide for Georgia growers. Redesigning the carton boxes to contain more vent space for increased airflow and comparing them to the current setup will be future steps of this research.

*Corresponding authors. Email: ramsey.corn@uga.edu; adeltsidis@uga.edu

Proc. Fla. State Hort. Soc. 136:160–161. 2023.

—Scientific Note—

Evaluating the Effect of Various Picking Intervals on the Storability and Postharvest Quality of Georgia-grown Blueberries

Amit Godara*, Karoline Ponder, Ramsey Corn, Camille Esmel McAvoy, Zilfina Rubio Ames, and Angelos Deltsidis*

Department of Horticulture Sciences, University of Georgia, Tifton, GA

Keywords. firmness, titratable acidity, total anthocyanin, total soluble solids, *V. corymbosum*

Blueberry (*V. corymbosum*) production and consumption have increased in the US in the last ten years (Rodriguez et al., 2019), with Georgia being one of the top blueberry producing states in the country. The industry's ability to remain economically viable depends on maintaining high yields and excellent fruit quality (Hajihassani et al., 2018). Blueberries are a highly perishable commodity, making postharvest management crucial to maintain quality (Steynberg et al., 2022). Blueberry fruit quality is determined by several factors, like firmness, color, sugar/acid ratio, flavor, and aroma (Talcott, 2007). One of the most important factors that can affect the storability and postharvest quality of blueberries is the picking interval or harvest schedule. The shift from manual to mechanical harvest has increased picking intervals, which may negatively impact fruit quality. The objective of this study was to examine the impact of delayed harvests on the fruit quality of two Georgia-grown blueberry cultivars. A randomized complete block design was used with 10 plots in which 5 subplots were harvested on May 17 or June 2 (H1—early harvest) blueberry cultivars, 'Meadowlark' and 'Brightwell' respectively. The remaining 5 subplots were harvested 7 days later (H2—delayed harvest: May 24 and June 9, Meadowlark and Brightwell respectively). Blueberries were stored at 1°C with 85% RH for up to 21 days. Quality parameters: firmness, fruit size, total soluble solids (TSS), titratable acidity (TA), and total anthocyanin (TAC) were evaluated after 1, 7, 14, and 21 days of storage (DOS). Statistical analysis was performed using JMP pro software by two-way ANOVA and Tukey's HSD test was used for mean separation.

The results of this experiment were significantly different across harvests which demonstrates significant interactions between harvest timing and length of storage. 'Meadowlark' fruit from the delayed harvest had lower firmness and bigger fruit size values after 7, 14, and 21 days in storage. The early harvest had firmer fruit even after 21 days of storage. The delayed harvest resulted in higher TSS, but lower TA (Fig. 2). Fruit from the delayed harvest had a higher anthocyanin content, and the degradation of total anthocyanin was also higher in fruit from H2 from day 1 to day 21 of storage (Fig. 3).

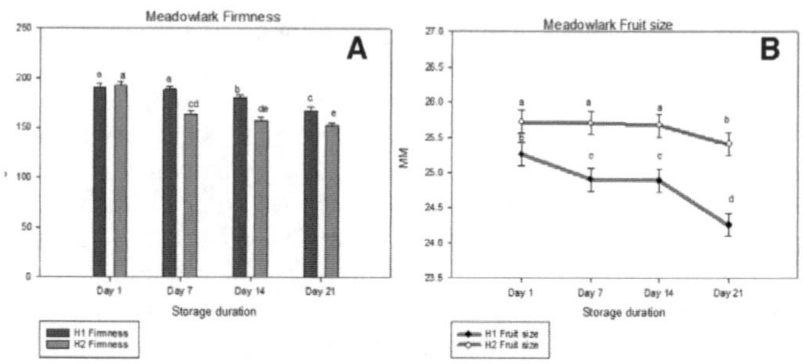

Fig. 1. Firmness (**A**) and fruit size (**B**) of fruit from harvest one (H1) and harvest two (H2—delay harvest) during storage (1, 7, 14 and 21 days of storage). Bars represent standard error. Bars with different letters indicate the values are significantly different ($P \leq 0.05$) based on Tukey's HSD test.

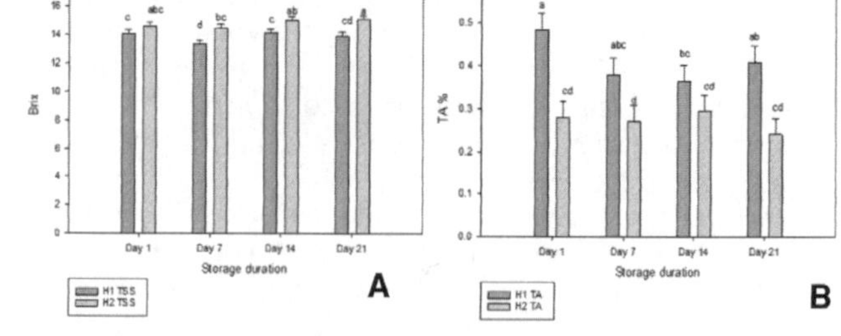

Fig. 2. Total soluble solids (**A**) and Titratable acidity (**B**) of fruit from harvest one (H1) and harvest two (H2— delay harvest) during storage (1, 7, 14 and 21 days of storage). Bars represent standard error,

A similar pattern was found for 'Brightwell' fruit firmness, size, TSS, TA, and anthocyanin. The early harvest had significantly higher firmness on day 21 compared to the delayed harvest on day 1 of assessment (Fig. 4). Fruit size was bigger in H2 but lower TA, and TSS remained not significant throughout storage (21 days) (Fig. 5). Fruit from the delayed harvest had higher anthocyanin content and the degradation of monomeric anthocyanin was also higher in fruit from harvest 2 for day 1 to day 21 of storage (Fig. 6).

In summary, our results indicated that delayed harvests diminished blueberry's postharvest qualities and attributes. Even

*Corresponding authors. Email: amit.godara@uga.edu; adeltsidis@uga.edu

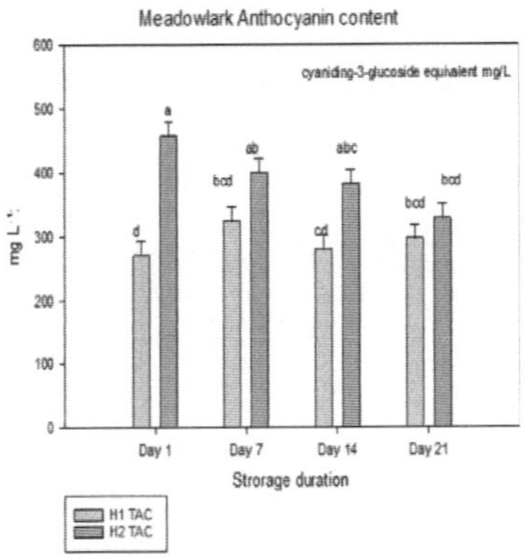

Fig. 3. Total Anthocyanin content of fruit from harvest one (H1) and harvest two (H2— delay harvest) during storage (1, 7, 14, and 21 days of storage). Bars represent standard error

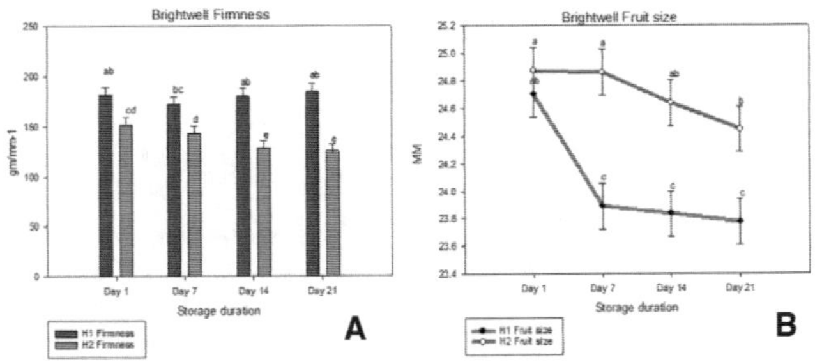

Fig. 4. Firmness (**A**) and fruit size (**B**) of fruit from harvest one (H1) and harvest two (H2— delay harvest) during storage (1, 7, 14 and 21 days of storage). Bars represent standard error,

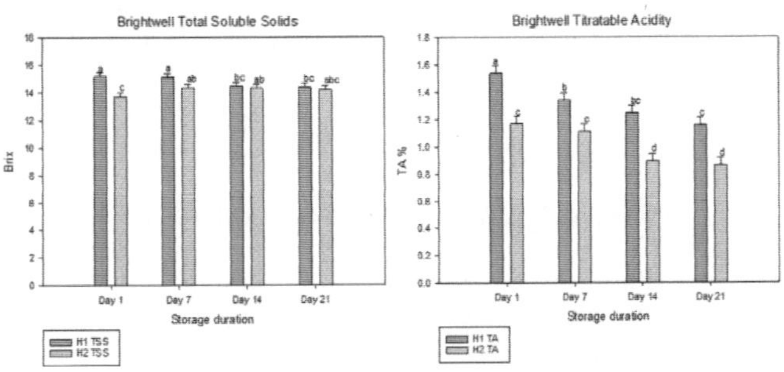

Fig. 5. Total soluble solids (A) and Titratable Acidity (**B**) of fruit from harvest one (H1) and harvest two (H2— delay harvest) during storage (1, 7, 14 and 21 days of storage). Bars represent standard error,

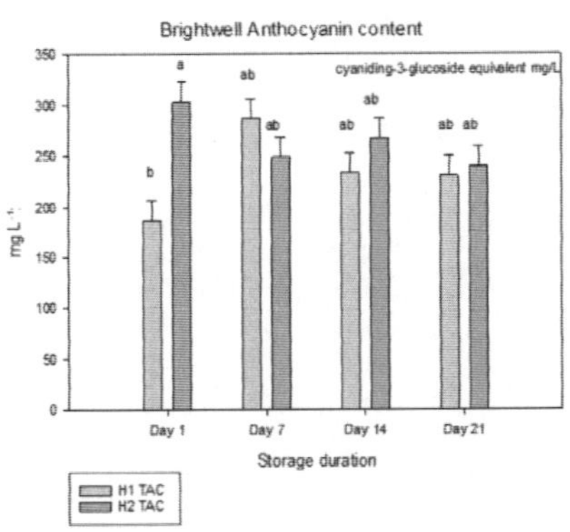

Fig. 6. Total anthocyanin content of fruit from harvest one (H1) and harvest two (H2—delay harvest) during storage (1, 7, 14, and 21 days of storage). Bars represent standard error

though fruit was placed at the proper temperature during storage, a delayed harvest reduced fruit storability since it negatively impacted fruit firmness and other quality characteristics. Therefore, timely harvests are recommended to maintain fruit quality and maximize storage potential.

Future Research

Based on the results from this study, a new experiment has been initiated using a more frequent picking schedule. The new experiment includes three consecutive harvests with 2, 3, and 7 days picking intervals in Rabbiteye blueberries. The results of this study will be useful to optimize and understand the effect of different harvest regimes on the storability and postharvest quality of blueberries.

Literature Cited

Hajihassani A., Lawrence K.S., & Jagdale G.B. 2018. Plant parasitic nematodes in Georgia and Alabama. p. 357–391. In: Subbotin S. & Chitambar J. (eds). Plant parasitic nematodes in sustainable agriculture of North America. Sustainability in plant and crop protection. Springer. Cham. https://doi.org/10.1007/978-3-319-99588-5_14

Rodriguez-Saona C., Vincent C., & Isaacs R. 2019. Blueberry IPM: Past successes and future challenges. *Annual Review of Entomology*. 64, 95–114. https://doi.org/10.1146/annurev-ento-011118-112147

Steynberg P., Goedhals-Gerber L.L., & van Dyk E. 2022. An analysis of the impact of logistics processes on the temperature profile of the beginning stages of a blueberry supply chain. *Horticulturae*. 8(12):1191. https://doi.org/10.3390/horticulturae8121191

Talcott, S.T. 2007. Chemical components of berry fruits. Ch. 2. In: Zhao Y. (ed.) Berry fruit: Value-added products for health promotion. CRC Press, Boca Raton, FL https://doi.org/10.1201/9781420006148

Proc. Fla. State Hort. Soc. 136:162. 2023.

—Scientific Note—

Evaluation of Surface Water Treatment Efficacy Protocol Using Calcium Hypochlorite Against *Salmonella* Spp. in Florida Water

LaTaunya Tillman* and Michelle D. Danyluk*

Citrus Research and Education Center, University of Florida IFAS, Lake Alfred, FL

Keywords. chlorine, peracetic acid, water treatment

Microbial contamination of surface water used in produce production has been implicated as a source of produce contamination resulting in outbreaks. Growers are under market-driven and regulatory pressure to treat surface water before uses which bring it in contact with produce. Currently there are no chemical treatments labeled by the United States Environmental Protection Agency to reduce foodborne pathogens in preharvest agricultural water. The Food and Drug Administration (FDA) released a protocol (Reg. No. 94151PA7) for the industry to use in testing claims for the effectiveness of a product for inactivating foodborne bacteria in preharvest agricultural water. The protocol implements the creation of a laboratory test water composed with Arizona Test Dust and controlled physiochemical parameters. The study aimed to evaluate the applicability of the FDA's revised water treatment efficacy chemical labeling protocol using a chlorine [calcium hypochlorite; (Cl)] sanitizer in actual Florida agricultural water.

Breakpoint chlorine (to 4 ppm free chlorine) was measured for an agricultural surface water source from a West Central Florida farm pond (98 mL). The pond water was inoculated with 1 mL of a Rifampicin (80 μg/mL) resistant seven-strain American Type culture collection (ATCC) *Salmonella* cocktail, with starting inoculation populations of 6.83 ± 0.57 log CFU/mL. Experiments were completed in triplicate for each contact time (1, 5, or 10 min), and temperature (12 °C or 32 °C) combination (n = 9). Following the addition of 1 mL of the *Salmonella* inoculum into agricultural water (99 mL), it was equilibrated to 12 °C or 32 °C for at least 30 minutes. After temperature equilibration, 1 mL of a Cl solution was added to inoculated pond water (100 mL) for contact times of 1, 5, or 10 minutes. Salmonella populations were determined by serial dilutions in a sodium metabisulfite solution (28 g/L) to neutralize any residual sanitizer, plated onto non-selective media (Brain Heart Infusion Agar) with rifampicin, and incubated at 35 ± 2 °C for 24 ± 2 h. Colonies were counted by hand and expressed as log CFU/mL. The limit of detection (LOD) was 2.0 log CFU/mL. When counts fell below the LOD, < 2.0 log CFU/mL was used for all calculations. Student t-tests and a three-way analysis of variance (ANOVA) were performed.

The amount of chlorine needed to achieve 4 ppm free chlorine twice consecutively, over a minimum of 30 minutes, was recorded (n = 3). The average amount of chlorine needed for the pond to reach the 4 ppm breakpoint target was 6.33 ± 1.46 ppm, and took 241.33 ± 80.41 minutes. *Salmonella* populations following treatments for up to 10 min ranged from 3.34 ± 0.54 to 3.55 ± 0.62 log CFU/mL. There were no significant differences ($P = 0.12$) between the 1 and 5 min contact times. At the 10 minutes, counts were non-detectable across all replicates. At 32 °C, counts were non-detectable at all contact times. Treatment contact time had a significant ($P < 0.05$) impact on Salmonella population recovery.

Calcium hypochlorite at concentrations of 2–4 ppm free Cl can be an effective surface water treatment to reduce *Salmonella* populations for growers using ponds in West Central Florida. Variability between breakpoint ranges and reductions achieved across treatment trials suggests that relationships between water sources and natural environmental interactions can influence treatment efficacy. Treatment effectiveness should be determined by the water source, and physiochemical attributes that can impact different treatment chemistries should be monitored. The practice of chemical treatment of surface water in growing areas has raised some questions about the long-term impacts on the soil and plant health in preharvest environments. More research is needed to evaluate the long-term use and impacts in growing environments.

*Corresponding authors. Email: tillmanl@ufl.edu; mddanyluk@ufl.edu

Proc. Fla. State Hort. Soc. 136:163. 2023.

—Scientific Note—

Simulation of the Risk of Microbial Contamination for Dropped and Drooping Grapefruits and Strawberries With Ink

Claudia A. Pegueros-Valencia*, Loretta M. Friedrich, and Michelle D. Danyluk*

University of Florida, IFAS, Citrus Research and Education Center, Lake Alfred, FL

Keywords. dropped, drooping, grapefruits, microbial contamination, simulation, strawberries

The Produce Safety Rule (PSR) on the Food Safety Modernization Act (FSMA) regulated by the Food and Drug Administration (FDA) establishes mandatory science based minimum standards for the safe growing, harvesting, packing, and holding of fruits and vegetables grown for human consumption to reduce foodborne illnesses. This rule prohibits the growers from distributing dropped covered produce (FDA, 2015). Nevertheless, the PSR does not mention drooping produce; a drooping produce item is one which is in contact with the ground but still attached to the plant. Dropped produce represents a food safety risk since produce damage or ground contact makes it more susceptible to pathogen contamination (Duffy & Schaffner, 2002). There is a need to better understand the risks associated with harvesting both dropped or drooping covered produce under various conditions. The objective of this study was to investigate the risk of contamination associated with drooped and dropped strawberries and grapefruits. A total of 64 strawberries (n = 64) and 80 grapefruits (n = 80) were dropped and drooped over black ink pads under different conditions. Four of each fruit were used in each of the conditions. The strawberries were drooped over the inkpad for 0 (touch), 0.16, 1, and 24 hours and the grapefruits were drooped for 0 (touch), 1, 3, 7, and 14 days. For the dropping treatments strawberries were dropped over the ink pad through a polyvinyl chloride (PVC) pipe with a diameter of 7.62 cm from a height of 15.24, 30.48, 60.96, and 121.96 cm and the grapefruits were dropped through a polyvinyl chloride (PVC) pipe with a diameter of 15.24 cm from a height of 15.24, 30.48, 60.96, 121.92, and 182.88 cm. After each treatment, each fruit was placed in a light box and pictures were taken next to a ruler for image scaling proposes. The total area (TA) and inked area (IA) were measured using the Java-based image processing software, Image J; finally, percentage of inked area (PIA; area with risk of microbial contamination) by dropped and drooped produce over the ink pad, the force (F) exerted over the ink pad by the drooping produce, and the impact force (IF) over the ink pad by dropping produce were calculated. For the statistics analysis, a one-way analysis of variance (ANOVA) was used to describe the effect of time and height over drooped and dropped fruits respectively. A *t*-test was used to compare the effect of drooped versus dropped for both fruits and linear regressions was used to evaluate correlations between, the fruit weight, ink transfer rates, and the force (F) or impact force (IF).

In the results, it was observed in general that grapefruit had a higher PIA (16.9 ± 9.8 to 34.3 ± 12.0% by dropping versus 1.8 ± 0.6 to 17.5 ± 2.4% by drooping) than the strawberries (8.5 ± 2.7 to 18.9 ± 10.11% by dropping and 2.8 ± 2.2 to 4.4 ± 1.2% by drooping). In both fruits, a tendency of the PIA to increase as the height increased was observed. Height had a significant effect of ($P < 0.05$) on the PIA for both fruits. The tendency of PIA to increase can occur due the impact force of the ink pad which increases with height. This force can deform the fruits bringing them in greater contact with the ink pad which simulates the ground (Celik et al., 2011). Regarding drooping, there was not a significant difference ($P \le 0.05$) between time and PIA for drooping strawberries over 24 hours. For grapefruits, PIAs increased significantly ($P < 0.05$) as contact time increased after 72 hours. No correlation was found between the weight or force and the PIA, during drooping or dropping for either fruit. There was a weak correlation between impact force and PIA for strawberries ($R^2 = 0.7005$). No correlation was found for the grapefruit.

The significance of this study is that evaluating the area of fruits impacted by drooping or dropping is an important first step in understanding the difference in food safety risks between drooped and dropped produce.

Literature Cited

Celik K., Rennie A., & Akinci I. 2011. Deformation behaviour simulation of an apple under drop case by finite element method. *J. Food Eng.* 104(2):293–298.

Duffy S. & Schaffner D.W. 2002. Monte Carlo simulation of the risk of contamination of apples with *Escherichia coli* O157:H7. *Int. J. Food Microbiol.* 78:245–255.

U.S. Food and Drug Administration (FDA). 2015. Food safety modernization act (FSMA). Public Law 111-353. U.S. Food and Drug Administration, Washington, DC.

*Corresponding authors. Email: claudiapeguewosv@ufl.edu; mddanyluk@ufl.edu

Proc. Fla. State Hort. Soc. 136:164. 2023.

—Scientific Note—

Preharvest-applied Fungicides for Postharvest Diplodia Stem-end Rot Control on Grapefruit

Jiuxu Zhang[1], Jiaqi Yan[2], Monty Myers[1], Cuifeng Hu[1], Edinson A. Diaz[3], Liliana M. Cano[3], Jinhe Bai[4], Anne Plotto[4], Wei Zhao[4], and Mark A. Ritenour[1]*

[1]*University of Florida, IFAS, Indian River Research and Education Center, Department of Horticultural Sciences, Fort Pierce, FL*

[2]*College of Horticulture, China Agriculture University, Beijing, China*

[3]*University of Florida, IFAS, Indian River Research and Education Center, Department of Plant Pathology, Fort Pierce, FL*

[4]*USDA–ARS, U.S. Horticultural Research Laboratory, Fort Pierce, FL*

Keywords. *Lasiodiplodia* spp., postharvest decay, postharvest disease management

Diplodia stem-end rot (SER), caused by *Lasiodiplodia* spp., is an important postharvest decay pathogen on fresh citrus in Florida. The epidemic of Huanglongbing (HLB) in Florida causes citrus tree decline resulting in more dead wood that significantly increases *Lasiodiplodia* populations and preharvest fruit infection, leading to increased postharvest Diplodia SER. New or improved Diplodia SER control methods are needed to sustain Florida's citrus industry and growers' economic returns. Evaluations of foliar-applied chemicals to suppress preharvest fungal populations and fruit infection to reduce subsequent postharvest Diplodia SER were conducted in multiple grapefruit groves over four seasons (2019–2023).

In these tests, each season we used the best-performing materials from previous season(s) plus new materials for successive field studies. Better performing products were repeated over multiple seasons to verify their decay control efficacy and repeatability. Chemical products tested over the four seasons included: Topsin 4.5 FL (a reference fungicide with excellent Diplodia SER control; Ritenour et al, 2004; Zhang & Timmer, 2007), Graduate A+, Headline, Quadris Top, Mentor EC, Mertect 340F, Switch 62.5WG, Miravis Prime, Miravis Top, Citrus Fix (2,4-D), Cannonball WG, Abound, Luna Sensation, Veltyma and Goldshield. A total of 12 separate field experiments were conducted in grapefruit groves in the Indian River Citrus District and near Frostproof, FL. Application rates were based on label instructions or, for materials with no preharvest citrus label, rates were based on their labels for other crops or their postharvest uses. Trees sprayed with water served as negative controls. Fruit were harvested 2 and 14 days after application. Harvested fruit were exposed to up to 5 days of degreening [5 ppm ethylene, 29 °C, and 90% relative humidity (RH)], then incubated at 24 °C with 90 to 95% RH for up to three weeks. Diplodia SER was recorded weekly, and both disease incidence and severity were calculated. During the 2022–23 fruit season, preharvest-treated fruit were also evaluated for decay control efficacy under simulated commercial postharvest handling practices (washed, coated with carnauba wax, and stored at 10 °C).

*Corresponding author. Email: ritenour@ufl.edu

Findings

Overall, results obtained from the multiple groves over four seasons showed that:

1) Graduate A+ (fludioxonil + azoxystrobin), effectively and consistently reduced Diplodia SER (reduced decay by 38.5 to 82.8%) when used at postharvest rates, which are higher than what would be allowed for preharvest applications.
2) Abound (azoxystrobin), Headline (pyraclostrobin), Cannonball WG (fludioxonil), Mertect 340F (thiabendazole), Quadris Top (azoxystrobin + difenoconazole), Luna Sensation (flupyram + trifloxystrobin) and Veltryma (pyraclostrobin + mefentrifluconazole) moderately reduce Diplodia SER but were not consistent.
3) Topsin 4.5 FL (thiophanate-methyl) consistently showed the best Diplodia SER control in all tests (reduced decay by 38.7 to 90.3%); however, it is not available for commercial use since efforts by the registrant to obtain an Environmental Protection Agency (EPA) label for citrus were abandoned in 2009.
4) Miravis Prime (fludioxonil + pydiflumetofen) consistently reduced Diplodia SER incidences in all tests (reduced decay by 27.5 to 78.8%), and it also showed good activity for decay control during the simulated commercial postharvest fruit cold storage.

Suggestions for Future Direction

Preharvest application of Miravis Prime is a promising product for postharvest Diplodia SER control. However, it is not currently registered for use on most citrus. The registrant has been contacted in efforts to obtain a U.S. Environmental Protection Agency label for use on grapefruit and other citrus not currently on the label to provide an additional tool for Diplodia SER control.

Literature Cited

Ritenour M.A., Pelosi R.R., Burton M.S., Stover E.W., Dou H., & McCollum, T.G. 2004. Assessing the efficacy of pre-harvest fungicide applications to control postharvest diseases of Florida citrus. *HortTechnology*. 14:58–62.

Zhang J. & Timmer L.W. 2007. Preharvest application of fungicides for postharvest disease control on early season tangerine hybrids in Florida. *Crop Protection*. 26:886–893.

Proc. Fla. State Hort. Soc. 136:165. 2023.

—Scientific Note—

Florida's Extension Programs Prepare Produce Growers for Produce Safety Rule Inspection

Clara Diekman[1]*, Micah Gallagher[2], Matthew Krug[3], Kirby Quam[4], Chelsea Peebles[4], Keith Schneider[2], Renee Goodrich[2], Michelle Danyluk[1], and Taylor O'Bannon[1]

[1]*University of Florida, IFAS, Citrus Research and Education Center, Lake Alfred, FL*

[2]*University of Florida, IFAS, Department of Food Science and Human Nutrition, Gainesville, FL*

[3]*University of Florida, IFAS, Southwest Florida Research and Education Center, Immokalee, FL*

[4]*Florida Department of Agriculture and Consumer Services, Bartow, FL*

Keywords. extension, inspections, produce safety, trainings

The Food Safety Modernization Act (FSMA) Produce Safety Rule (PSR) was the first legislation to establish minimum standards for growing, harvesting, packing, and holding fresh produce. The regulation meant Florida produce growers would require education and technical assistance to meet the requirements of the rule. The University of Florida, Institute of Food and Agricultural Sciences (UF/IFAS) and the Florida Department of Agriculture and Consumer Services (FDACS) have collaborated, since 2018, to provide education and outreach through Produce Safety Alliance (PSA) grower training courses and On-Farm Readiness Reviews (OFRR) to assist growers in meeting the requirements of the PSR. Objectives include determining 1) if the outreach and education programs were successful in improving the level of knowledge of foundational food safety principles and of the PSR and 2) the level of farm preparedness for FSMA PSR compliance.

The PSA grower trainings are conducted using a standardized curriculum developed by the Produce Safety Alliance and are delivered in-person across the state or remotely via Zoom. Pre- and post-tests were given at these trainings (n = 1448) to measure knowledge increase during the class. Post-test score means (21.73/25), were significantly higher than pre-test score means (18.23/25), indicating an increase in knowledge after participation in the training ($t = -9.21$, $P < 0.05$). Since 2020 there have been 10 in-person PSA trainings (n = 130) and 23 remote trainings (n = 384). Average post-test scores (22.24/25, 20.8/25) were significantly higher than average pre-test scores (20.02/25, 15.7/25) for both in-person and remote trainings, respectively. These results indicate that the remote delivery of PSA grower trainings is just as effective as in-person trainings at increasing participant knowledge. There is also increased attendance among specific demographics for remote trainings. Specifically a 13% increase in attendance from Hispanic/Latino growers, 11% increase in participants between the ages of 41 and 55, and a 12% increase in attendance from small farmers growing 1–10 acres. A followup survey to determine practices changed or implemented was sent four months post training. The most common cange implemented following the training was the creation or modification of record keeping systems.

Once requested, UF/IFAS and FDACS personnel travel to the farm to conduct an OFRR reviewing their growing, harvesting, and packing practices; water sources; and other related topics. These farm visits were an educational opportunity for growers to receive one-on-one feedback on areas to improve prior to a FSMA PSR inspection. Following the completion of each OFRR, surveys (n = 66) are completed to evaluate farm readiness for rule implementation, and highlighted areas where more education was needed to meet minimum requirements. These OFRR surveys indicated sanitation (35%), preharvest water (16%), and worker training (24%) required the most improvements, and that 47% of farms met the PSR requirements, 41% needed minor improvements, and 12% needed significant improvements. The major areas of concern with sanitation are cleaning and sanitation (42%) and record keeping (36%). For worker training the major areas identified are worker training (63%), record keeping (15%), and visitor policy (14%). Reviewers identified time (40%), technical support (21%), and facility upgrades (15%) as the assistance required by farms to meet minimum rule requirements.

The PSA grower trainings have positively impacted the level of PSR and foundational food safety knowledge among Florida growers. Remote delivery of the PSA grower training is effective at increasing knowledge gain, and increasing accessibility of extension education to minority populations, likely due to the reduction of barriers like time, travel, and money. The results from OFRRs show that the majority of farms visited have already met minimum rule requirements or need only minor improvements to meet those requirements. As inspections continue throughout the state of Florida, education and outreach programs are ongoing to ensure Florida farms are prepared for implementation of the FSMA PSR.

*Corresponding author. Email: cdiekman97@ufl.edu

Proc. Fla. State Hort. Soc. 136:166–167. 2023.

—Scientific Note—

Tissue-specific Proteome Profile Analysis Reveals Regulatory and Stress Responsive Networks in Passion Fruit during Storage

Ellen Garcia[1], Jin Koh[2], Xingbo Wu[3], Ali Sarkhosh[1]*, and Tie Liu[1]*

[1]*Horticultural Sciences Department, University of Florida, IFAS, Gainesville, FL*

[2]*The Interdisciplinary Center for Biotechnology Research, University of Florida, Gainesville, FL*

[3]*Research and Education Center, University of Florida, IFAS, Homestead, FL*

Keywords. fruit, postharvest, proteomics

Passiflora edulis, commonly known as passion fruit, is a crop with a fragrant aroma and refreshingly tropical flavor that is a valuable source of antioxidants. It offers a unique opportunity for growers because of its adaptability to tropical and subtropical climates. Passion fruit can be sold on the fresh market or used in value-added products, but its postharvest shelf life has not been well-researched, nor have superior cultivars been well-developed. Understanding the proteins expressed at the tissue level during the postharvest stage can help improve fruit quality and extend shelf life. In this study, we carried out comparative proteomics analysis on four passion fruit tissues (the epicarp, mesocarp, endocarp, and pulp), using multiplexed isobaric tandem mass tag (TMT) labeling quantitation.

Findings

Purple passion fruits from mature vines of *Passiflora edulis* (Sims) were obtained at full maturity from the University of Florida Tropical Research and Education Center in Homestead, FL. Fruits were kept at room temperature (RT), and the tissue layers were separated using a scalpel. Samples were frozen in aluminum foil in liquid nitrogen and stored at –80 °C until protein extraction. Four replicates were used for the mesocarp, endocarp, and pulp and two replicates were used for the epicarp. Each tissue was processed for protein extraction according to Fujiki et al. (1982), with modifications. Epicarp extracts were labeled with TMT tags 126, and 128C, mesocarp extracts were labeled with TMT tags 127N, 129N, 130C, and 132N, endocarp extracts were labeled with 127C, 129C, 131N, and 132C, and pulp extracts were labeled with 128N, 130N, 131C, and 133N. Aliquots of all sample mixtures were labeled with TMT tags 133C and 134N as the control group. The peptides were fractionated using an Agilent HPLC 1260 with a Zorbax Extended C18 bRP column (2.1 × 150 mm, 5 µm, Agilent, Santa Clara, CA). Proteome Discoverer v2.4 (Thermo Fisher Scientific, Waltham, MA), ShinyGO v0.77, Kyoto Encyclopedia of Genes and Genomes (KEGG), Search Tool for the Retrieval of Interacting Genes/Proteins (STRING) database, and Position-Specific Iterative Basic Local Alignment Search Tool (PSI-BLAST) was used to analyze the data.

The results indicated a total of 3352 proteins identified of which 295 proteins were determined to be differentially expressed in at least one tissue with a P-value < 0.05. Proteins shared between tissue types included lipid transfer proteins (LTP1, LTP10, LTP5), heat shock protein (HSP22), plastid lipid-associated protein (PAP4) and senescence associated gene 12 (SAG12) with different expression levels in each tissue (Fig. 1). This indicates the proteins involved in the stress response of the fruit and subsequent mechanisms used to counteract lipid accumulation, heat, protein degradation and oxidation that lead to apoptosis and decay. Some key differentially regulated processes included cuticle wax formation, hormone signaling pathways (ethylene, jasmonic acid, and gibberellic acid), heat shock proteins, stress tolerance proteins, antioxidant activity, cell wall metabolism, lipid metabolism and secondary metabolite biosynthesis. The downregulated proteins in the epicarp were related to glycosphingolipid biosynthesis, carotenoid biosynthesis, carbon fixation, pyruvate metabolism, cysteine and methionine metabolism, carbon metabolism, protein processing in the endoplasmic reticulum, and biosynthesis of secondary metabolites. Interestingly, glycosphingolipid and carotenoid biosynthesis were downregulated, while the transporters and oxidative damage proteins were upregulated. The roles

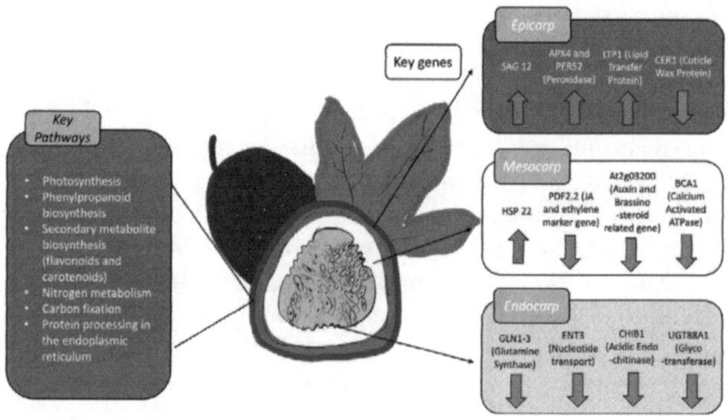

Fig. 1. A schematic of passion fruit tissue-specific key up- and down-regulated genes.

*Corresponding authors. Email: sarkhosha@ufl.edu; tieliu@ufl.edu of glycosphingolipids in senescence have been widely studied in animals, but not in plants. In yeast and animal cells, sphingolipids are involved in cellular responses, including aspects of signaling, apoptosis, and senescence. Moreover, we used bioinformatics tools to characterize 36 hypothetical proteins based on sequence and structure analyses. These proteins represent potential novel candidate genes in passion fruit, implicated in crucial processes such as photosynthesis, immunity, homeostasis, cell structure, protein metabolism, RNA and DNA repair, and senescence pathways. These are potential pathways to study further in passion fruit when it comes to working toward extending the shelf life.

Suggestions for Future Direction

In conclusion, our study on tissue-specific proteome profile analysis has provided valuable insights into the regulatory and stress-responsive networks operating in passion fruit during storage. The consumer acceptance and shelf life of fresh passion fruit are contingent on the preservation of an intact and firm epicarp without pathogenic blemishes. By investigating the underlying biological mechanisms governing texture transformation during ripening, we have shed light on crucial factors that can influence the storability of fresh fruit. The realistic and practical applications of this work are targeting tissue specific proteins through inhibitors, gene editing, conventional breeding, or custom packaging to enable transportation across distant locations and prolong fresh quality.

Literature Cited

Fujiki F.Y., Hubbard A.L., Fowler S., & Lazarow P.B. 1982. Isolation of intracellular membranes by means of sodium carbonate treatment: Application to endoplasmic reticulum. *J Cell Biol.* 93(1):97–102. https://doi.org/10.1083/jcb.93.1.97

Handling & Processing Section

Proc. Fla. State Hort. Soc. 136:168–173. 2023.

Maintaining Visual Quality and Fresh, Firm Texture during Countertop Ripening of Peaches, Guava, and Passionfruit With Vented Containers

Sonya Leonore Stahl* and Steven A. Sargent

Horticultural Sciences Department, University of Florida, IFAS, Gainesville, FL

Keywords. climacteric, room temperature, relative humidity, ripening, storage, *Prunus, Psidium, Passiflora*

Peaches, guavas, and passionfruits ripen best at room temperature (68 to 73.4 °F/20 to 23 °C) and high relative humidity (RH) (> 85%), however, the RH of an air-conditioned home is typically < 50% which accelerates desiccation. This project explored ripening these fruits under a passive system to generate high RH to investigate potential quality benefits. The high RH environment was generated by trapping water vapor from the ripening fruit inside a lidded, 5-L plastic container, vented with 225 holes of 3.4 mm² evenly distributed across the lid in a 15 × 15 grid. Fruit were stored in these lidded containers or on open trays from 3 to 8 days at 69.8 °F (21 °C) and 50% RH, simulating an air-conditioned kitchen environment. Once the fruit stored in open air were ripe and ready to eat, fruit from both treatments were evaluated. RH in the vented containers for guava and passionfruit was maintained at 92 to 94% and that in containers storing peach was 96 to 99%. Storage of peaches, passionfruits, and guavas in the vented containers prevented visually-evident shriveling and delayed ripening-associated softening compared to those that were stored under low RH conditions in ambient air. Peaches stored in high RH were firmer (20 N), than those in low RH (8–10 N), as were guavas (30–35 N and < 20 N), respectively. Weight loss for all fruit stored in high RH was much less (2 to 4%), compared with fruit stored in open air (13 to 25%), depending on the crop. Other benefits of ripening in the vented containers were that guava fruit maintained peel sheen, whereas that of fruit ripened under low RH became matte and dull; the peel of container-ripened peaches remained taut and fresh while that of peaches ripened on the simulated countertop became shriveled and visually undesirable. Therefore, storage in a passive system capable of maintaining a high RH, non-condensing environment can prevent shrivel and maintain visual and tactile quality of peaches, guavas, and passionfruits under typical air-conditioned conditions.

Peach, passionfruit, and guava are typically sold to consumers at the mature, firm ripeness stage and require ripening after purchase. During home ripening, refrigerator storage is not advised because typical temperatures can induce chilling injury, slowing or preventing metabolic reactions necessary to promote normal ripening. Room temperature (20 to 25 °C) is recommended for best fruit ripening, however, the relative humidity (RH) in many air-conditioned homes is usually < 50%, far below the 80-90% ideal for ripening produce (Kader, 1996). Lower RH environments create a high vapor pressure deficit between the 100% RH of the interstitial air within the fruit tissue and the surrounding ambient air, accelerating dehydration from fruit, thereby reducing fruit quality both visually and texturally (Wu, 2010). Often, consumers complain of fruit spoilage before it completes ripening. The creation of a passive countertop-ripening chamber capable of establishing a high-RH headspace would enable fruit to ripen at the appropriate temperature while retaining quality.

This concept employs the principle already employed in the refrigerator's high-humidity drawer. The countertop ripening unit would serve as a 'humidity drawer' intended for room temperature storage. The objective of these tests was to determine the efficacy of a container designed as a 'room-temperature humidity drawer' that would utilize venting to optimize RH during ripening while preventing condensation, in order to preserve quality and extend shelf life.

The authors would like to thank Adrian D. Berry for assistance in these studies.

*Corresponding author. Email: cocoplum@ufl.edu

Materials and Methods

Throughout several trials, peaches, guavas, and purple passionfruits at a firm, unripe stage were purchased by the case from local supermarkets and maintained at 20 to 22 °C during transport back to the University of Florida in Gainesville. Since they were refrigerated at the point of origin, all fruit were held on the lab benchtop to allow internal temperatures to equilibrate to 22 °C room temperature. Fruit were sorted for firmness, freedom from mechanical damage, and uniform visual ripeness stage: yellow ground color in peach, no yellow coloration in guava, and no rind shrivel in passionfruit.

Once sorted, fruit were placed either on an open tray to simulate consumer countertop ripening, or in one of three vented plastic containers. Preliminary tests compared number of vent holes (3.4 mm² in diameter) and distribution in the lids to determine the total vent area necessary for full containers of fruit to produce high RH (> 85%) while avoiding condensation on the inner surfaces or fruit. These tests determined that RH > 80% was not achieved when the total vent area in the lid was 2.4 in² (1639 mm²) (data not shown). However, lids with ~1.2 in² (768 mm²) achieved > 85% RH and this area was selected for the subsequent tests.

Uniform distribution of vents across the lid was found to be critical, otherwise condensation formed on the container walls beneath the unperforated lid areas.

The lids with uniformly distributed perforations were as follows, based on container volume:

- 5.0 L: 225 perforations totaling 1.19 in^2 (768 mm^2) in vent area
- 2.5 L: 116 perforations totaling 0.61 in^2 (394 mm^2) in vent area
- 1.2 L: 57 perforations totaling 0.3 in^2 (194 mm^2) in vent area

The number of fruit per container was determined by the maximum number of fruit that would fit comfortably in a single layer so that fruit would not be crushed by the weight of other fruit during ripening (Fig. 1). Ripening of single fruit was also evaluated, either in the 1.2-L container, or in the 5-L container with fewer vent holes in the lids to obtain 33% of the total vent area of the other 5-L lid (0.4 in^2/255 mm^2).

Fruit from each replicate container (n = 3) was weighed before sealing and control fruit (n = 3 groups) were weighed before being stored on the open tray (Fig. 2). Temperature and RH were recorded by data loggers (Hobo, Onset Computer Corporation, model U-12, Bourne, MA) placed into each container and logged every 30 minutes. The storage rack in the cold room was enclosed with a plastic cover (open bottom) to restrict air movement from the cold room fans (Fig. 3). The cold room was set to 21 °C and RH was set to 50% (± 5%) to simulate an air-conditioned home. Ambient room temperature and RH were also tracked in three locations: immediately next to the fruit on the trays, inside the rack cover but away from the control fruit, and outside the rack cover.

Fig. 1. Fruit placement before sealing the vented containers.

Fig. 2. Fruit placement on open trays.

Fig. 3. Fruit stored in vented containers and on open trays underneath enclosed rack cover.

We previously reported on trials for passionfruit and guava stored under conditions to simulate "room-temperature" ripening in countries that both have hot climates and do not customarily employ air conditioning in the home. Ripening in hot, arid environments was simulated for passionfruit (35 °C) and guava (30 °C), both at 50% RH. Ripening in hot, humid environments was simulated for passionfruit (30 °C/80%) RH (Stahl & Sargent, 2022). Some of these results are included here, along with new results from 21 °C, for ease of comparison.

Fruit ripening was tracked using the control fruit stored in open air; each test was terminated when control fruit were considered fully ripe and ready for consumption. Fruit were then reweighed and subjectively rated for appearance. Peaches and guavas were considered ripe when flesh responded to gentle manual pressure between the thumb and forefinger. Passionfruit held under ambient conditions were considered ripe when the usually spherical rind was shriveled and collapsed.

Guava and peach fruit firmness was determined with a Texture Analyzer (TA HD Plus, Texture Technologies, Hamilton, MA). A flat plate (75 mm diameter) attached to a 500-kg load cell was driven into opposite sides of fruit equators at 0.8 mm/sec; maximum force at 2.5-mm deformation was recorded in Newtons (N). For peaches, the equatorial surfaces evaluated were perpendicular to the "seam" of the fruit.

Statistics were calculated using standard error of the mean (n = 3).

Results and Discussion

Storage conditions. Setpoint storage conditions were maintained during the tests, where room temperature averaged 21 °C and RH averaged 50%. Upon sealing, RH inside the containers quickly rose to >90% in full, 5-L containers and to 80 to 90% for a single fruit stored in a 1.2-L container. RH from the data loggers placed next to guavas on the open tray matched RH in the room

(50%), while RH recorded in close proximity to peaches on the open tray was higher, approximately 60%. Grierson & Wardowski (1978) report that RH in refrigerated storage is altered by many environmental factors such as air movement, small variations in temperature, and proximity to absorbent materials, which supports this variation in microclimate.

Appearance. Visual appearance, weight loss, and firmness were the main limiting quality factors for fruit ripened in open air (21 °C/50% RH). Peaches became shriveled and soft to the touch after 4 days, while those ripened in vented containers remained firm, taut, and visually appealing (Fig. 4). Guavas on the open tray ripened in 6 days, by which time they lost their glossy peel sheen and exhibited a matte appearance; those ripened in vented containers remained glossy (Fig. 5).

All passionfruit ripened in open air at 21 °C/50% RH evidenced collapsed, wrinkled rinds associated with ripening, whereas 100% of those ripened in vented containers remained smooth and spherical (Figs. 6 and 7).

For passionfruit ripened under high ambient temperature/high RH (30 °C/80% RH) conditions, all fruit stored in vented storage containers were turgid and unripe after 6 days, whereas 55% of control fruit were considered ripe (Figs. 8, 9). However, ripening environment did not affect development of decay, where fruit stored either in open air or in vented containers had approximately 15% decay after 6 days. Storage under low or high ambient RH demonstrated the effectiveness of storage in vented containers to extend shelf life of passionfruit: vented containers, even under low ambient RH, were more effective than higher ambient RH alone.

Weight loss. Fruit ripened in vented containers, and therefore in a high-RH environment, retained more weight than those ripened

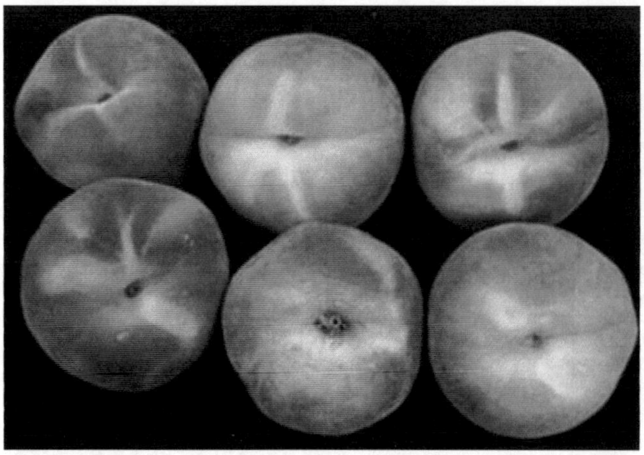

Fig. 4. Peaches after 4 days at 21 °C, ripened in vented containers (top row) or in open air (bottom row).

Fig. 5. Guava after 6 days storage at 21 °C in vented containers (left) or in open air (right).

Fig. 6. Passionfruit after 6 days at 21 °C: Top row—full containers with 100% open vents; middle row—single fruit stored in 33% open vents; and bottom row—fruit stored in open air.

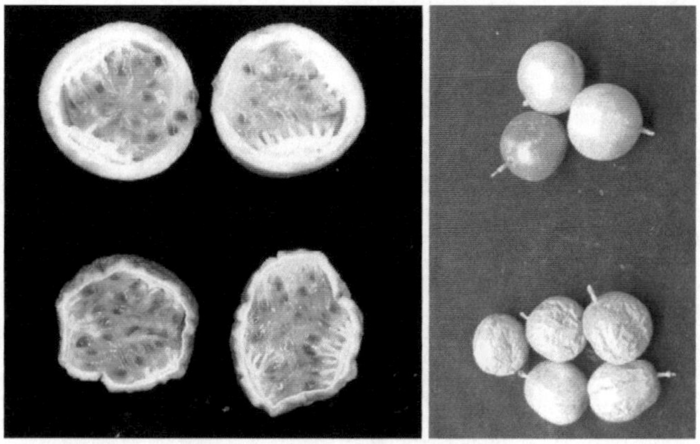

Fig. 7. Passionfruit appearance after 6 days under high temperature/high relative humidity (RH) conditions (30 °C/80% RH). In vented containers: turgid/unripe (top); in open air: shriveled/ripe (bottom).

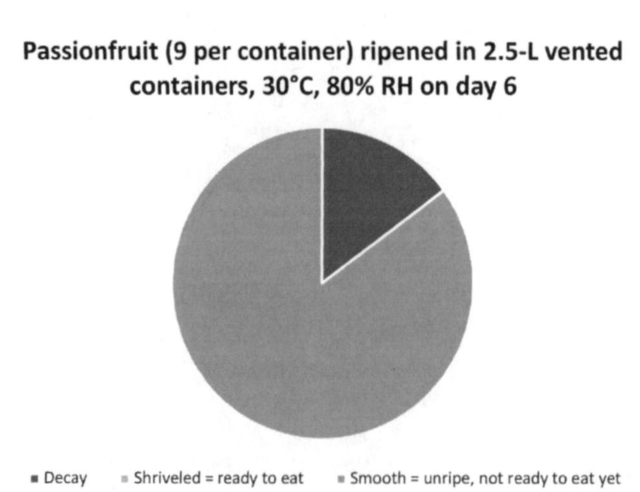

Fig. 8. Appearance of passionfruit (Decay, Round/smooth, Shriveled/ripe) after 6 days in vented containers at 30 °C/80% RH.

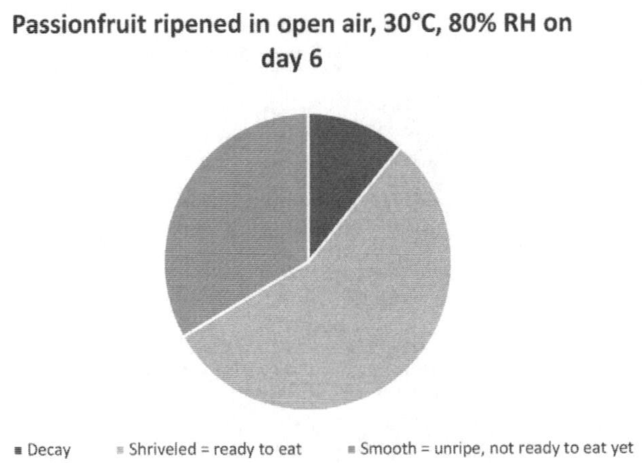

Fig. 9. Appearance of passionfruit (decay, round/smooth, shriveled/ripe) after 6 days in ambient air at 30 °C/80% RH.

in open air during storage at 21 °C/50% RH, across all trials. In four separate tests, peaches lost 13 to 18% of initial weight when stored in open air, depending on variety and duration of storage. In contrast, peaches ripened in 5-L, vented containers lost only 2 to 3% (Fig. 10). This confirms results reported by Whitelock et al. (1994) in which weight loss of stored peaches tripled from 0.43 to 0.73% per day to 1.28 to 2.30% per day as RH decreased from 98% to 67%, with variations dependent on cultivar. Additionally, the removal of fuzzy trichomes on the peach epidermis, a common practice in modern commercial peach production, accelerates moisture loss. When trichomes were removed prior to ripening, peaches lost 13% of initial weight following 2 days of storage at 24°C and 40% RH, whereas intact peaches lost only 5% (Fernández et al (2011).

Different quantities of fruit in the vented containers (in other words, whether the container held 1, 2, 4, or 9 peaches) all maintained weight loss at less than 3.5%; those held in ambient (50% RH) air lost >18% fresh weight (Fig. 10 d).

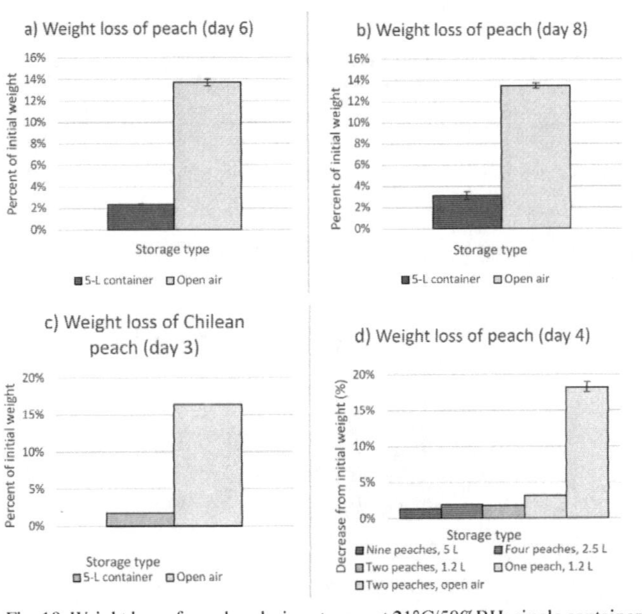

Fig. 10. Weight loss of peaches during storage at 21°C/50%RH: single container or open storage (a, b; separate tests); imported fruit (c). Effect of vented container size and number of fruit per container (d). (vertical bars represent standard error ($P \leq 0.05$)

Weight loss in both pink and white guava types was even more dramatically suppressed by vented container ripening (Fig. 11). Fruit ripened in open air at 21 °C/50% RH lost from 17% (pink) to 25% (white) of the initial weight; in contrast, guavas stored in a full 5-L container lost 2 to 4% of the initial weight. Pink guava fruit ripened in open air at 30 °C also lost 17% weight when stored in open air but only 4% when stored in vented 5-L containers (Fig. 12).

Passionfruit stored at low or high ambient temperatures also evidenced more weight loss when ripened in open air compared with those ripened in vented containers. Weight of passionfruit decreased 13% when stored in open air at 21 °C, compared with 3% when stored in full, 5-L vented containers (Fig. 13). However,

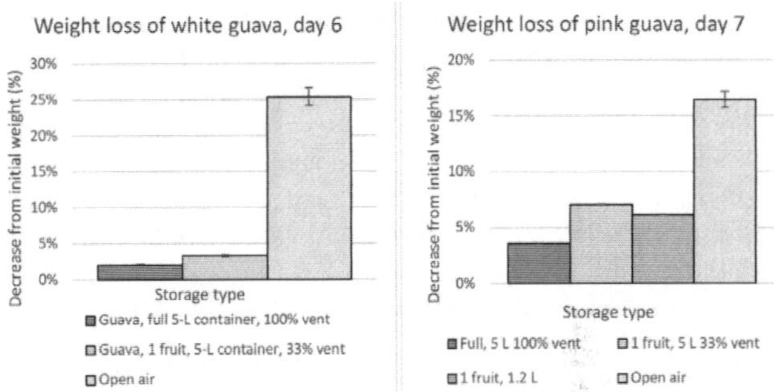

Fig. 11. Weight loss of white guava (left) and pink guava (right) after storage at 21°C/50% RH as affected by vented container size, vent opening and number of fruit per container (separate test.).

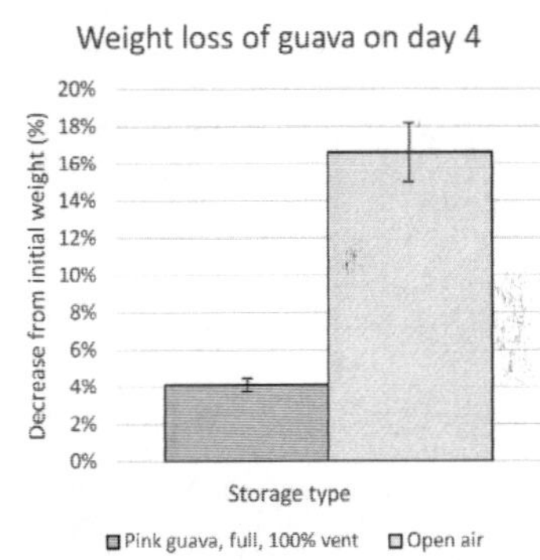

Fig. 12. Weight loss of pink guava in vented containers or open storage at 30°C/50% RH.

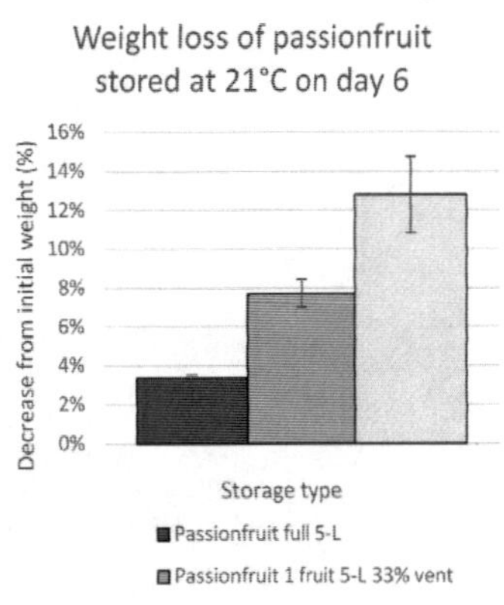

Fig. 13. Weight loss of passionfruit in full or single-fruit vented containers during storage at 21°C/50% RH.

a single fruit in the container lost roughly double the weight of fruit stored in the full container. Fresh weight of passionfruit decreased 15 to 16% when stored in open air at 35 °C/50% RH, compared with 5 to 6% when stored in full, 2.5-L or 5-L vented containers (Fig. 14). This occurred after 3–4 days compared with 6 days for those held at 21 °C, indicating the positive correlation between higher storage temperatures and weight loss. Weight of passionfruit decreased 14 to 16% when stored in open air at 30 °C and 80% RH, compared with 3 to 4% when stored in full, 2.5-L vented containers (Fig. 15). Note that fruit stored in 80% RH in open air still experienced more weight loss similar to those stored under 50% RH, indicating that elevated ambient RH alone is not sufficient to minimize weight loss.

A wide variety of other fruit stored in vented containers in related trials (usually climacteric fruit such as banana, mango, and avocado) also benefited during storage from 21 °C to 35 °C. Some fruits were more susceptible to weight loss than others. For example, at 21 °C pears ripened in open air had double the weight loss of those ripened in vented containers, whereas apricots ripened in open air lost five times the weight when stored under the same conditions (data not shown). Mangoes ripened in vented containers at 30 °C and 35 °C under 50% RH had 1/3 to 2/3 the weight loss of those ripened in open air (Stahl & Sargent, 2022). Grierson & Wardowski (1978) report that the effect of weight loss on marketability is commodity-dependent; for example, as little as 5% weight loss renders apples unsaleable.

Firmness. Peach fruit ripened in vented containers for 4 days at 21 °C/50% RH were more than twice as firm (16–21 N) as those ripened in open air (8 N) (Fig. 16). A single peach ripened in the vented container lost significantly more than those containing two or more peaches, but much less than those stored on the open tray. Similar results were also obtained in a separate trial with a 6-day storage period, in which firmness of peaches ripened in full containers was 20 N compared with those ripened in open air (10 N) (Fig. 17) Containers with a single fruit softened more than those that were half-full of fruit; reducing the vent area to 33% compensated for the lower number of fruit by maintaining firmness.

Guava firmness during storage at 21 °C/50% RH also benefited from the high-RH environment of container storage. Firmness for fruit stored in open air for 6 and 7 days at 21 °C was 19 N and 23 N, for white and pink types, respectively; firmness in vented containers was maintained for full containers (34 N) or single fruits when vent opening area was reduced to 33% or when container size was reduced to approximately 25% volume (Fig. 18).

Firmness for guavas stored for 4 days at 30 °C/50% RH, simulating a hot, dry climate (no air conditioning), was 60 N after ripening in vented containers, but only 28 N after ripening in open air (Fig. 19). However, high rates of decay in one of the two trials under these conditions indicate that open-air ripening for guava should not take place above certain ambient temperatures; in Pakistan, for example, guava is only harvested in cooler seasons (Dr. Faisal Shah, personal communication).

Passionfruit was not evaluated for firmness due to the external surface being a husk rather than a peel. However, other commodities such as Bosc pear and Fuyu persimmon were also firmer when ripened at 21 °C/50% RH in vented containers (data not shown).

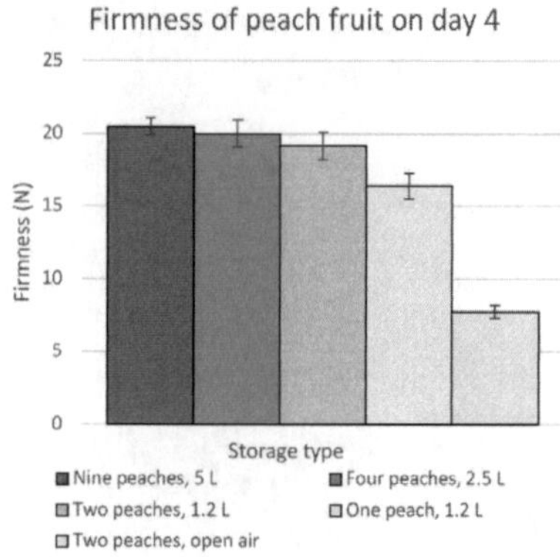

Fig. 16. Maximum compression force at 2.5 mm for peach fruit after 4 days of storage at 21 °C/50% RH.

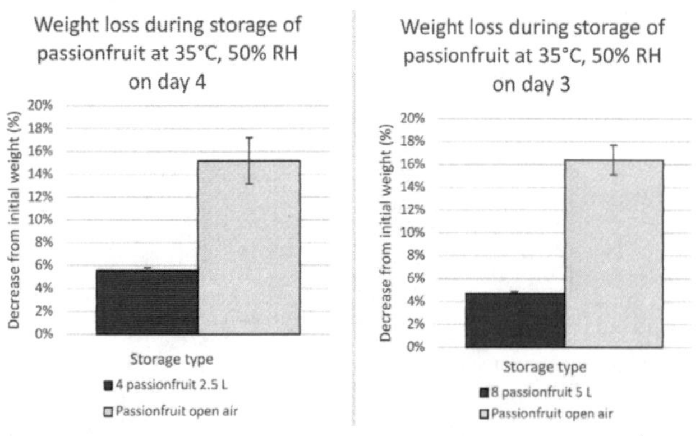

Fig. 14. Weight loss of passionfruit after storage at 35°C/50% RH. (Separate tests.)

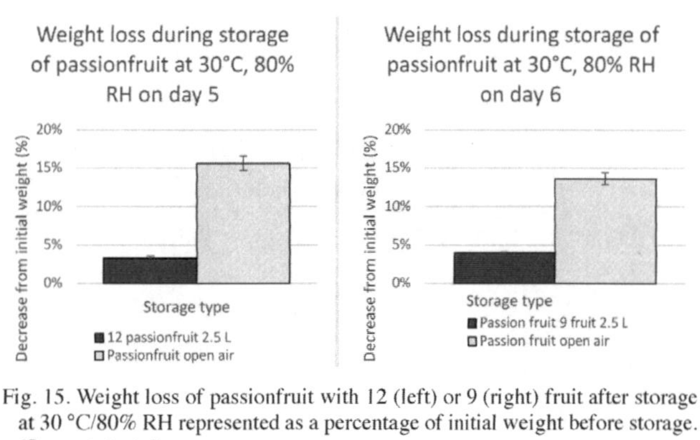

Fig. 15. Weight loss of passionfruit with 12 (left) or 9 (right) fruit after storage at 30 °C/80% RH represented as a percentage of initial weight before storage. (Separate tests.)

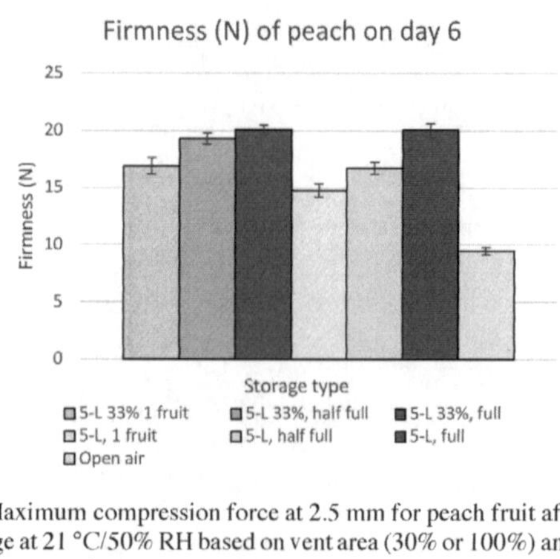

Fig. 17. Maximum compression force at 2.5 mm for peach fruit after 6 days of storage at 21 °C/50% RH based on vent area (30% or 100%) and number of fruit per container.

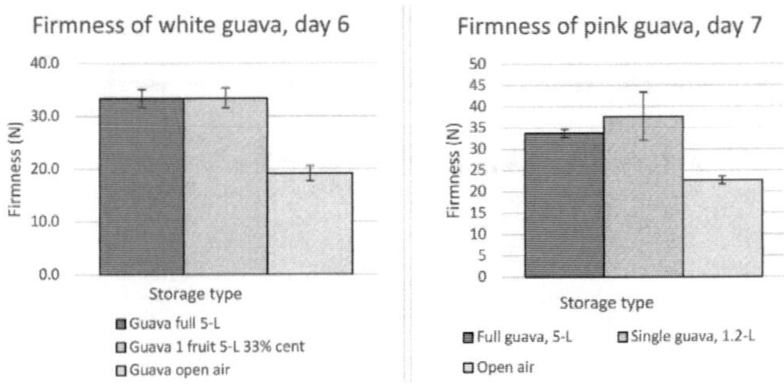

Fig. 18. Maximum compression force at 2.5 mm for white (left) and pink (right) guava after storage in full or single-fruit containers at 21 °C/50% RH. (Separate tests.)

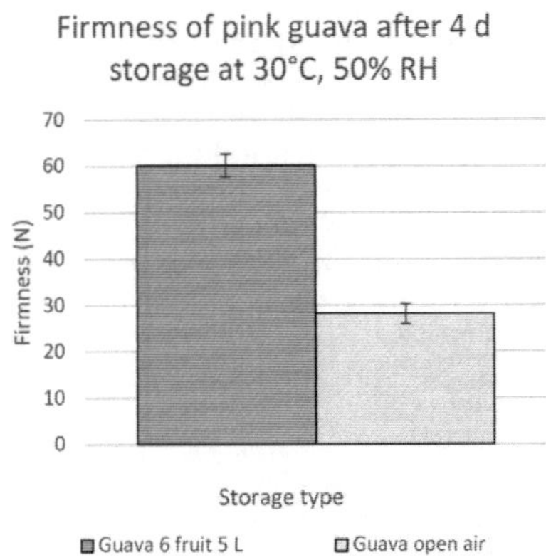

Fig. 19. Maximum compression force at 2.5 mm for pink guava after 4 d storage at 30 °C/50% RH (From: Stahl and Sargent, 2022).

Conclusion

Climacteric fruit ripen to best quality under high-RH conditions. The low RH typically found in air-conditioned homes interferes in this process. An environment of 80 to 90% RH was established and maintained by employing a vented countertop system properly designed to modulate the water vapor released by the fruit itself. Maintaining the RH in this range avoided condensation of free water inside the container, thus minimizing mold development. These conditions increased fruit quality by maintaining favorable conditions for fruit ripening, which in turn, extended structural integrity and visual quality. The most dramatic effects were seen on peach, guava, and passionfruit and have been reported here.

Literature Cited

Fernández V., Khayet M., Montero-Prado P., Heredia-Guerrero J.A., Liakopoulos G., Karabourniotis G., Del Río V., Domínguez E., Tacchini I., Nerín C., Val J., & Heredia A. 2011. New insights into the properties of pubescent surfaces: Peach fruit as a model. *Plant Physiology*. 156(4):2098–2108.

Grierson W. & Wardowski W.F. 1978. Relative humidity effects on the postharvest life of fruits and vegetables. *HortScience*. 13(5):570–574

Kader A., 1996. Recommendations for maintaining postharvest quality. Department of Plant Sciences, University of California, Davis. Accessed August 2023. https://postharvest.ucdavis.edu/Commodity_Resources/Fact_Sheets/Datastores/Fruit_English/?uid=9&ds=798

Stahl S.L. & Sargent S.A. 2022. Ripening rate and quality of selected tropical fruits under extreme home ambient conditions. *Proc. Fla. State Hort. Soc.* 135:144–150.

Whitelock, D.P., Brusewitz, G.H., Smith, M.W., & Zhang, X. 1994. Humidity and airflow during storage affect peach quality. *HortScience*. 29(7):798–801.

Wu C.T. 2010 An overview of postharvest biology and technology of fruits and vegetables. In: Technology on reducing post-harvest losses and maintaining quality of fruits and vegetables. Proceedings of 2010 AARDO Workshop.

Proc. Fla. State Hort. Soc. 136:174–178. 2023.

Can Mandarins Be Blended with Orange Juice and Still Maintain Orange Juice Flavor?

Anne Plotto*, Jinhe Bai, Kristen Jeffries, Gabriela Olmedo, Elizabeth Baldwin, John Manthey, Matthew Mattia, and Ed Stover

USDA–ARS, U.S. Horticultural Research Lab, Fort Pierce, FL

Keywords. citrus, Code of Federal Regulations (CFR), Food and Drug Administration (FDA), huanglongbing (HLB), orange juice

Since the citrus greening disease (aka Huanglongbing, HLB) was detected in 2005, orange (*Citrus sinensis*) juice production has decreased by more than 80% in Florida. Over the years, some mandarins (*C. reticulata*) and mandarin-like hybrids with *Poncirus trifoliata* in their pedigree have shown tolerance to the disease and have the potential to make up the deficit in the volume of oranges. However, current FDA rules allow for only 10% mandarin juice in orange juice. If citrus, other than *C. sinensis,* could be added to orange juice in amounts greater than 10%, it would give more flexibility to processors under the current production environment where oranges are declining due to citrus greening. The objective of this study was to evaluate the effect of blending mandarins and other citrus hybrids with orange juice on overall orange flavor. In the first year, a 9-member panel was trained to distinguish orange from mandarin flavor using reference standards, rate sweetness, sourness and bitterness, and to describe potential off flavors. Fruits were washed, sanitized, and manually juiced. Blends were made with 60:30 'Valencia':'Hamlin' juice, and 10% mandarin juice including 'Superna', FF-5-51-2, and a *Poncirus* introgressed mandarin hybrid, FF-1-89-11. These blends were compared with the standard 60:40 'Valencia':'Hamlin' juice. Panelists found no differences in orange flavor among samples. Mandarin flavor was slightly higher in blends containing FF-5-51-2, and sweetness and sourness were higher in blends containing FF-1-89-11, in comparison with the reference 60:40 'Valencia':'Hamlin' juice. A similar comparison was performed between 60:40 'Valencia':'Hamlin' juice and 60:20:20 'Valencia':'Hamlin':mandarin juice, the mandarin being FF-5-51-2, 'Superna' or Sugar Belle®. The trained panelists could perceive the mandarin flavor in those blends, but the orange flavor remained unchanged. In the following year, blends were prepared with juice that had been commercially processed and pasteurized and presented to untrained panelists in a difference test (Tetrad). Most panelists could not detect differences between control base juice (60:40 'Valencia':'Hamlin') and base juice to which 10%, 15% or 20% of FF-5-51-2 had been added. The test was also performed with commercial juice, and again, untrained panelists could not detect differences between blends and control. From these results, it is unlikely that untrained consumers would perceive mandarin or other hybrid added to orange juice at a level up to 20%.

The combination of a devastating disease, citrus greening (a.k.a. Huanglongbing, HLB) and multiple hurricane events has brought Florida citrus production to an all-time low in 2023 (15.8 million boxes in 2022–23, versus over 200 million boxes in 2000). Less than ten processing plants were operating in 2022, as opposed to over 40 in the early 2000s (Munch, 2023). Nevertheless, some growers are still planting and have become highly technical to bring trees to production (Archer & Albrecht, 2023; Li et al., 2020; Qureshi et al., 2019; Zhang et al., 2019; 2021). Furthermore, both the U.S. Department of Agriculture (USDA) in Fort Pierce, and the University of Florida in Lake Alfred and in Gainesville, have been active in plant improvement and there is hope for new hybrids that are tolerant to HLB (Stover et al., 2015; Ramadugu et al., 2016). The current breeding program at the USDA has inherited hybrids of *Citrus* × *Poncirus*, which appear to be tolerant to HLB. Likewise, some mandarin hybrids, such as 'LB8-9' Sugar Belle® from the University of Florida, are showing tolerance to HLB in the field. While those hybrids and mandarins cannot be used to make "orange" juice because of an obvious difference in flavor, their juice could be blended with the juice of oranges to complement the loss in production and to improve the poor quality of orange juice affected by HLB. However, a current FDA regulation limits the amount of mandarin or other citrus hybrids that can be added to orange juice to 10% (21 C.F.R. §146.140 – Pasteurized orange juice; and §146.141; §146.146; §146.151). Above that amount, the juice cannot be named "orange juice".

During a 6-year study, our laboratory has classified over 50 accessions (unnamed mandarins and *Poncirus*-introgressed man-

The authors thank JBT Corporation for assisting in processing oranges and mandarins in the year 2022–23.

*Corresponding author. Email: anne.plotto@usda.gov

darin hybrids as well as mandarin and orange named cultivars) into fruit having orange, mandarin or "poncirus" flavor by sensory evaluation using a trained panel (Baldwin et al., 2019; Fan et al., 2024). Some of these accessions with high sugar or sugar/acid ratio and with mandarin or close-to-orange flavor were chosen to be blended with orange juice. The selected accessions also showed tolerance to HLB. The objective of the present study was to measure the effect of adding the juice of selected mandarin or *Poncirus*-introgressed mandarin hybrids into orange juice on the sensory quality of such juice blends. The sensory quality was measured by descriptive analysis using a trained panel, as well as by difference testing using an untrained panel.

Materials and Methods

Fruit of 'Hamlin', 'Superna', Sugar Belle®, FF-5-51-2 and FF-1-89-11 were harvested from the experimental farm at the A.H. Whitmore Foundation Farm, Groveland, FL, or obtained from a local processor ('Valencia'). In 2021-2022, fruit were juiced manually using a reamer-type juicer (Vinci™ Hand Free Juicer, Vinci® Housewares, La Mirada, CA, USA), and frozen at -20 °C until used. In 2022-23, 'Hamlin', 'Valencia' and FF-5-51-2 fruit were processed using a JBT 391 extractor and pasteurized (Baldwin et al., 2018), with "standard" settings for 'Hamlin', and "premium" settings for 'Valencia' and FF-5-51-2. The "premium" setting is used with fruit that produce more peel oil when squeezed. In a separate experiment, commercial orange and mandarin juice was obtained from local processors. The commercial orange juice was an undisclosed blend of 'Hamlin' and 'Valencia' juice without any oil added. Fruit identification, harvest date, titratable acidity (TA), soluble solids concentration (SSC), and SSC/TA ratio are listed in Table 1.

Blends were prepared in the following way. In 2021-22, the orange juice was a mix of 'Valencia' and 'Hamlin' at a 60:40 ratio. The ratio became 60:30 or 60:20 of 'Valencia':'Hamlin' when the mandarin hybrids were blended at 10% or 20%, respectively (Tables 2 and 3). In 2022–23, the 'Valencia':'Hamlin' orange juice base remained at a 60:40 ratio, and the mandarin juice (FF-5-51-2 or commercial mandarin) were blended with that base at 10%, 15% or 20% (Tables 4 and 6).

In 2021-22, samples were evaluated by descriptive sensory analysis using a trained panel. Nine panelists were trained to rate intensities of sweetness, sourness, bitterness, orange and mandarin flavor, and off flavor in citrus juice using reference standards and a line scale with anchors at 0 (none), 7.5 (medium) and 15 (high) (Baldwin et al., 2019; Deterre et al., 2023). At each session, panelists evaluated three samples [45 mL juice at 14 °C in 118 mL (4 oz) plastic soufflé cups (Solo® Cups Co., Urbana, IL)] presented in a randomized order. Samples were tasted in duplicate, with two sessions per day. In 2022–23, difference testing was used to see if untrained panelists could perceive the difference between 100% orange juice and orange juice blended with 10%, 15% or 20% mandarin juice. A tetrad test was conducted. Panelists were presented with four sample cups and they were instructed that the cups were two pairs of the same sample. Panelists were asked to group the paired samples together. If they chose the correct answer (i.e. they could discriminate between the two samples presented as two pairs), they were asked to select which pair they preferred, which was sweeter, and which had more orange flavor and then comment on their selection. If they chose the incorrect answer, no further question was asked. There were 31 to 41 panelists who participated in the tests. All taste panels took place in individual booths under red lighting. Compusense© Cloud (West Guelph, ON, Canada) was used for data acquisition.

Juice and blends were analyzed for SSC and TA using previously described methods (Fan et al., 2024).

Instrumental data were analyzed by analyses of variance (ANOVA) using XLSTAT v 2010.1.1 [Addinsoft (2024), XLSTAT statistical and data analysis solutions, Boston, MA, USA]. Sensory trained panel data were analyzed by ANOVA using SenPAQ version 5.01 (Qi Statistics©, Reading, UK) (Deterre et al., 2023).

Table 1. Orange, mandarin and mandarin × *Poncirus* hybrids quality before blending: juice titratable acidity (TA), and soluble solids concentration (SSC).

Genotypes	Harvest date	TA (%)	SSC (°Brix)	SSC/TA
	2021–22 season			
Valencia	10-Apr-22	0.82 b[z]	10.6 cd	13.0 d
Hamlin	5-Jan-22	0.54 c	9.7 d	18.1 c
Superna	20-Dec-21	0.35 d	10.2 d	29.2 a
SugarBelle	5-Jan-22	0.86 b	13.4 b	15.5 d
FF-5-51-2	14-Dec-21	0.53 c	11.6 c	21.8 b
FF-1-89-11	10-Dec-21	1.54 a	15.9 a	10.4 e
	2022–23 season			
Valencia	6-Apr-23	0.99 a	10.4 b	10.5 c
Hamlin	10-Jan-23	0.43 c	11.2 a	26.2 a
FF-5-51-2	13-Dec-22	0.50 b	11.0 a	22.0 b

[z]Values followed by a different letter within a column and a season are statistically different by the Tukey test ($P < 0.05$).

Results and Discussion

Among hybrids tested in 2021–22, FF-1-89-11 had the highest TA and SSC (1.54% and 15.9 °Brix) but the lowest SSC/TA ratio (10.4), while 'Superna' had the highest SSC/TA ratio, followed by FF-5-51-2, both hybrids having low TA (Table 1). When added at 10% in a 'Valencia':'Hamlin' (60:30) base, FF-1-89-11 was the only hybrid which changed the SSC/TA ratio to a lower level than the other blends (Table 2). For all hybrids added to orange juice, orange flavor was unchanged, mandarin flavor was slightly higher than the orange juice base, but only significantly higher for FF-5-51-2. Sweetness and sourness were higher in the blend containing FF-1-89-11 (sourness only different from the blend containing 'Superna'), and bitterness and off-flavor were unchanged (Table 2). Some panelists commented on a slight "poncirus" off flavor in samples containing FF-1-89-11, but it was only a hint that was not quantified as an off flavor. However, for that reason, FF-1-89-11 was not included in the next trial.

When mandarin juice was added at 20% to the 'Valencia'/'Hamlin' orange juice, the SSC/TA ratio was different for each blend, but remained in a range of 14.3 to 15.7 (Table 3). The blend with 'Superna' had the highest SSC/TA ratio, and the blend with Sugar Belle® the lowest (Sugar Belle® also had the highest TA). Orange flavor was unchanged when mandarin juice was added at 20%, however, mandarin flavor increased by at least 1.5 points and was significantly different from the orange juice base for all blend types (Table 3). Sweetness was highest in the juice containing Sugar Belle® (although not different from juice with 'Superna') and sourness was higher in all juices containing a mandarin juice. Juices containing FF-5-51-2 and 'Superna' were perceived as having a slight pumpkin off flavor reflected in the off-flavor rating; that flavor had been found in our trials to be characteristic of some mandarin juice with low TA, such as 'Superna' and FF-5-51-2 in this trial.

Table 2. Titratable acidity (TA), soluble solids concentration (SSC), SSC/TA ratio, and sensory descriptors (scale 0–15) for orange juice blended with mandarins or mandarin × *Poncirus* hybrid at 10%.

Blends	TA (%)	SSC (°Brix)	SSC/TA	Orange	Mandarin	Sweet	Sour	Bitter	Off-flavor
60:40_Val:Ham[z]	0.65 b[y]	9.7 a	14.8 a	7.3 a	2.5 b	6.1 b	4.8 ab	0.5 a	0.5 a
60:30:10_Val:Ham:Superna	0.69 ab	10.4 a	15.1 a	7.3 a	3.1 ab	5.9 b	4.5 b	0.5 a	0.6 a
60:30:10_Val:Ham:FF-5-51-2	0.70 ab	10.4 a	15.0 a	7.1 a	3.9 a	6.1 b	5.0 ab	0.6 a	0.6 a
60:30:10_Val:Ham:FF-1-89-11	0.76 a	10.4 a	13.7 b	7.8 a	3.1 ab	6.6 a	5.2 a	0.5 a	0.4 a

[z]Cultivar abbreviations: Val = 'Valencia'; Ham = 'Hamlin'.

[y]Values followed by a different letter within a column are statistically different by the Tukey (instrumental) and Fisher's LSD (sensory) tests ($P < 0.05$). TA and SSC: n = 4. Sensory data: n = 9.

Table 3. Titratable acidity (TA), soluble solids concentration (SSC), SSC/TA ratio, and sensory descriptors (scale 0–15) for orange juice blended with mandarins or mandarin × *Poncirus* hybrid at 20%.

Blends	TA (%)	SSC (°Brix)	SSC/TA	Orange	Mandarin	Sweet	Sour	Bitter	Off-flavor
60:40_Val:Ham[z]	0.66 b[y]	9.7 b	14.8 c	7.3 a	2.5 b	6.1 b	4.8 b	0.5 a	0.5 b
60:20:20_Val:Ham:FF5-51-2	0.69 b	10.6 ab	15.2 b	6.5 a	4.6 a	6.2 b	5.7 a	0.7 a	1.0 ab
60:20:20_Val:Ham:Sugar B.	0.79 a	11.3 a	14.3 d	7.3 a	4.3 a	6.8 a	5.5 a	0.6 a	0.7 b
60:20:20_Val:Ham:Superna	0.66 b	10.3 ab	15.7 a	6.9 a	4.0 a	6.3 ab	5.0 ab	0.8 a	1.3 a

[z]Cultivar abbreviations: Val = 'Valencia'; Ham = 'Hamlin'; Sugar B. = Sugar Belle®.

[y]Values followed by a different letter within a column are statistically different by the Tukey (instrumental) and Fisher's LSD (sensory) tests ($P < 0.05$). TA and SSC: n = 4. Sensory data: n = 9.

Table 4. Titratable acidity (TA), soluble solids concentration (SSC,) and SSC/TA ratio for 60:40 'Valencia': 'Hamlin' orange juice (OJ base), and FF-5-51-2 mandarin blends at 10%, 15% and 20% in OJ base.

Blends	TA (%)	SSC (°Brix)	SSC/TA
60:40_Val:Ham (OJ base)	0.72 a[z]	11.0 a	15.3 a
10% FF-5-51-2 in OJ base	0.71 a	10.9 a	15.5 a
15% FF-5-51-2 in OJ base	0.70 a	10.7 a	15.4 a
20% FF-5-51-2 in OJ base	0.72 a	10.9 a	15.2 a

[z]Values followed by a different letter within a column are statistically different by the Tukey test ($P < 0.05$). n = 4.

In summary of the 2021-22 trials, a trained panel could perceive and quantify flavor differences when mandarin juice was added to a blend of 'Valencia'/'Hamlin' juice at 20% mandarin but not 10%. However, in these experiments, the juice was hand-squeezed and not pasteurized, the latter step significantly changes orange juice flavor in a commercial process (Perez-Cacho & Rouseff, 2008). Furthermore, blends were made by changing the 'Valencia':'Hamlin' ratio from 60:40 to 60:30 with the mandarin juice added, therefore they were not true comparisons with the initial juice. The ratio of 60:40 was arbitrarily chosen as a starting basis used by processors, but it may vary depending on the quality of each cultivar. In the next experiments, juice was processed and pasteurized in a pilot plant to produce a juice closer to a commercial extraction process, and the mandarin was added to the orange juice base to keep the 'Valencia':'Hamlin' ratio at 60:40.

Fruit harvested in 2022–23 were 'Valencia', 'Hamlin' and FF-5-51-2. 'Hamlin' had the highest SSC/TA ratio and 'Valencia' the lowest, with FF-5-51-2 over 20.0 as was 'Hamlin' (Table 1). It is to be noted that 'Hamlin' fruit were harvested from young (7–year–old) trees in a rootstock trial, and were quite healthy at that time, explaining their high SSC/TA ratio. 'Valencia' fruit, on the other hand, were purchased from a local processor and sourced from commercial groves across the State of Florida and were more representative of the quality of fruit affected by HLB (Dala Paula et al., 2017), with an SSC/TA ratio barely exceeding 10.0.

TA, SSC and SSC/TA were unchanged by adding FF-5-51-2 to the orange juice base at 10%, 15% or 20% (Table 4). Likewise, untrained panelists could not detect any difference among samples when each of the blends were compared to the orange juice base (Table 5). The tetrad test is a difference test that has gained popularity among sensory scientists because it is more powerful than the well accepted triangle test, and therefore differences between two products can be demonstrated with fewer panelists. While it cannot be used with products that have a strong carry-over, such as spicy foods, it was demonstrated to be appropriate to test orange juice (Ennis, 2012; Ishii et al. 2014). In our study, those who could perceive a difference between samples (12 out of 34, 13 out of 36, and 15 out of 31 panelists for each test, respectively) either preferred the juice with FF-5-51-2 when it was added at 10 or 15%, or liked it less at 20% (Fig. 1). Juice with 10% FF-5-51-2 was perceived as having more orange flavor than the 100% orange juice; however, with 15% or 20% FF-5-51-2, 100% orange juice was perceived as having more intense orange flavor.

For commercial juice, orange juice had lower TA, SSC and SSC/TA than mandarin juice, and all the blends of mandarin in orange juice (10% to 20%) had intermediate values, except for 10% mandarin in orange juice, which had the same SSC/TA as

Table 5. Tetrad test results for comparisons of orange juice (OJ) base versus 10%, 15%, or 20% mandarin FF-5-51-2 juice in OJ base.

Blends	N	Correct	Incorrect	d'	*P*-value	Significant
10% FF-1-51-2 vs. OJ base	34	12	22	0.3	0.47	NO
15% FF-5-51-2 vs. OJ base	36	13	23	0.4	0.42	NO
20% FF-5-51-2 vs. OJ base	31	15	16	1	0.06	NO

N = number of panelists participating in the test.

Table 6. Titratable acidity (TA), soluble solids concentration (SSC), and SSC/TA ratio for commercial orange juice (OJ), mandarin juice (M) and OJ and M blends at 10%, 15%, and 20% M in OJ.

Blends	TA (%)	SSC (°Brix)	SSC/TA
Orange (OJ)	0.67 c[z]	10.1 c	14.9 b
Mandarin (M)	0.81 a	12.3 a	15.3 a
90:10 OJ:M (10% M)	0.71 b	10.7 b	14.9 b
85:15 OJ:M (15% M)	0.72 b	10.7 b	15.0 ab
80:20 OJ:M (20% M)	0.71 b	10.7 b	15.2 ab

[z]Values followed by a different letter within a column are statistically different by the Tukey test ($P < 0.05$). n = 4.

orange juice (Table 6). As with experimental samples, panelists could not differentiate between 100% orange juice and orange juice with 10%, 15% or 20% mandarin juice (Table 7). We added one more comparison, 20% mandarin in orange juice versus 10% mandarin in orange juice, similarly as to what is allowed commercially, and again, no difference between the two samples was perceived by the untrained panelists. For those panelists who could detect the difference between the two samples, pure orange juice was preferred in all 3 tests (Fig. 2). Orange juice was also perceived to be sweeter when 10% or 15% mandarin juice was added but not 20%, and only the juice with 15% was perceived to

Table 7. Tetrad test results for comparisons of commercial orange juice (OJ) versus 10%, 15%, or 20% mandarin (M) juice in OJ.

Blends	N	Correct	Incorrect	d'	*P*-value	Significant
10% M vs. 100% OJ	41	15	26	0.4	0.38	NO
15% M vs. 100% OJ	36	15	21	0.7	0.19	NO
20% M vs. 100% OJ	32	14	18	0.8	0.14	NO
20% M vs. 10% M in OJ	36	17	19	0.9	0.06	NO

N = number of panelists participating in the test.

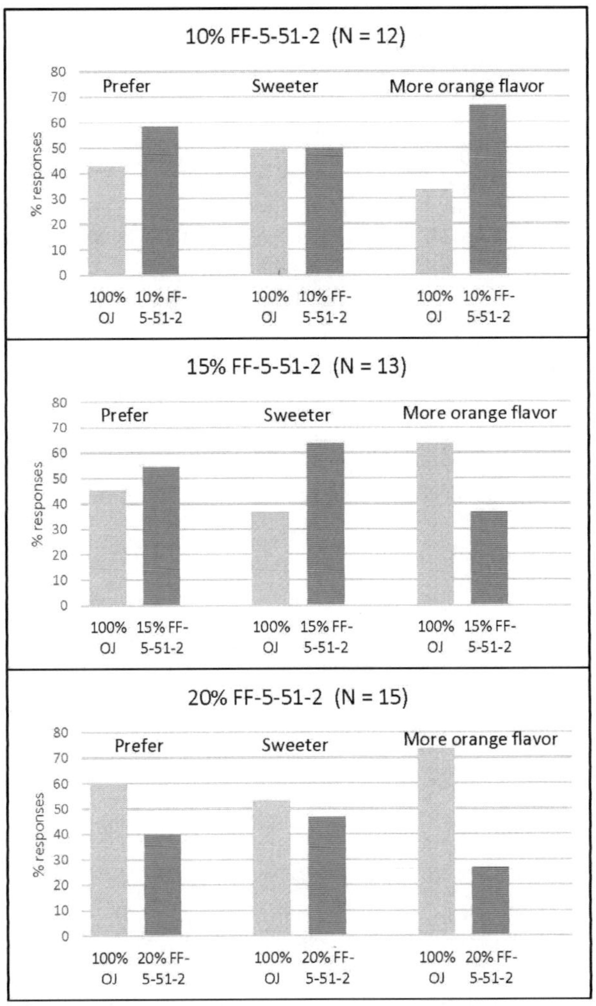

Fig. 1. Distribution of responses for juice preference, sweetness and orange flavor for panelists who could distinguish among the two samples (correct response in the Tetrad tests. N = 12, 13, 15 when the juice of mandarin FF-5-51-2 was blended at 10%, 15% or 20% with orange juice)

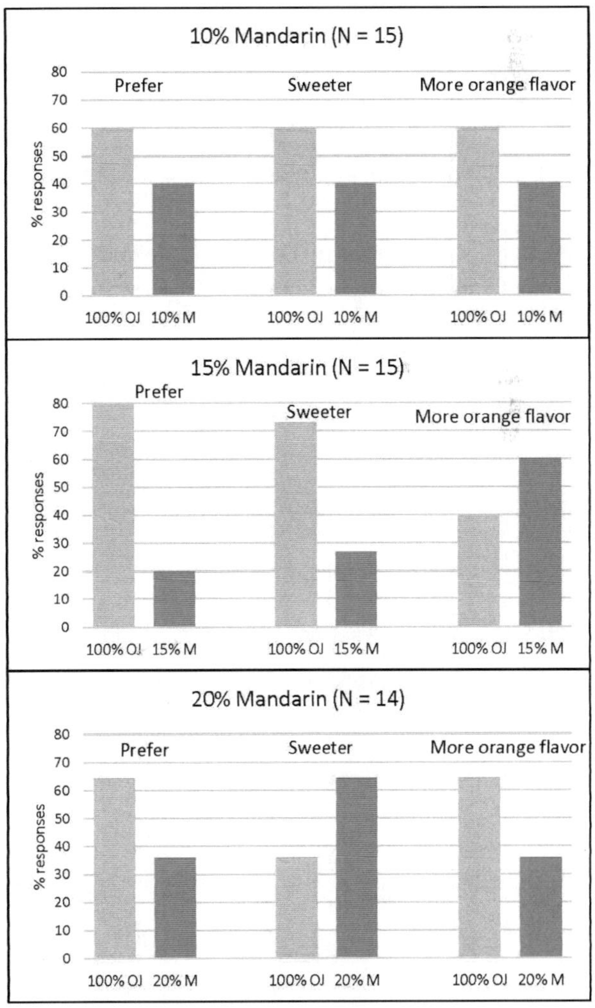

Fig. 2. Distribution of responses for juice preference, sweetness and orange flavor for panelists who could distinguish among the two samples (correct response in the Tetrad tests. N = 15, 15, 14 when commercial mandarin juice was blended at 10%, 15% or 20% with commercial orange juice).

have more orange flavor. However, these comparisons were with a small number of panelists (14–15), and some panelists made comments they did not really prefer one or the other sample, and that the differences were very slight.

Summary

Mandarin juice added to orange juice did not change the flavor of orange juice in a significant manner. Trained panelists could perceive an increase in mandarin flavor when the added mandarin juice was at 20% of the orange juice. The remaining flavors (sweetness and sourness) were slightly increased by mandarin in orange juice, but usually by less than one-point on a 15-point intensity scale. Untrained panelists could not detect differences between orange juice and orange juice with mandarin juice blended at 10%, 15% or 20%, except for a few very selective panelists. These data support the fact that mandarin juice could likely be blended with orange juice up to 20% without significantly changing the flavor of orange juice, although due to flavor variations among different cultivars, some might be perceived more easily at 15% or 20% in orange juice.

Literature Cited

Archer L. & Albrecht U. 2023. Evaluation of trunk injection techniques for systemic delivery of huanglongbing therapies in citrus. *HortScience*. 58(7):768–778. https://doi.org/10.21273/HORTSCI17172-23

Baldwin E., Plotto A., Bai J., Manthey J., Zhao W., Raithore S. & Irey M. 2018. Effect of abscission zone formation on orange (*Citrus sinensis*) fruit/juice quality for trees affected by Huanglongbing (HLB). *J. Agric. Food Chem*. 66:2877-2890. https://doi.org/10.1021/acs.jafc.7b05635

Baldwin E., Stover E., Manthey J., Sun X., Ference C., & Plotto A. 2019. Accelerating implementation of Huanglongbing-tolerant hybrids as new commercial cultivars for fresh and processed citrus. *Proc. Fla. State Hort. Soc*. 132:177–181.

Dala-Paula B.M., Plotto A., Bai J., Manthey J.A., Baldwin E.A., Ferrarezi R., & Gloria M.B.A. 2019. Effect of Huanglongbing or greening disease on orange juice quality, a review. *Frontiers in Plant Science*. 9:1976. https://doi.org/10.3389/fpls.2018.01976

Deterre S.C., Jeffries K.A., McCollum G., Stover E., Leclair C., Manthey, J.A., Bai J., Baldwin E.A., Raithore S., & Plotto A. 2023. Sensory quality of *Citrus* scion hybrids with *Poncirus trifoliata* in their pedigrees. *J. Food Science*. 88:1684–1699. https://doi.org/10.1111/1750-3841.16499

Ennis J.M. 2012. Guiding the switch from triangle testing to tetrad testing. *J. Sensory Studies*. 27(4):223–231. https://doi.org/10.1111/j.1745-459X.2012.00386.x

Fan Z., Jeffries K.A., Sun X., Olmedo G., Zhao W., Mattia M.R., Stover E., Manthey J.A., Baldwin E.A., Lee S., Gmitter F.G., Plotto A., & Bai J. 2024. Chemical and genetic basis of orange flavor. *Science Advances*. 10(9): eadk2051.

Ishii R., O'Mahony M. & Rousseau B. 2014. Triangle and tetrad protocols: Small sensory differences, resampling and consumer relevance. *Food Qual. Pref*. 31:49–55. http://dx.doi.org/10.1016/j.foodqual.2013.07.007

Li S., Wu F., Duan Y., Singerman A., & Guan, Z. 2020. Citrus greening: Management strategies and their economic impact. *HortScience*. 55(5):604–612. https://doi.org/10.21273/HORTSCI14696-19

Munch D., 2023. U.S. Citrus production – an uphill battle to survive. Market Intel. Apr 25, 2023. American Farm Bureau Federation. Accessed December 18, 2023. https://www.fb.org

Perez-Cacho P. R. & Rouseff R., 2008. Processing and storage effects on orange juice aroma: A review. *J. Agric. Food Chem*. 56(21):9785–9796. https://doi.org/10.1021/jf801244j

Qureshi J., Stelinski L., & Alferez F., 2019. HLB reduction strategies. *Citrus Industry*. 100(1).

Ramadugu C., Keremane M. L., Halbert S. E., Duan Y. P., Roose M. L., Stover E. & Lee R. F. 2016. Long-term field evaluation reveals Huanglongbing resistance in Citrus relatives. *Plant Disease* 100(9), 1858-1869. https://doi.org/10.1094/PDIS-03-16-0271-RE

Stover E., McCollum G.T., Driggers R., Lee R., Shatters Jr., R., Duan Y., Ritenour M., Chaparro J.X., & Hall D.G. 2015. Resistance and tolerance to Huanglongbing in citrus. *Acta Hortic*. 1065:899–903. https://doi.org/10.17660/ActaHortic.2015.1065.111

USDA, NASS, 2001. Citrus October Forecast. Maturity, test results and fruit size. [Accessed 18 Dec 2023]. USDA - National Agricultural Statistics Service - Florida - Historical Citrus Production Forecasts

USDA, NASS, 2023. Citrus December forecast. Accessed December 18, 2023. Citrus production forecast—December 2023. https://www.usda.gov

Zhang M., Karuppaiya P., Zheng D., Sun X., Bai J., Ferrarezi R.S., Powell C.A., & Duan Y. 2021. Field evaluation of chemotherapy on HLB-affected citrus trees with emphasis on fruit yield and quality. *Frontiers in Plant Science*. 12:611287. https://doi.org/10.3389/fpls.2021.611287

Zhang M., Yang C., Powell C. A., Avery P. B., Wang J., Huang Y., & Duan Y., 2019. Field evaluation of integrated management for mitigating citrus huanglongbing in Florida. *Frontiers in Plant Science*. 9:1890. https://doi.org/10.3389/fpls.2018.01890

Proc. Fla. State Hort. Soc. 136:179. 2023.

—Scientific Note—

Jicama as a Low-sugar Ingredient for Juice Production: The Impact of Soil Amendment and Harvest Maturity on Juice Quality

THAIS PÁDUA FREITAS, JINHE BAI, WEI ZHAO, JASON C. HONG, NANCY OWENS, MARLEE TRANDEL, AND ERIN ROSSKOPF*

USDA–ARS, Horticultural Research Laboratory, Ft Pierce, FL

Keywords. anaerobic soil disinfestation, juice, *Pachyrhizus erosus*, zeta potential

Jicama (*Pachyrhizus erosus*) tubers have potential to enhance insulin sensitivity, making it a great suitable for low-sugar juice. Limited information exists on Florida-grown jicamas.

Findings

In 2020, a study at the USDA Horticultural Research Laboratory in Fort Pierce, FL, compared jicama varieties 'Marde Ross' and '222' grown in a field treated with anaerobic soil disinfestation (ASD) to a control with conventional fertilizer. Tubers were harvested at 75, 105, and 147 days after seeding (DAS). The fleshy portion (FP), juice yield (JY), particle size (PS), zeta potential (ZP), soluble solids content (SSC), pH, titratable acidity (TA), and 2,2-diphenyl-1-picrylhydrazyl (DPPH) radical scavenging activity were measured. The fruit, flesh, and juice were white without discoloration. The FP accounted for 94.6%, and JY was 75.5%, with potential for improvement. The SSC ranged from 4.6 to 6.3%, TA about 0.03% with a pH of 6.4. Harvest time influenced the radical scavenging activity with values of 0.69, 0.55, and 0.36 mM Trolox equivalents on 75, 105, and 147 DAS, respectively (Table 1). 'Marde Ross' had higher FP, JY, and SSC. ASD-treated plot tubers had enhanced SSC and higher ZP value, indicating improved stability.

The cultivar, soil treatment, and harvest maturity influenced jicama yield and quality, but all samples met the standards for low-sugar juice.

Suggestions for Future Directions

Jicama tubers had over 90% water content and juice ratios varied between 72.5% and 81.4% in the 2020–21 season. This indicates that there is potential to improve juice yield. Consequently, future efforts should concentrate on investigating physical, chemical, and biological approaches to increase juice production.

Table 1. Effect of cultivar, soil treatment and harvest time on tuber and juice quality attributes in jicama.

	Cultivar		Soil		Harvest (DAS)		
Treatment	222	MR	ASD	Control	75	105	147
FP (%)	93.8	95.4*	94.1	95.1*	95.1	94.0	94.9
Water (%)	92.4	92.5	92.6	92.3	91.0 c	93.5 a	92.9 b
Pulp (%)	20.8*	18.5	19.7	19.6	17.8 b	17.7 b	23.6 a
JY (%)	72.5	78.5*	74.8	76.2	70.1 c	81.4 a	75.0 b
SSC (%)	5.3	5.8*	5.7*	5.4	5.8 a	5.0 b	5.9 a
TA (%) a	0.03	0.03	0.03	0.03	0.03	0.03	0.03
PS (nm)	362.3	418.8*	372.0	409.1	514.8 a	278.8 c	378.2 b
ZP (mV)	–20.2	–19.5	–20.7*	–19.0	–19.9	–20.5	–19.2
DPPH (%)	34.3	36.1	35.6	34.8	53.9 a	40.8 b	10.9 c

Data presented were the mean values of three biological replicates; * indicate treatment effect of cultivar and soil is significant at 5% level. Mean values in the same row that were not followed by the same letter are significantly different ($P < 0.05$) for harvest time comparison. FP = fleshy portion; JY = juice yield; SSC = soluble solids content; TA = titratable acidity; PS = particle size; ZP = zeta potential; DPPH = 2,2-diphenyl-1-picrylhydrazyl (DPPH).

*Corresponding author. Email: erinrosskopf@usda.gov

Proc. Fla. State Hort. Soc. 136:180–184. 2023.

Measuring Nitrate in Leachate During Greenhouse Production of *Philodendron* 'Green Princess': Comparing Analysis Methods and Evaluating Results

RORY CUSHEN[1], BRYCE CROSBY[2], MALCOLM M. MANNERS[3], ÖZLEM TUNCAY[4], SENTHIL B. GIRIMURUGAN[5], AND JOHN L. GRIFFIS, JR.[6]*

[1]*Florida Gulf Coast University, Integrated Studies Department, Fort Myers, FL*

[2]*Florida Southern College, Biology Department, Lakeland, FL*

[3]*Florida Southern College, Horticultural Sciences Department, Lakeland, FL*

[4]*Ege University, Horticulture Department, Bornova-Izmir ,Turkey*

[5]*Florida Gulf Coast University, Mathematics Department, Ft. Myers, FL*

[6]*Florida Gulf Coast University, Ecology and Environmental Studies Department, Ft. Myers, FL*

Keywords. Apogee MC-100, Hanna Nutrient Analysis Photometer, Horiba LAQUAtwin, Nix Pro Color Sensor, Opti-Sciences CCM-300

Nutrient runoff from excess fertilizer during commercial plant production may cause significant environmental damage while also posing a human health hazard. Determining effective and affordable methods of measuring nutrient leachate may help growers reduce nutrient waste while also reducing production costs. The main objective of this study was to compare the efficacy of several application rates of three different controlled release fertilizers during a six-month commercial production cycle for a *Philodendron* 'Green Princess' crop. Nitrogen is often considered to be the most important component of commercial fertilizers and we attempted to match the nitrogen sources and percentages in the fertilizers evaluated. Using the standard pour-through method, leachate was collected throughout the production cycle and evaluated for various nutrients. This portion of the project aimed to identify methods for measuring nitrate that were equally accurate and practical for growers to implement to reduce nitrate runoff. There were no visible symptoms of nitrogen deficiency on the *Philodendron* 'Green Princess' crop during the production cycle with any of the control release fertilizers tested. Numerous leachate samples contained substantial quantities of nitrate ions on some collection dates. Both destructive and non-destructive chlorophyll analyses were conducted on the crop and no chlorophyll deficiencies were discovered.

In the United States, Florida has been the leading producer of foliage plants since the 1960s (Chen et al., 2005). The most frequently produced tropical foliage plants can be found in a number of genera in the Araceae including *Aglaonema*, *Dieffenbachia*, *Epipremnum*, *Homalomena*, *Philodendron*, *Spathiphyllum*, and *Syngonium*. Numerous species of *Dracaena* and *Ficus* and several genera in the Arecaceae are also commonly produced by Florida growers (Chen et al., 2005). New varieties of foliage plants are released annually to the interior plant industry (Chen et al., 2002), but the fertilizer recommendations that are available to growers of foliage plants are usually general in nature or they were developed for different cultivars that were more popular some years ago (Conover & Poole, 1990).

Not surprisingly, fertilizer formulations have changed considerably over time and controlled release fertilizers (CRFs) of different types release various nutrients to plants differently under different growing situations (Incrocci et al., 2020, Merhaut et al., 2006, and Terlingen et al., 2016). Relatively recent CRF recommendations for the production of various foliage plants mention only a few cultivars and a few different types of CRFs; philodendrons were not mentioned (Stamps et al., 2008). For Florida growers, there are only general, non-specific fertilizer recommendations for commercial production of various philodendrons (McConnell et al., 2018). The main research objective of these experiments was to evaluate the efficacy of various rates of three different CRFs incorporated into a soilless medium (combined with two light levels in the greenhouses because of hurricane Ian damage to the greenhouse roofs) for producing a commercial crop of *Philodendron* 'Green Princess' based on both plant visual quality and fertilizer usage/loss by the end of a six-month production period. Additionally, the experiments were set up to train undergraduate students in both data collection and experimental design. Measuring nitrates in leachate is one of the more common methods used to measure and evaluate fertilizer loss during crop production (Stamps et al., 2008). The specific objectives of this portion of these experiments were: 1) to compare ease-of-use and accuracy of two different methods of measuring nitrates in leachate; and 2) to compare several non-destructive chlorophyll and color measurements taken from plants grown under various CRF treatments to determine if leaf chlorophyll content or leaf color was related to either rate of CRF applied or to nitrates in leachate.

*Corresponding author. Email: jgriffis@fgcu.edu

Materials and Methods

On September 7, 2022, at the Florida Southern College greenhouses (Lakeland, FL), 98 tissue-cultured liners of *Philodendron* 'Green Princess' (Oglesby Plants International, Altha, FL) were potted up into 15.24-cm diameter green plastic azalea pots (HC Companies, Twinsburg, OH) using Canadian peat:perlite:vermiculite (70:20:10 by volume) potting mix (Jolly Gardener Pro-line C/GP Mix, Oldcastle APG, Atlanta GA). There were seven different fertilizer treatments for these trials (as well as a biostimulant trial) and there were 12 replicates for each treatment for a total of 96 pots. For treatment #1 (which was used as the control for these trials), we added 2.4g/pot of a controlled release fertilizer (CRF) product, 20–6–9, 180-day formula Nursery Advantage (Diamond R Fertilizer, Winter Garden, FL) based on both existing fertilizer recommendations for philodendrons (Conover & Poole, 1990) and upon results from our recent trials with other varieties of philodendrons. (Griffis et al., 2022a; Griffis et al., 2022b). Treatment #2 had the same 2.4 g/pot of Diamond R Fertilizer, but the plants were treated at later dates with two biostimulant supplements (Haifa North America, Altamonte Springs, FL) For treatments #3, #4, and #5, we added either 2.0 g, 2.7 g, or 3.4 g/pot of another CRF product, 18–6–8 180 Total (Florikan, Sarasota, FL). This CRF had nitrate/ammoniacal nitrogen ratios similar to the Diamond R CRF product, but the total quantity of nitrogen in this CRF was less, about 90%, so treatment #4 (2.7 g/pot) would have about the same amount of nitrogen per pot as treatment #1; treatments #3 and #5 would have 25% less or 25% more nitrogen per pot than treatment #1 respectively. For treatments #6, #7, and #8, we added either 2.4 g, 3.2 g, or 4.0 g/pot of another CRF product, 15–7–15 + 2MgO (8) Haifa Multicote (Haifa North America, Altamonte, Springs, FL). This CRF had nitrate/ammoniacal nitrogen ratios similar to the Diamond R CRF product, but the total quantity of nitrogen in this CRF was less, about 75%, so treatment #7 (3.2 g/pot) would have about the same amount of nitrogen per pot as treatment #1; treatments #6 and #8 would have 25% less or 25% more nitrogen per pot than treatment #1 respectively. We also supplemented the soilless growing medium in each pot with 14g dolomite and a soluble trace elements mix (S.T.E.M.—Peters Professional S.T.E.M. G99900, ICL Fertilizers, Dublin, OH) drench. Pots were half-filled with potting mix only, moistened slightly with tap water, then the upper half of the pots were filled with potting mix, fertilizer and dolomite and stirred a bit before moistening again. In six of the pots, HOBO MX2202 Bluetooth dataloggers (Onset Computer Corp., Bourne, MA) were switched on and placed at the interface of the 2 potting medium layers before the upper half of the pot was filled, so that they would be about 7 cm deep in the filled pots. After the pots were filled and tagged for each treatment, they were moved to the greenhouse where they were placed on the bench in a pre-determined complete randomized block design with four blocks each containing 3 replicates of each of the 8 treatments. Pots were evenly moistened and a dibble was used to make a hole for the liners. The liners were inserted into the potting mix and all pots were then saturated with tap water. Afterwards, 50 mL of STEM solution (5.6 g S.T.E.M./3.78 L water) was poured into each pot as a one-time, corrective application.

Unfortunately, on September 28, 2024, Hurricane Ian struck Florida's Gulf Coast, causing significant damage to many areas. The roof of the greenhouses at Florida Southern College was badly damaged, with whole sections of the roof torn off by wind. Surprisingly, none of the plants were toppled over, but they were soon exposed to direct sunlight. Since most foliage plants are grown under shade conditions and many foliage plants may be damaged by direct sunlight, the experiment was covered with two shade structures previously used in earlier experiments (Griffis et al., 2022a). This resulted in a modification of the trials to a split plot design (30% shade and 60% shade) (Envirotech Greenhouse Solutions, Richmond CA) with two randomized blocks within each plot with 3 replicates per treatment per block and 16 total treatments. Initially the pots were hand watered as needed. Since hand watering can easily lead to unequal amounts of water being applied to individual pots (and uneven watering can become an unintended variable in an experiment), an emitter system (Maxijet, Winter Haven, FL) was inserted in the pots when the shade cloth system was installed, so that all pots could be watered simultaneously for the same length of time, with approximately the same amount of water applied to each pot. Several emitters were first placed in plastic beakers and the irrigation system was run for one minute several times. On average, the emitters dispensed 420 mL ± 5% per minute.

Non-destructive chlorophyll measurements were taken throughout the growing period for this crop, starting about one month after the initial planting date. Leaf chlorophyll was measured using two different chlorophyll content meters, an MC-100 (Apogee Instruments, Logan, UT) that works by transmitting light through the leaf to a sensor and a CCM–300 (Opti-Sciences, Hudson, NH) that works by reading light reflected from the leaf to a sensor. For the measurements, the newest fully expanded leaf (determined by feel, appearance, and position on the plant) on each plant was selected and marked. For each chlorophyll meter, five non-destructive chlorophyll measurements were taken from the selected leaf between visible leaf veins. All readings were taken within the same marked area on the chosen leaf to ensure consistency in the measurements and the five chlorophyll measurements were averaged by the chlorophyll meter software (Sweeney et al., 2019). Monthly, for six months, the newest fully expanded leaf was selected on each plant and these non-destructive chlorophyll measurements were taken. Additionally, leaf color was determined, also on a monthly basis using a Nix Pro Color Sensor (Nix Sensor Ltd, Hamilton, ON, Canada). For the Nix Pro Sensor, five measurements were taken and the measurements were averaged by the sensor software. Chlorophyll measurements and leaf color measurements were evaluated to determine if a relationship could be determined between the measurements by Pearson's Correlation test (Tuncay, 2011). We also attempted to determine if either chlorophyll or color measurements were related to the different fertilizers and fertilizer rates applied to the crop or to the different light levels of the growing environment. Other crops as different as parsley and rocket (Tuncay, 2011), garden cress (Tuncay et al., 2011), maize (Rhezali & Lahlali, 2017; Xu et al., 2010), and apples (Ye et al., 2020) have also been evaluated to see if a relationship between chlorophyll development and leaf color could be correlated to crop nutrition. A "student buyer" survey of the crop was conducted at the end of February 2023 to determine if prospective buyers preferred plants grown under any of the 16 treatment regimens (Crosby et al., 2023). For all the statistical analyses, IBM SPSS Statistics for Windows, Version 25.0. (IBM Corp., Armonk, NY) was used.

To determine the release rates of the various fertilizers, leachate samples were collected approximately every four weeks (October 16, November 15 and December 12, 2022, and January 21 and February 17, 2023) throughout the six-month growing period beginning leachate collection 39 days after the potting-up date. Leachate samples were collected using the pour-through method (Wright, 1986; Blythe & Merhaut, 2007; Torres et al., 2010), where all plants were irrigated for several minutes until saturation occurred and water was flowing out from the bottom of the pots. After 30 minutes to allow pots to drain, plastic saucers (20.3 cm) were placed under all pots and 50 mL of tap water was poured into each pot. These "pour-through" samples (about 50 mL) were collected in the saucers, and along with four samples of greenhouse tap water, were poured into 50 mL conical centrifuge tubes. These samples were transported to Florida Gulf Coast University and were stored at –20 °C until analyzed. Fertilizers were reapplied to all pots on December 12, 2022 at the same rates as they were initially applied.

Electrical conductivity (EC) and pH measurements of leachate were determined using a Hanna HI9814 GroLine pH/EC/TDS/temperature meter (Hanna Instruments, Woonsocket, RI). For this portion of the experiment, both nitrate and nitrate-nitrogen content of the leachate were first determined using a Horiba LAQUAtwin NO3-11 nitrate ion meter (Kyoto, Japan). Since no reagents were required and sample aliquot size was only 0.5 mL, all measurements with the Horiba meter were conducted in triplicate to check for variation, and averaged. No dilutions of the original samples were required for measurements using this Horiba nitrate meter. For comparison, nitrate and nitrate-nitrogen were also determined using the Hanna HI83325 Nutrient Analysis photometer (Smithfield, RI) following the nitrate test procedures (an adaption of the cadmium reduction method) included with this device. Since there were 100 samples for analysis every four weeks and the Hanna analysis system required larger sample aliquot volumes and additional reagents and considerable time, these Hanna nitrate and nitrate-nitrogen content measurements were not replicated. For the Hanna meter measurements, if the Horiba meter indicated that the nitrate content in the samples was outside the Hanna meter measurement range, the sample aliquots were diluted 1:10 or 1:20. The nitrate levels (according to the Horiba meter) were sometimes quite high and out of range for the Hanna devices without a sample dilution. Various other chemicals in the leachate may have interfered with the readings from either meter.

Results and Discussion

The amounts of nitrate-nitrogen lost in the leachate of the various fertilizer treatments were fairly high early on in the trials but they declined quickly each month (as did other nutrients), such that the same fertilizer treatments had to be added again to the pots about 90 days after the initial fertilizer application. All fertilizers in the trials were rated for 180 days or longer, but under our growing conditions the fertilizers did not last nearly that long. Nitrate-nitrogen was not the limiting nutrient with any of the CRFs trialed; both the potassium and phosphate levels declined in the leachates even more rapidly than the nitrate-nitrogen levels (data not shown).

The Horiba nitrate meter nitrate-nitrogen readings for all leachates tended to be higher than the readings from the Hanna nutrient analysis photometer, indicating the possibility of some level of interference with the nitrate-nitrogen ion measurements by some other component within the leachate. The Horiba meter does has a ± 10% margin for error, but so does the Hanna photometer for the nitrate-nitrogen test method. In any case, the readings from both meters show strong positive correlations to each other (Table 1). We did not discover any significant differences among treatments with regard to chlorophyll production, leaf color, or any of the growth parameters we measured. Correlation tests between nitrate concentrations in treatment leachates and different color parameters (*L, *a, *b) of leaves were performed. Test results showed that all the color parameters correlated significantly with nitrate concentration. The buyer survey did not discover any significant differences in marketability of plants produced with any of the fertilizer treatments or with either of the light level treatments (Figs. 1, 2, and 3) (Crosby, et al., 2023).

Conclusions and Summary

We did not find any significant differences among any of the fertilizer treatments with regards to the quality or marketability of the *Philodendron* 'Green Princess' plants. Therefore we can recommend that this crop grown under similar Central Florida growing conditions should be produced using the lowest rate of each CRF fertilizer trialed, with reapplication of the fertilizer

Table 1. Pearson Correlation test comparing nitrate-nitrogen measurements (all dates combined) from Horiba nitrogen ion meter (Horiba) and Hanna nutrient analysis photometer (Hanna) with non-destructive chlorophyll measurements of *Philodendron* 'Green Princess' taken with the Opti-Sciences CCM-300 (Chl ref) and with the Apogee MC-100 chlorophyll content meter set for either generic measurements (Generic) or for SPAD measurements (SPAD). Sig. = significance level.

Correlations		Horiba	Hanna	Chl ref	Generic	SPAD
Horiba	Pearson correlation	1.0000	0.779354	0.267825	–0.09607	0.259502
	Sig. (2-tailed)		0.0000	0.0000	0.0356	0.0000
Hanna	Pearson correlation	0.779354	1.0000	0.135683	0.135683	0.135683
	Sig. (2-tailed)	0.0000		0.0029	0.9803	0.0002
Chl ref	Pearson correlation	0.267825	0.135683	1.0000	0.135683	0.70129
	Sig. (2-tailed)	0.0000	0.0029		0.0000	0.0000
Generic	Pearson correlation	–0.0961	0.0011	–0.5336	1.0000	–0.3772
	Sig. (2-tailed)	0.0356	0.9803	0.0000		0.0000
SPAD	Pearson correlation	0.2595	0.1692	0.7013	–0.37716	1.0000
	Sig. (2-tailed)	0.0000	0.0002	0.0000	0.0000	

Fig. 1. Philodendron 'Green Princess' grown under lower light conditions (60% shade), 5 ½ months after planting tissue culture plugs.

Fig. 2. Philodendron 'Green Princess' grown under higher light conditions (30% shade), 5 ½ months after planting tissue culture plugs.

after 90 days. This should reduce the amount of nitrate (and other nutrients in the CRF) lost from the pots without reducing the quality of the plants or extending the time needed to produce the crop. The higher light level of the two tested in these trials might produce a marketable crop one to two weeks sooner than the lower light level.

We can recommend either the Horiba LAQUAtwin nitrate meter or the Hanna Nutrient Analysis Photometer for measuring nitrates and/or nitrate-nitrogen in leachate based on accuracy, but the Horiba LAQUAtwin is considerably less expensive to use and is much quicker to provide measurements in a greenhouse situation.

We can recommend either the Opti-Sciences CCM-300 or the Apogee MC-100 for measuring leaf chlorophyll content. However, the CCM-300 has the advantage when plants have thick leaves, since the CCM-300 can read reflectance from a leaf when the MC-100 light source cannot penetrate through the leaf to allow the meter to determine light absorbance. Even the newer leaves of *Philodendron* 'Green Princess' can be thick and sometimes several extra readings had to be taken with the MC-100 to get a measurement. The software provided with

Fig. 3. Philodendron 'Green Princess' set up with student buyer survey in progress, 6 months after planting tissue culture plugs. Plants grown under either higher or lower light levels were distributed equally among the plants for evaluation.

the chlorophyll meters is adequate and the data can be easily transferred to a spreadsheet. We can also recommend the Nix Pro Color Sensor for determining color measurements of plant leaves. The device is light weight, easy to use, and the software (app) provides a large amount of information that can be used in color analysis.

Literature Cited

Blythe E.K. & Merhaut D.J. 2007. Testing the assumption of normality for pH and electrical conductivity of substrate extract obtained using the pour-through method. *HortScience* 42(3):661–669. https://doi.org/10.21273/HORTSCI.42.3.661

Chen J., Henny R.J., & McConnell D.B. 2002. Development of new foliage plant cultivars. p. 466-472. In: J. Janick & Whipkey, A. (eds.) Trends in new crops and new uses. ASHS Press Alexandria, VA.

Chen J., McConnell D.B, Henny, R.J., & Norman D.J. 2005. The foliage plant industry. *Hortic. Rev.*, 31:45–110. https://doi.org/10.1002/9780470650882.ch2

Conover C.A. & Poole R.T. 1990. Light and fertilizer recommendations for production of acclimatized potted foliage plants. CFREC-A Research Report RH-909-1. University of Florida, IFAS, Mid Florida Research and Education Center, Apopka, FL. https://mrec.ifas.ufl.edu/foliage/resrpts/rh_90_1.htm

Crosby B., Manners M.M., Griffis Jr. J.L., & Serrano S.A. 2023. Performance of *Philodendron* 'Green Princess' grown under various light levels and fertilizer regimes. *Proc. Fla. State Hort. Soc.* 136:187–190.

Griffis Jr J.L., Manners, & Barnes Q. 2022a. Controlled release fertilizer and soilless medium temperatures significantly interact during greenhouse production of *Philodendron* 'Imperial Green'. *Acta Hortic*. 1334:289–296.

Griffis Jr. J.L., Manners M.M., Bennett K., & Tuncay Ö. 2022b. Unusual variegation pattern in *Philodendron* 'Birkin' related to leaf chlorophyll content and leaf color. *Proc. Fla. State Hort. Soc*. 135:187–190.

Incrocci L., Maggini R., Cei T., Carmassi G., Botrini L., Filippi F., Clemens R., Terrones C., & Pardossi A. 2020. Innovative controlled-release polyurethane-coated urea could reduce N leaching in tomato crop in comparison to conventional and stabilized fertilizers. *Agronomy* 10(11):1827. https://doi.org/10.3390/agronomy10111827

McConnell D.B., Chen J., Henny R.J., & Everitt K.C. 2018. Cultural guidelines for commercial production of interiorscape *Philodendron*. Extension Publication ENH893. University of Florida, Gainesville, FL. https://doi.org/10.32473/edis-ep150-2003

Merhaut D.J., Blythe E.K., Newman J.P., & Albano J.P. 2006. Nutrient release from controlled-release fert,ilizers in acid substrate in a greenhouse environment: I. Leachate electrical conductivity, pH, and nitrogen, phosphorus, and potassium concentrations. *HortScience* 41(3):780–787. https://doi.org/10.21273/HORTSCI.41.3.780

Rhezali A. & Lahlali R. 2017. Nitrogen (N) mineral nutrition and imaging sensors for determining N status and requirements of maize. *Journal of Imaging* 3(4):51.

Stamps R.H., Chen J., Natarajan S., & Popenoe J. 2008. Evaluations of new controlled-release fertilizers for use in production of interiorscape and landscape plants. *Proc Fla. State Hort. Soc*. 121:346–349. https://journals.flvc.org/fshs/article/view/87437

Sweeney K.J., Jones M., Cruz A., Cureton D., Manners, M.M., & Griffis Jr. J.L. 2019. Relationships between leaf chlorophyll content, fertilizer application rates and visual grades of greenhouse-grown Homalomena 'Emerald Gem'. *Proc. Fla. State Hort. Soc*. 132:205–209.

Terlingen J.G.A., Radersma S., Out G.J.J., Hernández-Martínez J., & Raemakers-Franken, P.C. 2016. Current developments in controlled release fertilisers. In: Proceedings International Fertiliser Society (No. 781:1–24). International Fertiliser Society, Colchester, U.K.

Torres A.P., Mickelbart M.V., & Lopez R.G. 2010. Leachate volume effects on pH and electrical conductivity measurements in containers obtained using the pour-through method. *HortTechnology* 20(3):608–611. https://doi.org/10.21273/HORTTECH.20.3.608

Tuncay O. 2011. Relationships between nitrate, chlorophyll and chromaticity values in rocket salad and parsley. *African Journal of Biotechnology* 10(75):17152–17159.

Tuncay O., Esivok D., Yagmur B., & Okur B. 2011. Yield and quality of garden cress affected by different nitrogen sources and growing period. *Africa J. Agricultural Research* 6(3):608–617.

Wright R.D. 1986. The pour-through nutrient extraction procedure. *HortScience*. 21(2):227–229. https://doi.org/10.21273/HORTSCI.21.2.227

Xu Y., Wang X., Sun H., & Wang H. 2010. Study of monitoring maize leaf nutrition based on image processing and spectral analysis. p. 465–468. In: 2010 World Automation Congress. IEEE.

Ye X., Abe S., Zhang S., & Yoshimura H. 2020. Rapid and non-destructive assessment of nutritional status in apple trees using a new smartphone-based wireless crop scanner system. *Computers and Electronics in Agriculture* 173:105417.

Proc. Fla. State Hort. Soc. 136:185. 2023.

—Scientific Note—

Promoting Home Gardens and Healthier Eating through Garden to Table Workshops

ASHLEY STONECIPHER* AND TERRI KEITH

University of Florida, IFAS, Jackson County Extension, Marianna, FL

There is a high interest among our residential community in producing sustainable herb and vegetable gardens and connecting the harvest directly to healthier home cooked meals. There is an economically motivated desire to grow food at home to achieve cost savings. We developed a series of workshops with the objectives to educate and encourage home gardening, cooking with herbs and vegetables, develop an understanding of health benefits associated with eating fresh produce and safe food preservation.

Each garden to table workshop was 2–3 hours long and was designed and delivered with informational PowerPoint presentations and food tasting. Additionally, participants took home seeds, planting trays, and small transplants for their future garden and containers.

Materials developed and or used in these workshops included various extension publications with best management practices, seed and transplant information, a PowerPoint that used references from the University of Florida, other land grant universities, the National Center for Home Food Preservation and the US Department of Agriculture. Factsheets were created to use at home on incorporating herbs and fruits into everyday cooking. At each workshop the participants were provided with all the copies of the materials used and seeds and/or plants to encourage them to garden at home.

The program was marketed with flyers that were sent to extension advisory boards, volunteer groups and posted at local stores. Facebook posts were created on the pages for Family and Consumer Sciences, Master Gardener Program, the County Extension and Gardening in the Florida Panhandle and shared with additional groups. Press releases were sent to 10 local news and radio outlets and a community newsletter that reached 21,000 people. An additional promotion effort used email blasts sent via the University of Florida, IFAS Subscription Management Systems to a multi-county mailing list of 1668 people. These promotional efforts resulted in 242 registrants who attended the workshops in the series from 2021–23.

To evaluate each Garden to Table workshop, a paper survey was provided to each participant at the end of the class. We used Qualtrics to enter our survey data and summarize the impact of all the classes that were conducted. Evaluations from the Garden to Table Series provided these following results:

- 99% of participants responded that they increased their knowledge or gained useful ideas at one of the workshops. Examples include food storage, planting depth, new recipes, requirements for growing fruit and how to use herbs in cooking.

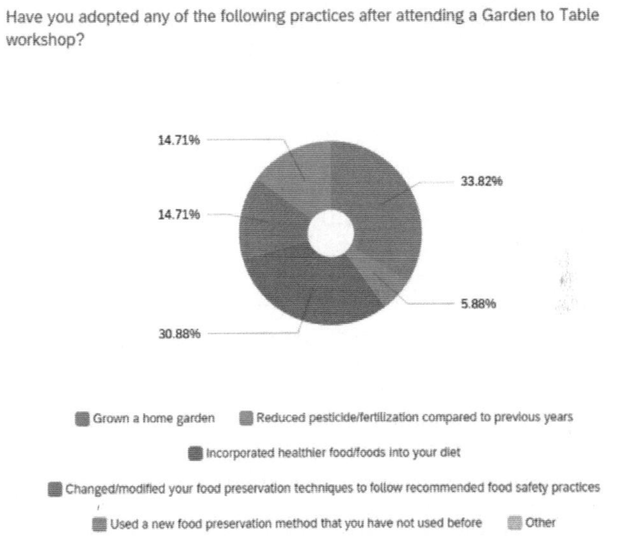

Fig. 1. Garden to Table Workshop participant's reported practices after completing the workshop.

- 98% of the attendees intended to use information from the workshops, including using herbs when cooking to make healthier meals (34%), modifying at least one recipe by using fruits to make it healthier (25%) and to grow one or more vegetables they have not grown before (23%).

A follow-up survey has been conducted after each workshop to determine actual behavior change (Fig. 1).

By combining gardening and cooking, residents are more motivated to start growing herbs and vegetables at home. This is a start to reach the community with information on how to grow a garden and how to use its harvest products in the kitchen. Demand for information about residential small-scale vegetable production techniques continues to grow. This program was well received, and we will continue to reach out to more residents in Jackson County. Programs such as this Garden to Table series enable and encourage gardeners to utilize their garden space and grow their own vegetables to use in their home cooking. In Summer 2022, we conducted a pilot Garden to Table one-day children's camp for ages 8–12. In the future, this series could be adapted into a more comprehensive youth camp in partnership with the 4-H agent.

*Corresponding author. Email: ams2904@ufl.edu

Ornamental, Gardening, & Landscaping Section

Proc. Fla. State Hort. Soc. 136:186. 2023.

—Scientific Note—

Encouraging Homeowner Citrus

DAVID A. AUSTIN*

University of Florida, IFAS, Highlands County Extension, Sebring, FL

Keywords. Asian citrus psyllid, homeowner citrus, Huanglongbing

Florida's citrus industry has been devastated by a bacterial disease called Huanglongbing (HLB) (*Candidatus* Liberibacter asiaticus). The disease is vectored by an insect, the Asian citrus psyllid (*Diaphorina citri*). The bacterium lives within the phloem of the citrus and hinders carbohydrate movement downward through the tree. Presently, many dooryard citrus trees are infected with HLB and will succumb to or already have succumbed to the disease. Their owners have spent money to care for their trees while experiencing small to no returns on their investment. Many homeowners, not educated on this disease, have become disenchanted with growing citrus due to the lack of success they have experienced. Although no citrus is resistant to HLB, there are tolerant varieties.

The Highlands County Extension agent has been encouraging residents to grow citrus in their yards by offering classes on the proper use of cultural practices to successfully maintain a tree that would supply quality fruit for a longer duration. These classes introduce participants to varieties of HLB-tolerant citrus, such as Sugar Belles®, and SunDragons, that can be more easily grown and maintained. The classes are conducted using a two-hour PowerPoint presentation on growing citrus. The presentation teaches participants how to identify HLB and other common citrus pests and diseases, gives them a better understanding of citrus nutrition, soil health, and irrigation practices and introduces them to individual plant covers (IPC).

The two citrus varieties described are made available for the participants to purchase on the day of the class. Administering pretests and posttests to participants has shown a knowledge gain of 50% based on pre- and post-assessments on the day of the class. Eighty percent of participants reported success growing their citrus trees. Eighty percent also reported keeping the IPC on their trees at the time of the survey. These last two objectives will be ascertained with a survey sent 9 months after the class. Another follow-up survey will be sent two years after the class to check the progress of the purchased trees.

Eighty-two participants attended the Growing Citrus at Home program. During this program, 92 trees were purchased and taken home, each with an IPC. Seventy-five participants took a pre- and post-assessment to ascertain knowledge gain. Knowledge gain, based on a 10-question pre-and post-assessment, went from 35.5% to 67.9% which is an increase of 91.2 % (Table 1). This was 41.2% higher than the initial objective. Twenty participants responded to the post-class survey. One hundred percent of the respondents had left the IPC on the tree which had been emphasized during the class as a best management practice. Based on the early success of the program, it will be continued again for a second year and annual data can be obtained. Early information on Sugar Belle® and SunDragon citrus are showing much promise as homeowner citrus trees.

Table 1. Questions used to determine knowledge gained in the Growing Citrus at Home program administered at the Highlands County Extension, Sebring, FL. Calculating knowledge gain for this class was based on the following formula: Post-assessment score—Preassessment score = X; X/preassessment score = percent knowledge gain.

Pre- and post-assessment questions (N = 75)	Pre- % correct	Post-% correct	% change
Question 1: A major symptom of citrus canker is uneven discoloration of the citrus leaf known as blotchy mottle.	15.8%	22.2%	40.7%
Question 2: What specific vector moves citrus greening to and around a citrus tree?	51.3%	84.7%	65.1%
Question 3: Citrus greening is likely to infect nearly what percent of homeowner trees?	34.2%	98.6%	188.2%
Question 4: An individual protective cover protects trees from all damaging insects by preventing them from getting to the tree.	35.5%	55.6%	56.4%
Question 5: Citrus canker is currently the most damaging disease for commercial citrus.	36.8%	68.1%	84.7%
Question 6: A general yellowing of older leaves with lighter-colored veins on citrus is what type of deficiency?	31.6%	90.3%	185.9%
Question 7: There are no citrus trees that are resistant to HLB.	51.3%	75.0%	46.2%
Question 8: Micronutrient nutrient deficiencies are more likely to show up in what type of leaf growth?	26.3%	56.9%	116.4%
Question 9: Deformed new leaf growth on a citrus tree accompanied by brown to greying tunneling appearance is probably caused by what organism?	47.4%	68.1%	43.7%
Question 10: A thick black film or layer on plant leaves or fruit that can be removed by rubbing your fingers over it is an indicator of what type of problem?	25.0%	59.7%	138.9%
Knowledge gain results.	35.5%	67.9%	91.2%

*Corresponding author. Email: davidaustin@ufl.edu

Proc. Fla. State Hort. Soc. 136:187. 2023.

Serving Up Bite-sized Information from Plant to Plate

Cynthia Nazario-Leary*, Martha Maddox, and Kevin Korus

University of Florida, IFAS, Alachua County Extension, Gainesville, FL

—abstract—

A collaborative effort between Alachua County and the University of Florida, IFAS Extension created the Plant of the Month (POTM) program to provide residents with easy-to-understand information on how and when to grow edible plants. Each month, information on how to plant, care for, harvest, and prepare a selected edible plant is curated by the Environmental Horticulture Extension agent, Master Gardener volunteers, and the Family & Consumer Sciences Extension agent. Plants are selected based on their ability to grow successfully in the area and their popularity with home gardeners. Plants are matched to a month based on when the plant should be planted or harvested. The Plant of the Month program is shared via public service announcements (PSA), posted on a POTM website, Master Gardener social media sites, and complimented monthly on the Alachua County Extension podcast. Podcasts are a great opportunity to extend the educational outreach of the POTM program while building a connection with public audiences. The podcast is distributed and stored on Spotify® for Podcasters and can be listened to on various podcast platforms. The POTM podcast began airing in January 2022 and currently has 15 episodes with a total of 1048 plays. Ninety-five percent of our listeners are from within the United States, but the podcast has been downloaded in 24 other countries. The audience is comprised mostly of 45- to 59-year-olds (62%), followed by 35- to 44-year-olds (26%) with male and female listeners make up 59.6% and 40.4% of the audience, respectively.

*Corresponding author. Email: cnazarioleary@ufl.edu

Proc. Fla. State Hort. Soc. 136:188–189. 2023.

—Scientific Note—

Designing Climate-smart Extension for the Commercial Horticulture Industry in Florida

HENRY MAYER[1]*, YVETTE GOODIEL[2], DON RAINEY[3], MIKE ORFANEDES[4], HAMUTAHL COHEN[5], RYAN KLEIN[6], AND JOHN ROBERTS[7]

[1]*University of Florida, IFAS, Miami-Dade County Extension, Homestead, FL*

[2]*University of Florida, IFAS, Martin County Extension, Stuart, FL*

[3]*University of Florida, IFAS, Regional Specialized Agent, Water Resources, Plant City, FL*

[4]*University of Florida, IFAS, Broward County Extension, Davie, FL*

[5]*University of Florida, IFAS, Collier County Extension, Naples FL*

[6]*University of Florida, IFAS, Environmental Horticulture Department, Gainesville, FL*

[7]*University of Florida, IFAS, Palm Beach County Extension, West Palm Beach, FL*

Keywords. climate change, green industry

Urban areas currently account for approximately 70% of global fossil-fuel carbon dioxide (CO_2) emissions (Gu et al., 2015). Lawns dominate urban landscape in the U.S. and require intensive management, including frequent mowing, fertilization and irrigation which may influence CO_2 emissions from both biogenic and anthropogenic sources (Lerman & Contosta, 2018).Climate change refers to long-term shifts in weather patterns, rising sea levels and increased global temperatures. These changes have far-reaching environmental consequences, including impacts for the landscaping industry. Extreme weather events, shifting growing seasons, accelerated coastal erosion and increased pest and disease pressures are some of the challenges green industry personnel now face.

Furthermore, climate change affects the function and suitability of plant species for specific regions. Plant hardiness zones which provide guidance on adapted species ranges, are beginning to shift due to changing temperatures. This means that landscapers will need to manage different plant pallets as the impacts of climate change become more apparent. Water management is another crucial aspect influenced by climate change. Droughts and water scarcity are becoming more common in many regions, necessitating more efficient irrigation practices. Landscapers should know about water-conserving techniques, such as smart irrigation systems, rainwater harvesting and drought-resistant and salt-tolerant plant choices. By implementing these strategies, they can also help facilitate a carbon-capture landscape, while minimizing water usage. Warmer temperatures and altered rainfall patterns may create favorable conditions for pests and pathogens to thrive. Adhering to environmentally friendly solutions is vital for maintaining a healthy landscape without relying heavily on chemical pesticides.

University of Florida, IFAS faculty who work with the green industry in Florida want to understand how the industry perceives the effects of climate change. Extension agents are commonly viewed as trusted messengers because they reflect the values of the stakeholders they serve. As a result, they can organize discussions and facilitate conversations around a variety of contentious topics, including climate change (Monroe et al., 2018). However, in any given audience they are likely to encounter people from a variety of backgrounds who have a variety of attitudes about climate change.

In Summer 2022, we launched a needs assessment survey through Qualtrics to determine how green industry professionals perceive climate change (CC). The survey was sent via email, and direct contact through focus groups to people who work in landscape maintenance, tree care professionals, nursery operations, retail garden center, urban foresters, environmental consulting companies, pesticide applicators, fertilizer applicators and landscape architects. Of respondents, 84% were white, 39% had bachelor's degrees, and 42% had a license issued by the Florida Department of Agriculture and Consumer Services (FDACS).

The survey was completed by 615 people. Of the respondents, 85% stated that they have been working in the green industry (GI) >10 years, 84% have a management/supervisory role, 38% market their company/service as climate friendly, 12% of the people receive information from extension, and 13% from government agencies such as National Aeronautics and Space Agency (NASA) or the National Oceanic and Atmospheric Administration (NOAA). Additionally, 83% of respondents believe that human activities are the primary cause of climate change, 65% think that the green industry should be doing more to address the issue, 67% believe that climate change is relevant to their business, 69% are concerned about climate change, 75% want to learn more about climate change, 57% consider the carbon footprint of their equipment, and 60% believe their region will see increased incidence of extreme weather events such as hurricanes.

When asked about what they feel is the biggest obstacle to implementing climate-friendly practices, 42% mention cost of

*Corresponding author. Email: hmayer@ufl.edu

tools and equipment, 23% lack of knowledge about climate-friendly practices, and 17% lack of client interest and demand.

The results of the survey are encouraging. Thus, we are proposing state-wide extension programming around climate change, mitigation, and resiliency for the commercial horticulture industry. We believe it is important to provide professionals with both climate change education and tools to promote their livelihoods, and therefore we plan to develop a climate-change business certification program with online resources, training modules, and in-person demonstrations in English and Spanish. We are seeking funding for the development of a website with a carbon calculator that will act as a hub for the project.

Literature Cited

Gu C., Crane J., Hornberger G., & Carrico A. 2015. The effects of household management practices on the global warming potential of urban lawns. *Science Direct*. 151:233–242. https://www.sciencedirect.com/science/article/abs/pii/S0301479715000092

Lerman S. & Contosta A. 2018. Lawn mowing frequency and its effects on biogenic and anthropogenic carbon dioxide emissions. *Landscape and Urban Planning*. 182:114–123. https://www.sciencedirect.com/science/article/abs/pii/S0169204618304080?via%3Dihub

Monroe M., Megalos M., & Needham C. 2018. Strategies for communicating climate change to extension clientele. Extension Publication FOR333/FR402, University of Florida, Gainesville, FL. https://edis.ifas.ufl.edu/publication/FR402

Proc. Fla. State Hort. Soc. 136:190–193. 2023.

Performance of *Philodendron* 'Green Princess' Grown Under Various Light Levels and Fertilizer Regimes

Bryce Crosby[1]*, Malcolm M. Manners[2], John L Griffis, Jr.[3], and Susan A. Serrano[4]

[1]*Biology Department, Florida Southern College, Lakeland, FL*

[2]*Horticultural Science Department, Florida Southern College, Lakeland, FL*

[3]*Ecology and Environmental Studies Department, Florida Gulf Coast University, Ft. Myers, FL*

[4]*Mathematics Department, Florida Southern College, Lakeland, FL*

Keywords. Controlled release fertilizer, daily light integral, tropical foliage

***Philodendron* 'Green Princess' is a new hybrid variety with considerable commercial potential. There are no specifically published protocols for the production of a high-quality, marketable crop. Plants were grown in a Central Florida greenhouse at two light levels, with varying application rates of three commercially available controlled-release fertilizers. Soil *p*H and electrical conductivity were regularly measured throughout the production period. At the end of the 24-week production period, plant height, width, and number of shoots were measured. Florida Southern College students served as a "consumer" group, visually rating the plants for quality on a Likert scale, choosing the "best" and "worst" plants in each evaluation block—explaining why the plants were so rated, and indicating whether they would purchase each plant for a price of $9.99. Measurement and survey data were analyzed to determine if either production light levels or fertilizer application rates correlated with physical plant characteristics or qualities preferred by consumers. Student consumers preferred "larger/fuller" plants. Plants with greater numbers of visible shoots, that were also wider and taller, were rated as being higher quality. Such plants were more likely to be purchased for $9.99 and were consistently voted "best" plant per block. Further analysis revealed that none of the fertilizer rates or formulations were related to the superior plants, but there were significant correlations between crop production light level for both higher survey rankings and larger plant measurements. *Philodendron* 'Green Princess' can be produced satisfactorily under higher light levels using any of the fertilizers we evaluated.**

Growing plants with high consumer value and low input is beneficial to the commercial grower. Philodendrons have long been popular tropical foliage plants with historically high consumer value and low nutritional requirements. With the creation of new hybrid varieties every year there exists a lack of data to guarantee proper fertilizer treatment and light-level exposure to obtain the highest quality crop. *Philodendron* 'Green Princess' is a new self-heading (Henley et al., 1991) hybrid variety, with high potential commercial value. Due to 'Green Princess' being a new variety, there are no published protocols for ideal fertilizer amounts, release methods, and light levels, to produce a high-quality marketable crop most efficiently, with minimal inputs. Previous research (Tew et al., 2021) on Philodendron 'Birkin', another self-heading hybrid variety, found that none of the three rates of Diamond R Nursery Advantage 20-6-9, or light levels was significant in producing a "higher quality plant." Quality was determined through physical plant measurements and survey analysis by a "consumer" group. This research models previous research (Tew et al., 2021), while also testing two different commercially available control release fertilizers at nitrogen levels lower than, the same as, and higher than the lowest rate previously used (Tew et al., 2021).

Thank you to Florida Southern College and Florida Gulf Coast University for the joint effort in this research.

*Corresponding author. Email: Brycecrosby5@gmail.com

Materials and Methods

Experimental Design. Previous research (Tew et al., 2021) discovered the lowest concentration tested of Diamond R Nursery Advantage 180 20–6–9 produced *Philodendron* 'Birkin' plants of the same quality as the higher concentrations tested. Therefore, the low concentration, 2.4 g/pot of Diamond R Nursery Advantage 180 20–6–9 (release over 180 days) was used as the control group for this research regarding *Philodendron* 'Green Princess'. The two commercially available fertilizers tested were Florikan 18–6–8 Total (release over 180 days) and Haifa Multicote (8) 15–7–15 + 2MgO (release over 8 months). Ninety-six tissue culture plugs of *Philodendron* 'Green Princess' (Oglesby Plants International, Altha, Fla.) were potted in 6" azalea pots (HC Companies, Twinsburg, OH) with Jolly Gardener potting mix (60% Canadian peat, 40% perlite) on September 19, 2022. Eight replication groups of twelve plants each were formed to separate fertilizer treatments. The eight treatments were as follows: 2.4 g/pot of Diamond R Nursery Advantage 180 20–6–9 (control), 2.4 g/pot of Diamond R Nursery Advantage 180 20–6–9 with HaifaStim™ Wall-Up S and Vital (biostimulants), 75% of control nitrogen concentration for both Haifa Multicote (8) 15–7–15 + 2MgO and Florikan 18–6–8 Total, 100% of control nitrogen concentration for both Haifa Multicote (8) 15–7–15 + 2MgO and Florikan 18–6–8 Total, 125% of control nitrogen concentration for both Haifa Multicote (8) 15–7–15 + 2MgO and Florikan 18-6-8 Total. Within each 12-plant replicate group, 6 plants were placed at the high light level and 6 plants were placed at the low light level, placement orientation within each light level was randomized. Light levels were created by erecting shade boxes covering half the crop in a 30% shade cloth and the other half in a 60% shade cloth. Plants were then grown in a climate-controlled greenhouse over a 24-week production period being watered once a week with 300 mL of central Florida tap water.

Measurements. Light intensity in each level was measured by HOBO® Pendant® MX Temperature/Light Data Loggers (Onset Computer Corp., Bourne, MA) to calculate the average Daily Light Integral (DLI) (Torres & Lopez, 2010). Hurricane Ian damaged the roof tiles of Florida Southern College's greenhouse, forcing the crop to be relocated to a different greenhouse, where the plants sat without shade boxes (same DLI) from December 20, 2022 to January 13, 2023 (*ca*.24 days). Therefore, DLIs were calculated for the original location and new location following the hurricane (the period where all plants sat in the same DLI was removed from the calculation).

Leachate was collected monthly throughout the production period to monitor electrical conductivity and pH to provide insight on the amounts of available fertilizer. Electrical conductivity was corrected for temperature by the standard equation:

$$EC_t = EC_{25}\,[1 + a(t - 25)].$$

This equation is a linear model but has proven to be accurate within temperature ranges of 0 to 30 °C (Hayashi, 2003). The common groundwater value for the temperature compensation value a = 0.02 was used to obtain EC_{25} (Matthess, 1982). The formula

$$\text{TDS (ppm)=A*EC25(micro siemens)}$$

was used to determine total dissolved solids (TDS). A natural water value of 0.59 was used as the TDS factor=A (Hem, 1985). TDS present in the tap water used to collect leachate was subtracted out theoretically to provide the TDS only within the soil. Leachate collected on December 12, 2022, revealed that nitrogen levels had been exhausted in all treatments beyond the 150-ppm recommendation for foliage (Conover & Poole, 1990). Therefore, the crop was re-fertilized on December 12, 2022 (84 days after the original application). Plants were randomly separated into 4 individual tables for surveying on March 31, 2023, and surveyed by 22 Florida Southern College students on April 1, 2023 (193 days after planting). Surveys included ranking each plant by quality on a Likert (Likert, 1932) 1–5 scale (1 = very poor quality, 2 = below average quality, 3 = average quality, 4 = above average quality, and 5 = superior quality), "best" and "worst" plants per table, and would you buy for $9.99? (Yes, No, Maybe). Physical measurements of plants included: height (bench to tallest point), width (sum of widest side and side perpendicular to the widest), number of visible shoots, and chlorophyll (A+B) concentration. Chlorophyll determination was carried out by weighing exactly 100 mg ± 1 mg of leaf tissue from each plant. Tissue was incubated excluding light in 10 mL of DMSO for 14 hours at 40 °C, 23 hours at 65 °C, and 48 hours at 65 °C. Vials were lightly mixed before absorbance was measured by two Genesys 20 Ultraviolet-Visible Spectrometers (Thermo Fisher Scientific, Waltham, Mass.). Spectrometers were set to read two wavelengths which correlate to the absorbance of both Chlorophyll A (665 nm) and Chlorophyll B (649 nm). Absorbance values were converted to µg/mL (micrograms of Chlorophyll per mL of solution) using standard equations for a resolution range of 1–4 nm. (Wellburn, 1994).

Data analysis. Statistical analysis was performed using Minitab® statistical software. Statistical correlation was performed on survey quality ranking averages, would you buy for $9.99 averages, height, width, number of visible shoots, and Chlorophyll (A+B) concentrations to identify traits that increase the quality and marketability of a 'Green Princess' according to the consumer group. Ordinary Least Square Regression was performed at a 95% confidence interval to compare the effects of various fertilizer and light treatments on the measurements recorded, compared to contrast groups, Diamond R Nursery Advantage 180 20-6-9 (control) and high light level. Total dissolved solids were graphed over time to observe the change in available fertilizer for each brand.

Results and Discussion

All plant measurements were positively correlated with the survey criteria (Table 1) This observation means that as plants become wider, taller, possess more shoots, and produce more chlorophyll the perceived quality and likelihood of purchasing increases. Therefore "fuller" and "greener" plants are characteristics (Fig. 1, Fig. 2) that are of interest to commercial growers and should be targeted to produce a higher-quality *Philodendron*. Upon regression analysis of light levels (Table 2), it was observed that high light level significantly outperformed low light level in survey criteria, height, and Chlorophyll (A+B) concentration (Table 3). We recommend that the commercial grower expose their *Philodendron* 'Green Princess' to the high light level tested to produce a higher quality more marketable crop. Regression

Table 1. Correlations between measurements of *Philodendron* 'Green Princess' and consumer survey results.

	Quality ranking average	Would you buy for $9.99?
Would you buy for $9.99	0.956	–
Number of visible shoots	0.305	0.292
Plant width	0.336	0.357
Plant height	0.203	0.218
Chlorophyll (A+B) concentration	0.422	0.400

Fig.. 1. Philodendron 'Green Princess' plant with highest number of "Best" plant per block votes from a consumer panel.

Fig. 2. *Philodendron* 'Green Princess' plant with highest number of "Worst" plant per block votes from a consumer panel.

Table 2. Comparison of measurements between *Philodendron* 'Green Princess' plants within the low light level to measurements of plants within the high light level. Results representing a significant ($P < 0.005$) significant negative correlation (– Coefficient) are shaded.

Measurement:	Light level	Coefficient	*P*-value
Quality Ranking Average	Low	–0.322	0.004
Would You Buy for $9.99? Average	Low	–0.2279	0.007
Number of Visible Shoots	Low	0.208	0.481
Width	Low	–2.18	0.056
Plant Height	Low	–1.648	0.001
Chlorophyll (A+B)concentration	Low	–1.847	0.002

Table 3. Daily light integrals for both high/low light levels used to grow *Philodendron* 'Green Princess'.

Location	High light level	Low light level
Greenhouse #3	4.246 DLI	3.347 DLI
Greenhouse #4	3.077 DLI	1.975 DLI
Average	3.6615 DLI	2.661 DLI

analysis of fertilizer treatments (Table 4) revealed that three treatments were significant in producing larger plant measurements, but with analysis of survey criteria, no treatment outperformed the control group (Diamond R Nursery Advantage 180 20-6-9). This distinction shows that while some fertilizer treatments outperformed others, the consumer group did not recognize the difference. It is important to note that the control treatment with biostimulants (DR+BIO) performed significantly worse in survey criteria and Chlorophyll (A+B) concentration. Therefore, it is recommended that the grower apply the lowest rate of the most affordable fertilizer tested to decrease runoff and input costs. Observing the change in TDS over time (Fig. 3) in all three fertilizer brands, with and without biostimulants, it was discovered that the higher light level contained less available fertilizer than the low light level during the 5 months of testing. CRFs release at a faster rate as temperature increases. For that reason,

Table 4. Measurements of various fertilizer treatments tested on *Philodendron* 'Green Princess' compared to the measurements of control treatment (2.4g/pot of Diamond R Nursery Advantage 180 20-6-9). Results representing a significant ($P < 0.005$) positive (+ Coefficient) correlation are light shaded while significant negative correlations (–Coefficient) are dark shaded. **Key:** DR+Bio = 2.4 g/pot-Diamond R Nursery Advantage 180 20-6-9 (control) with HaifaStim™ Wall-Up S and Vital (biostimulants). F1 = 75% of control Nitrogen-Florikan 18-6-8 Total. F2 = 100% of control Nitrogen-Florikan 18-6-8 Total. F3 = 125% of control Nitrogen-Florikan 18-6-8 Total. M1=75% of control Nitrogen-Haifa Multicote (8) 15-7-15 + 2MgO. M2=100% of control Nitrogen-Haifa Multicote (8) 15-7-15 + 2MgO. M3 = 125% of control Nitrogen-Haifa Multicote (8) 15-7-15 + 2MgO.

	Quality Ranking Average (survey)	
Treatment	Coefficient	*P*-value
DR+Bio	–0.441	0.045
F1	0.076	0.728
F2	–0.115	0.597
F3	–0.324	0.14
M1	–0.042	0.848
M2	–0.017	0.937
M3	–0.005	0.982
	Would You Buy it for $9.99? Average (survey)	
Treatment	Coefficient	*P*-value
DR+Bio	–0.379	0.023
F1	0.059	0.718
F2	–0.027	0.871
F3	–0.267	0.107
M1	0.032	0.844
M2	–0.033	0.840
M3	0.031	0.850
	Number of Visible Shoots	
Treatment	Coefficient	*P*-value
DR+Bio	0.167	0.778
F1	0.417	0.481
F2	0.000	1.000
F3	–0.167	0.778
M1	–0.167	0.778
M2	–0.417	0.481
M3	0.000	1.000
	Plant Width	
Treatment	Coefficient	*P*-value
DR+Bio	1.42	0.531
F1	1.25	0.580
F2	2.58	0.254
F3	3.52	0.121
M1	3.85	0.090
M2	3.94	0.084
M3	4.79	0.036
	Plant Height	
Treatment	Coefficient	*P*-value
DR+Bio	0.792	0.421
F1	0.792	0.421
F2	1.396	1.58
F3	2.117	0.033
M1	2.017	0.042
M2	1.833	0.065
M3	1.771	0.074
	Chlorophyll (A+B) concentration	
Treatment	Coefficient	*P*-value
DR+Bio	–3.49	0.004
F1	0.57	0.628
F2	0.40	0.736
F3	–0.19	0.873
M1	–0.75	0.527
M2	–0.430	0.803
M3	–2.29	0.055

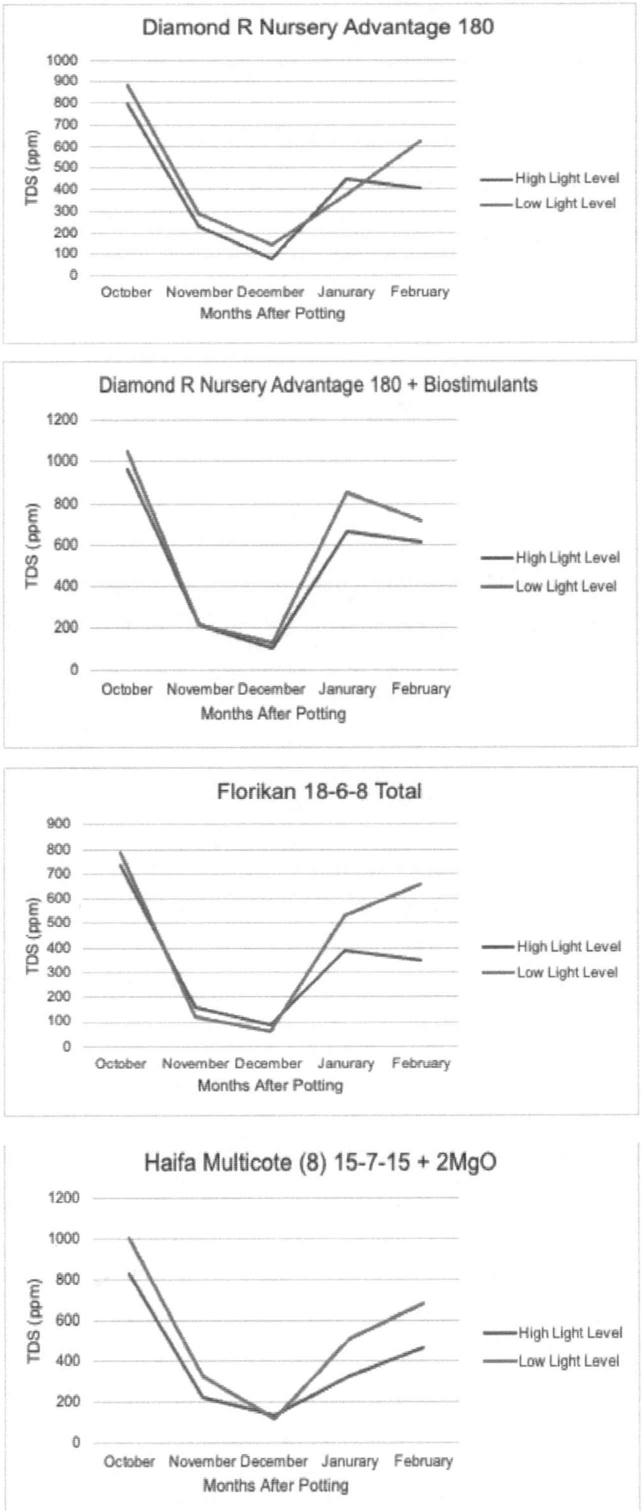

Fig. 3. Total dissolved solids in the growing media vs. time in a growing experiment with *Philodendron* 'Green Princess' (Note: fertilizer was re-applied in December.)

we would expect the higher light level to have greater levels of available fertilizer compared to the low light level in the months just following fertilization (October/December), but that was not observed. Therefore, the ability of the plants to utilize the available fertilizer is increased at high levels, explaining why TDS levels were normally lower and plant quality was significantly greater.

Summary

To produce the most attractive and marketable *Philodendron* 'Green Princess' with limited runoff and input cost, we recommend under the environmental parameters of this experiment to use the most affordable fertilizer we employed at the lowest rate tested, re-fertilizing every 90 days. While exposing the crop to the high light level tested for the duration of the 24-week production period.

Literature Cited

Tew M.M., Manners M.M., Griffis L.J, & Tuncay O. 2022. Effects of light level and controlled-release fertilizer level on greenhouse production and consumer perception of *Philodendron* 'Birkin'. *Proc. Fla. State Hort. Soc.* 135:195.

Torres A.P., & Lopez R.G. 2010. Measuring daily light integral in a greenhouse. Extension publication HO-238-W. Purdue University, Lafayette, IN.

Henley R.W., Chase A.R., & Osborne L.S. 1991. Philodendrons—Self heading types. Extension publication RH-91-27. University of Florida, Gainesville, FL.

Hayashi M. 2004. Temperature-electrical conductivity relation of water for environmental monitoring and geophysical data inversion. *Environmental Monitoring and Assessment*. 96:119–128. https://doi.org/10.1023/b:emas.0000031719.83065.68

Matthess G. 1982. The properties of groundwater. *Advances in Water Resources*. 6(2):71. https://doi.org/10.1016/0309-1708(83)90054-4.

Hem J.D., 1985. Specific electrical conductance, p. 66-67 in: Study and interpretation of the chemical characteristics of natural water. U.S. Geological Survey, Alexandria, VA.

Conover C.A. & Poole R.T. 1990. Light and fertilizer recommendations for production of acclimatized potted foliage plants. Extension publication RH-90-1. University of Florida, Gainesville, FL.

Likert R. 1932. A Technique for the Measurement of Attitudes. *Archives of Psychology* 22(140):1–55.

Wellburn A.R. 1994. The spectral determination of chlorophylls a and b, as well as total carotenoids, using various solvents with spectrophotometers of different resolution. *Journal of Plant Physiology*. 144(3):307–313.

Proc. Fla. State Hort. Soc. 136:194. 2023.

—Scientific Note—

Multi-agent Collaboration: Revitalizing a Native Landscape into an Educational Tool

Emily Marois*, Anna Mészáros, Kara Krueger, Frank Dowdle, and Matthew VanWeelden

University of Florida, IFAS, Palm Beach County Extension, West Palm Beach, FL

Keywords. demonstration gardens, extension

The goal of the Everglades Research and Education Center (EREC) landscape revitalization project was to increase awareness of the benefits of native plants, beautify the EREC campus, and to protect the original installation from being removed.

A multi-disciplinary collaboration between five Palm Beach County extension agents, the Palm Beach County Master Gardener Volunteer Program and the EREC Landscaping Committee resulted in a transformation of this overplanted landscape into a Florida-Friendly™ native plant demonstration garden. The garden consists of 28 species of native plants such as; state-endangered Florida bitterbush (*Picramnia pentandra*), redberry stopper (*Eugenia confusa*) and Bahama strongbark (*Bourreria succulenta*). Volunteers and extension agents participated in workdays to rehabilitate the garden (Fig. 1), and volunteers assisted in the creation of high-quality educational interpretive signage (Fig. 2) to inform EREC visitors about the common and rare native plants that reside in this landscape. QR codes as well as cultural requirements/wildlife value symbols are included on each of the 28 individual plant signs which link back to University of Florida publications or other reputable information sources. Additionally, the main garden sign includes a QR code that sends visitors to a survey recording knowledge gain and potential behavior change data. The QR code survey attached to the main garden sign has been employed and is currently collecting data on the visitors' experience, what they learned, and what practices they may implement in their own landscape. The EREC Garden has been transformed from a forgotten landscape into an educational tool that is accessible to EREC faculty, staff and visitors, schools, and all other community members.

This garden area serves as a resource for extension, offering a platform to involve the local community, schools, and decision-makers, inspiring them to take on the role of environmental stewards. Incorporating this demonstration garden into upcoming educational workshops, field trips, and field days will enhance the learning experience through hands-on activities.

Fig. 1. Volunteers and extension agents participated in workdays to rehabilitate the garden at the Everglades Research and Education Center, Belle Glade, FL.

Fig. 2. Volunteers assisted in the creation of high-quality educational interpretive signage to inform visitors about the common and rare native plants that reside in this landscape at the Everglades Research and Education Center, Belle Glade, FL.

*Corresponding author. Email: emarios@pbcgov.org

Proc. Fla. State Hort. Soc. 136:195. 2023.

—Scientific Note—

How Does Consumers' Knowledge of Native and Pollinator-friendly Plants Impact Their Plant Purchase Decisions?

XUAN WEI[1]*, MELINDA KNUTH[2], AND HAYK KHACHATRYAN[1]

[1]*Food and Resource Economics Department and Mid-Florida Research and Education Center, University of Florida, Apopka, FL*

[2]*Department of Horticulture Science, North Carolina State University, Raleigh, NC*

Keywords. generalized ordered logit, objective knowledge, physical attributes, plant care characteristics, plant sustainability features, subjective knowledge

Native and pollinator-friendly plants play a critical role in maintaining healthy pollinator populations and preserving biodiversity in urban landscapes. Many homeowners are becoming increasingly aware of the benefits of native and pollinator-attractive plants and seek out these species for their gardens and landscapes. The horticulture industry is shifting towards promoting these plants, but consumer knowledge gaps remain. To address this research gap, we surveyed 1788 participants using online and in-person experiments, to investigate how consumers' knowledge of native and pollinator-friendly plants may impact their plant purchase decisions. The national survey was administered online, using the Qualtrics XM survey software platform, from December 2017 to January 2018. Participants were pre-screened to ensure responses were collected from relevant audiences (e.g., people who have purchased plants in the past twelve months; lived in a home with a yard or garden).

The results showed that participants with relatively higher knowledge of native and pollinator-attractive plants weigh more on sustainable production methods relative to the plant's physical attributes (e.g., plant size, shape, etc.) and horticultural characteristics (e.g., plant health, easiness of care, etc.). Plant enthusiasts (participants who spent more money on plants) were more likely to prioritize sustainable production methods over the physical attributes of plants. In contrast, participants who primarily purchase plants from mass merchandisers or box stores tended to focus more on physical appearance or aesthetic characteristics. The findings emphasize the significance of subjective knowledge in prioritizing sustainability and shed light on the need to bridge the gap between perceived and actual knowledge. This study offers insights for policymakers, industry stakeholders, and consumers, contributing to the broader goal of supporting local biodiversity and sustainable urban development.

*Corresponding author. Email: wei.xuan@ufl.edu

Proc. Fla. State Hort. Soc. 136:196–198. 2023.

Viburnum Disease Management During Propagation

SHAWN T. STEED[1]*, WAEL ELWAKIL[1], AND GARY VALLAD[2]

[1]*University of Florida, IFAS, Hillsborough County Extension, Seffner, FL*

[2]*University of Florida, IFAS, Gulf Coast Research and Education Center, Balm, FL*

Keywords. *Cercospora* sp., *Corynespora* sp., cuttings, fungicide dips, fungicide drench, *Colletotrichum* sp., overhead-irrigation, *Phyllosticta* sp., sandankwa viburnum, sweet viburnum, *Viburnum odoratissimum*, *Viburnum suspensum*

***Viburnum* species are among the top-selling ornamental shrubs in Central Florida for placement in landscapes. Nurseries propagate viburnum by using landscape or production stock plants that may appear healthy but often harbor several foliar diseases prevalent in nurseries and landscape environments. These diseases then compromise production during the propagation stage. Research was conducted to determine if either drenching or dip-applying fungicides at the beginning of propagation would minimize disease severity. Two experiments using five treatments of fungicides were screened for efficacy. The pathogens *Colletotrichum* sp., *Cercospora* sp., *Corynespora* sp., and *Phyllosticta* sp. were isolated during the experiment. As a drench, both Postiva™ and Orkestra® significantly reduced final disease severity by > 50% compared to the untreated water control. For the dip-applied fungicide treatments, Postiva™, Orkestra®, Ryora®, and Omega® all significantly reduced final disease severity by 57% on average compared to the control. The same four treatments also prevented disease severity from increasing in a statistically significant manner over time, unlike the control. Root length, fresh and dry weight were only negatively impacted by Ryora® as a drench.**

The *Viburnum* genus of the family Adoxaceae is one of the top sellers among ornamental shrubs in Central Florida for placement in landscapes. This includes sandankwa viburnum (*Viburnum suspensum*) and sweet viburnum (*V. odoratissimum*). Nurseries propagate viburnum from vegetative cuttings using outdoor stock plants that may appear healthy but often harbor several foliar diseases prevalent in nurseries and landscape environments (Steed et al., 2021). These diseases then compromise production during the propagation phase. This is especially problematic for container nursery production that uses overhead irrigation, which is cheap to deploy but increases splash dispersal and leaf wetness, which is optimal for plant disease spread and development, respectively. Most growers and landscapers struggle with disease control that reduces the salability or aesthetics of plants. While effective fungicide applications can help manage foliar diseases during container production, making applications during or immediately following propagation may improve liner production and limit disease losses following transplanting. Some nurseries use hot water treatments on cuttings or fungicide sprays of stock plants to minimize disease problems before propagation. We propose that there might be a benefit of applying effective fungicides during cutting propagation as either a soil drench or as a dip-treatment for cuttings to minimize disease development early in the production cycle. The objective then, was to evaluate the benefit of applying fungicides during cutting propagation for liner production to minimize disease dispersal and development early in the production cycle.

Methods

Cuttings of sweet viburnum (*Viburnum odoratissimum*) were used to initiate two fungicide trials in Summer 2022. Both trials were conducted in July through August, using cuttings from naturally infected *V. odoratissimum* plants that were grown in 3-gallon (gal) containers at a commercial production plant nursery in Hillsborough County, FL. Semi-hardwood cuttings were about 3–4 inches long with one node and two leaves were taken from the mature growth. The trial was designed in completely randomized blocks with five replicates including an untreated water control and five fungicide treatments. Fungicide treatments were Postiva™ (benzovindiflupyr and difenoconazole) 28 oz/100 gal; Orkestra® (pyraclostrobin and fluxapyroxad) 10 oz/100 gal; Ryora® (Flutriafol) 14 oz/100 gal; Omega® (fluazinam) 16 oz/100 gal; Topsin4.5L® (thiophanate methyl) 20 oz/100 gal; and a water control. Fungicides were either applied as a soil drench (10 mL) to cuttings following the application of rooting hormone or as dip treatment to prepared cuttings followed by the application of rooting hormone and planting. Per standard nursery practice, cuttings (12 per replicate) were stuck in potting medium (50% peat: 50% perlite) on June 21, 2022, in 40-cell liner trays. Trays were immediately placed under an overhead mist system (4-second mist every six minutes) to keep leaves moist. After an initial period of two weeks, cuttings were rated for disease incidence and severity on a weekly basis. The percentage of symptomatic foliage was rated weekly for six weeks to calculate the area under the disease progression curve (AUDPC). Leaves were periodically sampled for disease identification. After six weeks, plants were removed, given a final assessment on week seven and uprooted to assess root length and fresh and dry root biomass. Data analysis was conducted using a generalized mixed model analysis (PROC GLIMMIX) within SAS (version 9.4) with blocking as a random variable and fungicide treatment as a fixed effect, with the Student Panel command in SAS to check for normality. Disease severity over time was analyzed using the repeated measures function within GLIMMIX and a first-order autoregression to model covariance structure. Means separations were performed using Fisher's protected LSD at a 95% level of confidence and degree-of-freedom

Special thanks to Rosales Nursery, J&R Nursery, ID Tree Farm and FNGLA for the research grant.

*Corresponding author. Email:ststeed@ufl.edu

contrasts to further analyze differences between multiple treatment means.

Results and Discussion

Identification of isolated fungi revealed the presence of multiple pathogens throughout both trials. Similar to previous nursery trials, *Colletotrichum* sp. was the most abundant, but others including *Cercospora* sp., *Corynespora* sp., and *Phyllosticta* sp. were observed causing symptoms of leaf spotting, blighting, and defoliation on propagated cuttings. Both drench-applied and dip-applied fungicides had a statistically significant effect on disease based on final disease severity and AUDPC measurements (Table 1 and Table 2). For drench-applied fungicide treatments, both Postiva™ and Orkestra® significantly reduced final disease severity by > 50% compared to the water control. In addition, both Postiva™ and Orkestra® prevented disease severity from increasing over time, relative to the control and other ineffective fungicide treatments (Table 1). For the dip-applied fungicide treatments, Postiva™, Orkestra®, Ryora®, and Omega® all significantly reduced final disease severity by 57% on average compared to the control. The same four treatments also prevented disease severity from increasing over time, unlike the control. However, none of the treatments were significantly different from the control based on AUDPC. This may be an artifact of how the measurements were made. If leaves fell off due to disease severity, they were not continually counted for the measurements of leaf areas with disease which may have impacted overall results.

Drench-applied fungicide treatments had a statistically significant effect on cutting vigor based on root length and root biomass (Table 3). Relative to the other fungicide treatments, plants treated with Ryora® had the shortest root length, and lowest fresh and dry root biomass compared to the control and the other fungicide treatments. Interestingly, dip-applied fungicide treatments did not have a statistically significant effect on cutting vigor (Table 4), and on average produced liner plants with roots that were on average 13% longer with 18% and 13% greater fresh and dry biomass, respectively, compared to liner plants that received drench applied fungicide treatments.

Summary

Fungicides applied to propagative cuttings as either a dip treatment prior to planting or as a drench treatment following planting benefited disease control during liner production. Similar to nursery trials previously conducted (Steed et al., 2021), Orkestra® and Postiva™ were the most effective at controlling the diverse diseases observed on the cuttings; whereas the other fungicide treatments were more variable depending on whether they were applied as a dip or drench. For example, Ryora and Omega®, both appeared to perform better as a dip, rather than a drench. This would make sense for Omega® (fluazinam), which is a contact fungicide, but Ryora® (flutriafol) is a systemic fun-

Table 1. Effect of drench-applied fungicides to *Viburnum* cuttings for liner production on final disease severity, disease severity over time, and area under disease progression curve (AUDPC).

Product	Active ingredient	FRAC	Rate/100 gal	P_{TIME}[z]	DS_{final}[y]	AUDPC[x]
Postiva™	benzovindiflupyr + difenoconazole	7+3	28 fl oz	0.0971	5.8 c[w]	145 c
Orkestra®	pyraclostrobin + fluxapyroxad	11+7	10 fl oz	0.3392	5.1 c	173 bc
Ryora®	flutriafol	3	14 fl oz	< 0.0001	16.3 a	289 ab
Omega®	fluazinam	29	16 fl oz	< 0.0001	16.8 a	317 a
Topsin 4.5L®	Thiophanate methyl	1	20 fl oz	0.0002	9.5 bc	239 ab
Control (water)	–	–	–	< 0.0001	12.7 ab	288 ab
P-value				NA	0.0001	0.0618*

[z]Statistical effect of fungicide treatments on disease severity over time, based on repeated measures.
[y]Final disease rating, based on percent symptomatic foliage, collected on August 19, 2022.
[x]Area under the disease progression curve (AUDPC), calculated using seven disease severity ratings.
[w]Means followed by the same letter are not significantly different at the 95% level of confidence.
**P-value* is of marginal significance, mean separation provided for guidance.

Table 2. Effect of dip-applied fungicides to *Viburnum* cuttings for liner production on final disease severity, disease severity over time, and area under disease progression curve (AUDPC).

Product	Active ingredient	FRAC	Rate/100 gal	P_{TIME}[z]	DS_{final}[y]	AUDPC[x]
Postiva™	benzovindiflupyr + difenoconazole	7+3	28 fl oz	0.8919	5.7 bc[w]	244 ab
Orkestra®	pyraclostrobin + fluxapyroxad	11+7	10 fl oz	0.9487	4.1 c	131 b
Ryora®	flutriafol	3	14 fl oz	0.9585	4.3 c	181 b
Omega®	fluazinam	29	16 fl oz	0.3415	7.0 bc	236 ab
Topsin 4.5L®	Thiophanate methyl	1	20 fl oz	0.0775	9.7 ab	325 a
Control (water)	–	–	–	0.0002	12.3 a	243 ab
P-value				NA	0.0011	0.03

[z]Statistical effect of fungicide treatments on disease severity over time, based on repeated measures.
[y]Based on final disease rating, based on percent symptomatic foliage, collected on August 19, 2022.
[x]Area under the disease progression curve (AUDPC), calculated using 7 disease severity ratings.
[w]Means followed by the same letter are not significantly different at the 95% level of confidence.

Table 3. Effect of drench-applied fungicides to propagated *Viburnum* cuttings on final root length and fresh root and dry root biomass.

Product	Active ingredient	FRAC	Rate/100 gal	Root length (cm)	Root biomass	
					Fresh (g)	Dry (g)
Postiva™	benzovindiflupyr + difenoconazole	7+3	28 fl oz	4.8 a[z]	24.8 ab	3.8 a
Orkestra®	pyraclostrobin + fluxapyroxad	11+7	10 fl oz	5.2 a	27.1 a	3.9 a
Ryora®	flutriafol	3	14 fl oz	3.9 b	14.1 c	2.3 b
Omega®	fluazinam	29	16 fl oz	5.0 a	19.0 bc	3.2 ab
Topsin 4.5L®	Thiophanate methyl	1	20 fl oz	4.8 a	20.2	3.2 ab
Water control	–	–	–	4.9 a	22.3 ab	3.6 a
P-value				0.0595*	0.0372	0.0872*

[z]Means followed by the same letter are not significantly different at the 95% level of confidence.
**P*-value is of marginal significance, mean separation provided for guidance.

Table 4. Effect of dip-applied fungicides to propagated *Viburnum* cuttings on final root length and fresh root and dry root biomass.

Product	Active ingredient	FRAC	Rate/100 gal	Root length (cm)	Root biomass	
					Fresh (g)	Dry (g)
Postiva™	benzovindiflupyr + difenoconazole	7+3	28 fl oz	5.7	30.2	4.4
Orkestra®	pyraclostrobin + fluxapyroxad	11+7	10 fl oz	5.4	21.4	3.3
Ryora®	flutriafol	3	14 fl oz	5.6	26.1	4.2
Omega®	fluazinam	29	16 fl oz	5.7	29.5	4.1
Topsin 4.5L®	Thiophanate methyl	1	20 fl oz	5.4	29.3	4.2
Water control	–	–	–	5.7	18.9	2.7
P-value				0.9479	0.1662	0.2210

gicide. Topsin4.5L® and Omega® were included in these trials as broad-spectrum fungicides that have shown success against similar fungal pathogens on other crops. On average, drench-applied fungicide treatments appeared to have a larger impact on root development, including the controls that were treated with water. Fungicides can have profound impacts on plant growth. However, the observed differences in the water-treated controls suggest that the effect was likely due to the drench applications diluting the rooting hormone applied to the propagated cuttings prior to planting. Whereas in the other trial, the rooting hormone was applied after the cuttings were dipped in a fungicide solution.

Overall, while fungicide applications during propagation seem promising to control disease, it is not curative. We did not follow these studies through container production since disease severity at the end of both trials was similar to disease levels observed at the initiation of prior nursery trials. Hence, we do not believe the observed gains in disease control during these propagation trials are biologically significant given the favorable environmental conditions during subsequent container production due to the use of overhead irrigation. We propose developing alternative methods to propagate cuttings without the use of overhead irrigation, combined with the use of drip-emitters and timely fungicide applications to minimize the impact on foliar diseases. In addition, the use of hot-water treatments could be evaluated as a non-chemical means to cure cuttings of any latent infections.

Literature Cited

Steed S.T., Elwakil W., Vallad L, & Vallad G. 2021. Viburnum foliar pathogen identification and fungicide efficacies. *Proc. Fla. State Hort. Soc.* 134:186–189.

Proc. Fla. State Hort. Soc. 136: 199–201. 2023.

Palm Fruit Management

A.D. Ali[1]*, Don R. Hodel[2], Manny Nassar[3], and Len Burkhart[4]

[1]*Davey Institute, The Davey Tree Expert Company, Fort Myers, FL*

[2]*Extension Advisor (retired), University of California, Seal Beach, CA*

[3]*Davey Institute, The Davey Tree Expert Company, St. Augustine, FL*

[4]*The Davey Tree Expert Company , San Marcos, CA*

Keywords. *Cocos*, fruit elimination, dikegulac-sodium, palms, *Phoenix*, *Syagrus*, *Washingtonia*

Four studies were conducted in South Florida and in Southern California to evaluate the growth regulator dikegulac-sodium (Atrimmec®) for palm fruit elimination. Species tested were queen palms [*Syagrus romanzoffiana* (Cham.) Glassman] and coconut palms *Cocos nucifera* L. in Florida; Mexican fan palms *Washingtonia robusta* H. Wendl. and date palms *Phoenix dactylifera* L. in California. Dikegulac-sodium was applied either as a trunk injection or basal soil drench. Evaluations were made of quantity of inflorescences, infructescences, and date yields. No significant differences were noted in any of the species tested. Future studies should investigate higher doses of dikegulac-sodium for palm fruit management.

Palms are arborescent monocots that comprise an integral part of urban landscapes in states such as Florida, California, and Hawaii. Of the approximately 2600 species known, two dozen are common in landscapes throughout southern and western US states. Popular landscape species in landscapes include Mexican fan palm *Washingtonia robusta* H. Wendl., Canary Island date palm *Phoenix canariensis* hort ex Chabaud, queen palm *Syagrus romanzoffiana* (Cham.) Glassman, coconut palm *Cocos nucifera* L. and date palm *P. dactylifera* L. The latter also is grown commercially in desert areas of Southern California for the edible dates, and the plantings serve as a reliable source of relatively inexpensive and uniform, mature palm trees for the landscape.

While palms offer a unique shape and stature among other plants in landscapes, they do pose some challenges. One is the nuisance fruit production with species such as queen palms and those in the genus *Phoenix*. Spent fruits fallen on sidewalks create a trip and fall hazard, while larger fruit such as mature coconuts can lead to injury to people or property. Thus, fruit reduction has been implemented mostly by pruning, which can be a labor-intensive practice.

Plant growth regulators (PGRs) have been studied for their effects on various palm species. Broschat and Donselman (1986) found that monthly sprays of ethephon retarded elongation and reduced the incidence of flowering in 1-year-old bamboo palms *Chamaedorea seifrizii* Burret. Similarly, Fisher (1980) reported leaf size reduction with ethephon sprays or leaf tip treatments in young fishtail palms *Caryota* L., *Chamaedorea* Willd., and areca palms *Chrysalidocarpus* H. Wendl. (formerly *Dypsis* H. Wendl.), *Phoenix* L., and lady palms *Rhapis* L. f. ex Aiton. Fisher and Theobald (1989) evaluated long term effects of trunk injection with gibberellin or cytokinin in mature coconut palms. Neither product induced vegetative proliferation intended for mass propagation of coconuts.

Ali and Bernick (2010) evaluated annual basal drenches of paclobutrazol (Cambistat®, Rainbow Ecosciences, Minnetonka, MN, USA) on royal palms [*Roystonea regia* (Kunth) O.F. Cook, nom. cons.]. They reported slight effects on vertical growth reduction, but no observable differences in flowering. Preliminary work by the senior author (unpublished data) showed that ethephon (Florel®, Monterey, Fresno, CA, USA) can be effective in aborting flowers but would require multiple applications during the season for most species.

Another PGR, dikegulac-sodium can be applied internally through trunk injection or as a soil drench to the root zone. The systemic activity offers advantages such as prolonged activity, less labor, and reduced drift and runoff when compared to spraying with ethephon. This paper reports on studies in Florida and California evaluating dikegulac-sodium (Atrimmec®, PBI/Gordon, Shawnee, KS, USA) for palm fruit elimination.

Materials and Methods

Florida Study. Mature coconuts growing in a well-maintained landscape setting in Boca Raton, FL. were selected for this study. The palms had approximately 5.5 m clear trunk and an average diameter breast height (dbh) of 25.4 cm. Treatment evaluations were made by counting the quantity of inflorescences (opened and unopened) and fruit for each palm at 11 weeks after treatment (WAT; May 3) and 19 WAT (June 26). The study was initiated on February 19, 2019 and completed on June 28, 2019.

The authors wish to express gratitude to PBI/Gordon for providing product, and to Marianne Waindle and Dawn Fluharty formerly from ArborJet for providing product and assistance in treatment applications in California. We appreciate the assistance of Mr. James Komen in analyzing certain parts of the California data. Gratitude also is expressed to Habcenter Nursery and to Ave Maria Home Owners Association for providing sites for the studies.

*Corresponding author. Email: ad.ali@davey.com

Mature queen palms growing in a well-maintained landscape setting in Ave Maria, FL. were selected for this study. The palms had approximately 4.2 m clear trunk and an average dbh of 25.4 cm. Treatment effects were determined by counting the quantity of inflorescences in bud (unopened flower clusters), inflorescences opened and fruit for each palm at 10 WAT (May 1), 19 WAT (July 31) and 24 WAT (June 25). The study was initiated on February 21, 2019 and completed on June 25, 2019.

Treatments for both species were as follows: 1) dikegulac-sodium trunk injection at 1 mL/2.54 cm dbh, 2) dikegulac-sodium basal drench at 90 mL/3.8 L water/palm, and 3) untreated check.

Trunk injections were made by drilling a single hole 1.4 m above grade, 5–7 cm deep then injecting the product with a Q-Gun injector (Rainbow Ecosciences, Minnetonka, MN, USA). Basal drenches were conducted by pouring the mixed product evenly around the base of the palm. Experimental design was randomized complete block with 10 replications (palms) per treatment. Data were analyzed with analysis of variance (ANOVA) and mean separation via Tukey's test at $P = 0.05$.

California Study. Date palms growing in a commercial date orchard in Indio, CA. were selected for this study. The palms had approximately 3 m clear trunk and an average dbh of 46 cm.

Inflorescence quantities were rated into three groups low (0–3 inflorescences), medium (4–6 inflorescences), and high (7 or more inflorescences) on each palm at 21 WAT (July 2nd). Additionally, the inflorescences were examined to see if fruit set had occurred. At 32 WAT (September 16), the weights of four infructescences in each cardinal direction were determined with a field scale and the quantity of infructescences was counted on each palm.

The study was initiated on February 6, 2019, and completed on September 16, 2019. Experimental design was randomized complete block with 10 replications (palms) per treatment. Fruit weight data were analyzed with ANOVA and mean separation via Tukey's test at $P = 0.05$.

Mexican fan palms growing as street trees in Long Beach, CA, were selected for this study. The palms had approximately 6.7 m clear trunk and an average dbh of 41 cm.

Inflorescence quantities were counted on each palm at 21 WAT (July 3rd). Additionally, inflorescences were examined to see if fruit set had occurred at 35 WAT (October 10). Infructescences were qualitatively rated low, medium, and high for fruit set on each palm. The study was initiated on February 6, 2019 and completed on October 10, 2019.

For the date palms and Mexican fan palms treatments were the same as the Florida study. Trunk injections were made by drilling a single hole 1.4 m above grade, 5–7 cm deep then injecting the product with a Quickjet injector (Arborjet, Woburn, MA, USA). Basal drenches were conducted by pouring the mixed product evenly around the base of the palm.

Results

Florida study. Effects of dikegulac-sodium on coconut palm inflorescence in bud, inflorescence opened and fruit formation are shown in Figs. 1–3, respectively. No significant reductions were observed on treated vs. untreated palms. Effects of dikegulac-sodium on queen palm inflorescence in bud, inflorescence opened and fruit formation are shown in Figs. 4–6, respectively. The drench treatment resulted in a slight (though not statistically significant) reduction in all three variables compared to the injection treatment or the untreated palms.

California study. During the July 3 rating, all date palms had high quantities of fruit bearing stalks in all treatments. Similarly, all fruit formed were of comparable size.

Date palm fruit yield and quantity of fruit clusters are shown in Table 1. No significant differences were noted among treatments with average weight ranging from 4.4 to 4.9 kg per harvest bag.

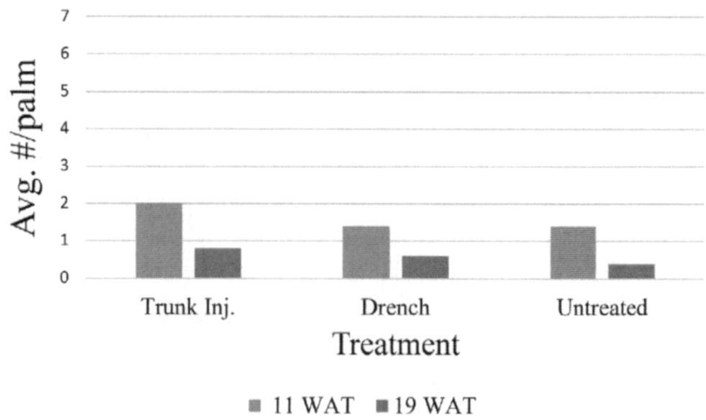

Fig. 1. Effects of dikegulac-sodium on coconut palms (*Cocos nucifera*) inflorescence in bud formation, South Florida.

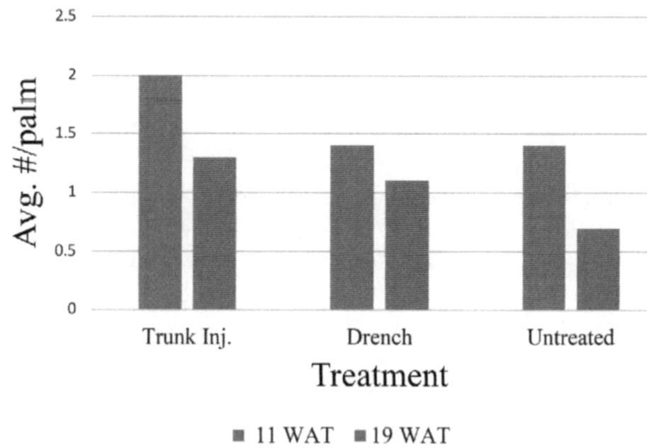

Fig. 2. Effects of dikegulac-sodium on coconut palms (*Cocos nucifera*) open inflorescence, South Florida.

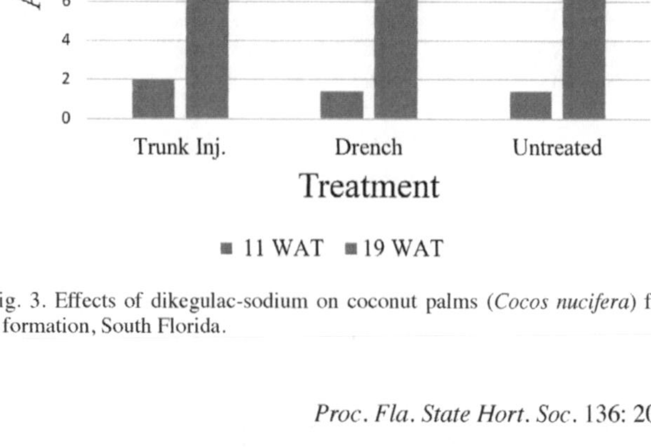

Fig. 3. Effects of dikegulac-sodium on coconut palms (*Cocos nucifera*) fruit formation, South Florida.

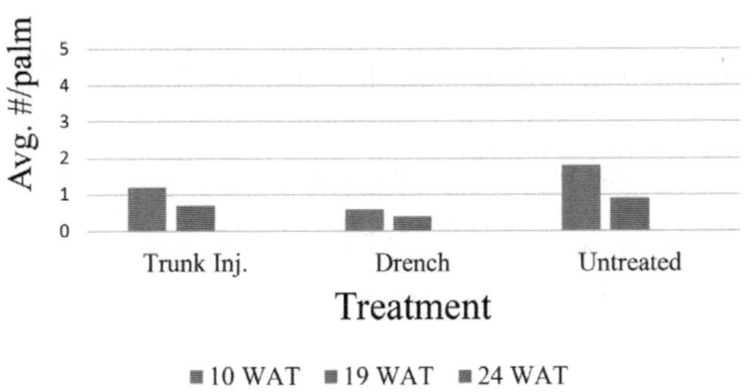

Fig. 4. Effects of dikegulac-sodium on queen palm (*Syagrus romanzoffiana*) inflorescence in bud formation, South Florida.

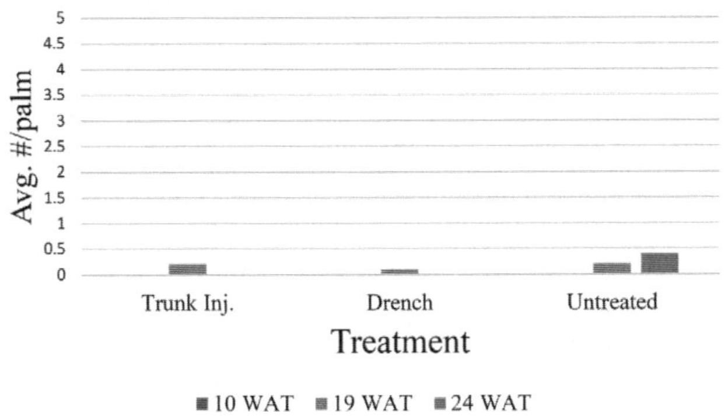

Fig. 5. Effects of dikegulac-sodium on queen palm (*Syagrus romanzoffiana*) open inflorescence formation, South Florida.

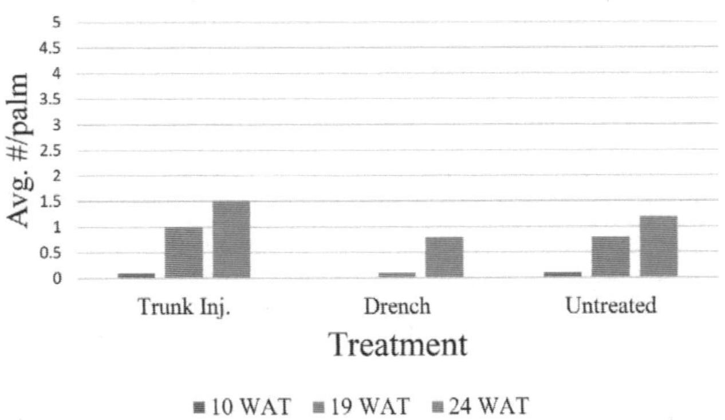

Fig. 6. Effects of dikegulac-sodium on queen palm (*Syagrus romanzoffiana*) fruit formation, South Florida.

Table 1. Influence of Atrimmec® on date palm (*Phoenix dactylifera* L.) fruit yield, Indio, Calif., September 18, 2019.

Treatment	Avg. wt. of harvest bag (kg)	Avg. no. fruit clusters/palm
Trunk injection	4.9	12.1
Basal drench	4.4	10.8
Untreated	4.7	10.7

No differences among treatment means, Tukey's test ($P > 0.05$).

Mexican Fan palms were visually evaluated for flower and fruit formation on July 2. No treatment effects were observed. Average number of flower and fruit stalks per palm were 9.6, 9.4, and 9.4 for the injection, drench and untreated, respectively. The palms also were evaluated visually for fruit set on October 10, on a scale of none, low, medium and high. Most palms were rated medium or high with no discernible treatment differences.

Discussion and Conclusion

Dikegulac-sodium is a product that has been safely and effectively used to suppress growth in approximately 50 tree and shrub species (https://www.cdms.net/ldat/ld60B008.pdf). It also has been used for suppression of flower and fruit formation in olive *Olea europaea* L., glossy-leaf privet *Ligustrum lucidum* W.T.Aiton, multiflora rose [*Rosa multiflora* Thunb. f. watsoniana (Crép.) Matsum.] and Japanese holly *Ilex crenata* Thunb.

In our studies, dikegulac-sodium did not provide acceptable reduction in flower and fruit formation of coconut, queen, date or Mexican fan palms. Additional studies should investigate the effects of higher doses. Due to the uniqueness of palm biology, soil drench applications may be more appropriate than trunk injections. The latter create permanent wounds that never cover over and they may serve as entry sites for pathogens and pests. Furthermore, cost/benefit analyses should be conducted to compare higher dose rates with traditional labor-intensive pruning.

Literature Cited

Ali A.D. & Bernick S.C. 2010. Effects of paclobutrazol on royal palm *Roystonea regia* trunk and height growth. *Arboriculture & Urban Forestry*. 36(5):221–223.

Broschat Timothy K. & Donselman H. 1986. Effects of several plant growth substances on height, flowering and lateral shoot development of *Chamaedorea seifrizii*. *Principes (Palms)*. 30:135–137.

Fisher Jack B. 1980. Morphogenetic effects of gibberellins and other growth regulators on palms. p. 21–32. In: Brickel C.D., Cutler D.F., & Gregory M. (eds.). Peraloid monocotyledons. Linn. Soc. Symp. Ser. 8, Academic Press, London.

Fisher J.B. & Theobald W.F. 1989. Long-term effects of gibberellin and cytokinin on coconut trees. *Principes (Palms)*. 33:5–17.

Proc. Fla. State Hort. Soc. 136:202. 2023.

—Scientific Note—

Measuring Phosphate, Calcium, and Magnesium in Leachate during Greenhouse Production of *Philodendron* 'Green Princess'—Comparing Analysis Methods and Evaluating Results

KARA FISCHER[1], RORY CUSHEN[2], BRYCE CROSBY[3], MALCOLM M. MANNERS[4], JOHN L. GRIFFIS, JR.[1]*, AND SENTHIL GIRIMURUGAN[5]

[1]*Florida Gulf Coast University, Ecology and Environmental Studies Department, Fort Myers, FL*
[2]*Florida Gulf Coast University, Integrated Studies Department, Fort Myers, FL*
[3]*Florida Southern College, Biology Department, Lakeland, FL*
[4]*Florida Southern College, Horticultural Sciences Department, Lakeland, FL*
[5]*Florida Gulf Coast University, Mathematics Department, Florida Gulf Coast University, Fort Myers, FL*

Keywords. controlled release fertilizer, CRF, foliage plants, Horiba

The main objective of this study was to compare the efficacy of several application rates of three different controlled-release fertilizers (CRF) during a 180-day commercial production crop cycle for *Philodendron* 'Green Princess'. The three fertilizers tested, Florikan CRF (18-6-8), Haifa Multicote™ (15-7-15+2MgO), and Diamond R Nursery Advantage, Powered by Purkote™ (20-6-9), were selected based upon their similar nitrate and ammoniacal nitrogen profiles, slow-release properties, and because they were readily available to Florida growers. Based on a number of measurements including overall plant visual quality, the larger experiment was designed to determine the optimal rate of fertilizer application for production of a high quality, saleable crop.

Over a 180-day period, various rates of three controlled-release fertilizers were evaluated under greenhouse conditions for successful production of a crop of *Philodendron* 'Green Princess'. On September 19, 2022, 96 individual tissue-cultured plugs (Oglesby International) were potted up into 15.24 cm diameter green azalea pots filled with a 70:20:10 peat:perlite:vermiculite potting mix. Pots were initially assigned 8 different fertilizer treatment groups with 12 pots per group. The control group received the Diamond R 180-day Nursery Advantage (20-6-9) fertilizer at 2.4g/pot (and a second group that would receive Haifa biostimulants also received 2.4 g/pot of the Diamond R fertilizer). The next 3 groups of pots received three different fertilizer rates of Florikan (18-6-8) 180-day Total = 2.0g/pot, 2.7g/pot, and 3.4g/pot. The final 3 groups of pots received three different fertilizer rates of Haifa Multicote™ 15-7-15 +2MgO 8-month = 2.4g/pot, 3.2g/pot, and 4.0g/pot. The middle rates for the Florikan and Haifa fertilizers were selected to match the available nitrogen rate of the Diamond R fertilizer. All CRF fertilizers were stirred into the potting mix before addition of the plugs; both dolomite (14 g/pot) and a S.T.E.M. (Soluble Trace Element Mix) drench were also added to the pots.

The plants were placed on the greenhouse benches at Florida Southern College, Lakeland, FL, in a generalized randomized block design, with four blocks each containing three replicates of each fertilizer treatment. Unfortunately, Hurricane Ian struck the greenhouses the next week and removed most of greenhouse roofs. Since this is an interior foliage crop that could be seriously damaged by full sun, our only choice was to erect shade structures over the plants. The available shade cloth (30% and 60% shade) resulted in an unplanned difference in light levels and we continued with 2 randomized blocks each containing 3 replicates under each of 2 different light regimes, resulting in 16 different treatments. Various direct measurements were taken throughout the trials. Using the standard pour-through method (Wright, 1986), leachate was collected monthly throughout the 180-day production cycle and evaluated for various nutrients. Leachates were evaluated at Florida Gulf Coast University, Fort Myers, FL, using a Hanna nutrient analysis photometer (using an adaptation of the calmagite method for determining calcium, the oxalate method for magnesium, and the amino acid method for phosphate, respectively (various reagents were required) as well as a Horiba LAQUAtwin calcium ion meter (no reagents were required). Both electrical conductivity (EC) and pH of leachates were also measured. Plants were also inspected for apparent nutrient deficiencies by looking for yellowing or purpling of the leaves, dead leaf margins and/or leaf tips, leaf expansion abnormalities, or any discoloration. No symptoms of calcium, magnesium, or phosphorus deficiency were apparent with any treatment. Both destructive and non-destructive chlorophyll analyses was conducted on the crop and no deficiencies were discovered. By December, all nutrient concentrations were found to be very low in the leachate; the EC was also low. These 90-day measurements showed that all fertilizers released nutrients quickly. An additional application of fertilizers at the original rates for all treatments was applied on day 90. A buyer survey of the plants was conducted after 180 days and the crop was deemed attractive and saleable. The lowest tested fertilizer application rates were sufficient to produce a quality crop of *Philodendron* 'Green Princess'.

Literature Cited

Wright Robert D. 1986. The pour-through nutrient extraction procedure. *HortScience* 21(2):22–229. https://doi.org/10.21273/hortsci.21.2.227

*Corresponding author. Email: jgriffis@fgcu.edu

Proc. Fla. State Hort. Soc. 136:203. 2023.

Profitability and Cost Estimates for Producing Tropical Hibiscus: A Partial Budget Analysis

Nicholas Haley[1]*, Hasmeen Abu[1], Trent Blare[2], and Fredy H. Ballen[2]

[1]*Food and Resource Economics Department, University of Florida, IFAS, Gainesville, FL*

[2]*Tropical Research and Education Center University of Florida, IFAS, Homestead, FL*

—abstract—

Tropical hibiscus (*Hibiscus rosa-sinensis*) is a bushy evergreen shrub or small tree that requires very high amounts of sunlight and is intolerant to temperatures below 50° F, so it is mostly grown in tropical and subtropical regions. Florida is an ideal climate for the commercial production of tropical hibiscus and is currently the largest domestic producer, with the majority occurring in South Florida. About 20% to 25% of plants sold from Miami-Dade County are hibiscus, where the market value of ornamental plants was $697 million (farmgate price) in 2017. Based on interviews with five hibiscus growers of different scales in South Florida, one of the largest impediments to optimal hibiscus production for Florida growers was pests. At the same time consumers are demanding plants that are grown with few pesticides, limiting growers' options to control these pests. To understand the cost effectiveness of biological controls, we estimated the costs and returns for starting and operating a hibiscus nursery, and the cost differences between using conventional pesticides and biological controls. Our results indicate that unskilled labor, fertilization, and pest control are the largest costs for operating a hibiscus nursery, and that there is potential for the use of biological controls to minimize insecticide costs.

*Corresponding author. Email: nhaley@ufl.edu

Proc. Fla. State Hort. Soc. 136:204–205. 2023.

—Scientific Note—

Measuring Potassium in Leachate During Greenhouse Production of *Philodendron* 'Green Princess'—Comparing Analysis Methods and Evaluating Results

Heidi Randall[1], Rory Cushen[2], Malcolm M. Manners[3], John L. Griffis, Jr.[1]*, and Senthil B. Girimurugan[4]

[1]*Florida Gulf Coast University, Ecology and Environmental Studies Department, Fort Myers, FL*

[2]*Florida Gulf Coast University, Integrated Studies Department, Fort. Myers, FL*

[3]*Florida Southern College, Horticultural Sciences Department, Lakeland, FL*

[4]*Florida Gulf Coast University Mathematics Department, Fort Myers, FL*

Keywords. controlled release fertilizer, CRF, foliage plants, Hanna nutrient analysis photometer

Potassium is one of the major components of commercial fertilizers. Plants utilize potassium to control cell membrane permeability. Potassium deficiencies can produce visual symptoms in various under-fertilized crops, but plant roots growing in soilless medium are good at capturing available potassium in leachate and visual deficiencies are rare in fertilized plants (White et al., 2021). Excess potassium might impact availability of other ions, but this condition would be unusual when plants are grown using recommended rates of commercial controlled release fertilizers (CRF). This study evaluated the impacts of several nutrients, including potassium, provided by the application of different rates of 3 commercially available CRF fertilizers on the production of the foliage crop *Philodendron* 'Green Princess' in a central Florida greenhouse environment.

On September 19, 2022, individual tissue culture plugs (Oglesby International) were potted in 15.24 cm diameter green azalea pots. These were then assigned 8 different fertilizer treatment groups. The control group (DR1) received a 2.4g/pot application of Diamond R 20-6-9 180 Nursery Advantage. A second group (DR1+Bio) also received a 2.4g/pot application of Diamond R 20-6-9 180 Nursery Advantage as well as drenches of Haifa biostimulants. The next three groups of plants received different rates of Florikan 18-6-8 180 Total. These rates were group (F1) 2.0g/pot, group (F2) 2.7g/pot, group (F3) 3.4g/pot. The final three groups of plants received different rates of Haifa Multicote 15-7-15 +2MgO (8 month). These rates were group (M1) 2.4g/pot, group (M2) 3.2g/pot, group (M3) 4.0g/pot. All CRF fertilizers were stirred into the 70:20:10 peat:perlite:vermiculite potting mix before addition of the plugs; both dolomite (14g/pot) and a S.T.E.M. (Soluble Trace Element Mix) drench were also added to the pots. There were 96 pots total, divided into 12 replicates of each treatment. These replicates were randomly divided into 4 blocks of 3 replicates per block. Fertilizers were reapplied at the same rates after 90 days.

*Corresponding author. Email: jgriffis@fgcu.edu

Using the standard pour-through method, leachate was collected monthly throughout the six-month production cycle and evaluated for various nutrients (Wright, 1986). Potassium measurements were made using two different devices: a Horiba LAQUA twin potassium ion meter (no reagents were required) and a Hanna nutrient analysis photometer (using a version of the tetraphenylborate method; reagents were required). This study indicated that there were no significant differences among the fertilizers and that they all produced similar excellent plant quality results. All fertilizer treatments were shown to release potassium rapidly under our growing conditions (Fig. 1). This rapid release may have been related to daytime soilless medium temperatures at or in excess of 30 °C. Although the medium

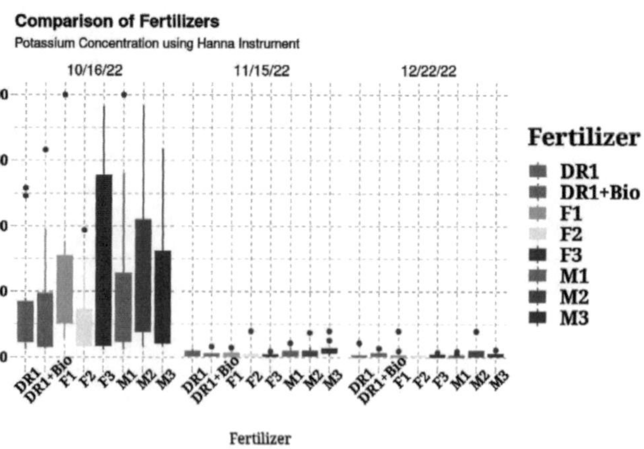

Analysis of Variance: Comparison of Fertilizers for K Concentration (Hanna Instrument)

Term	df	SS	MS	F	p-value
Fertilizer	7	5181.671	740.2388	0.8092609	0.5801925
Date	2	145016.654	72508.3270	79.2692252	0.0000000

Fig. 1. Comparison of potassium concentrations in leachate from container grow *Philodendron* 'Green Princess' determined by a Hanna photometer.

temperatures dropped to about 22 °C to 25 °C overnight, the high daytime temperatures accelerated the CRF fertilizer release of nutrients; relatively high quantities of potassium were present in the initial leachate measurements (Fig. 1). However, once the plants were well-established, the roots were able to rapidly absorb the available potassium; this resulted in a significant lowering of potassium levels in the subsequent leachate readings to less than 4 ppm (Fig. 1). Nutrient availably was also related to the pH of the water; many nutrients could be present, but not available to the plant because of the high pH levels > 8. No potassium deficiencies were observed in this study; however, in prior year experiments with philodendrons, when fertilizers were not reapplied after 90 days, nutrient deficiencies were observed. For this study, the lowest tested fertilizer application rates were sufficient to produce a quality crop of *Philodendron* 'Green Princess'.

Literature Cited

White P.J., Bell M.J., Djalovic I., Hinsinger P., & Rengel, Z. 2021. Potassium use efficiency of plants. p. 119–145 In: Murrell T.S., Mikkelsen R.L., Sulewski G., Norton R., & Thompson M.L. (eds.). Improving potassium recommendations for agricultural crops. Springer, New York, NY.

Wright R.D. 1986. The pour-through nutrient extraction procedure. *HortScience*. 21(2):22–229. htttps://doi.org/10.21273/hortsci.21.2.227

Proc. Fla. State Hort. Soc. 136:206. 2023.

How Do Different pH Meters Compare? A Look at Soil pH Meters: Store-bought, Portable Scientific Meters, and Analytical Lab pH Meters

Grantly Ricketts*, Kate Rotindo, and Amir Rezazadeh

University of Florida, IFAS, Extension, St. Lucie County, Ft. Pierce FL

—abstract—

Homeowners visit the University of Florida, Institute of Food and Agricultural Sciences (UF/IFAS) Extension St. Lucie County, seeking solutions for unhealthy plants. They are advised to send soil samples to the UF/IFAS Extension Soil Testing Laboratory (ESTL) for pH analysis. Clients informed Agents they intend to utilize store-bought pH meters. This experiment aimed to determine whether pH meters in ESTL and store-bought pH meters produced similar results. One hundred fifteen St. Lucie County residents were surveyed. Questions included: a) have you measured your soil pH, and if so, how?; b) have you ever used a store-bought soil pH meter, and c) do you feel that these types of pH meters are accurate? 64% (n = 74) of respondents believed store-bought pH meters are accurate, 32% (n = 37) stated store-bought pH meters may or may not be accurate, and 4% (n = 4) stated that store-bought pH meters are not accurate. To test the accuracy of store-bought pH meters, a soil sample was submitted to the University of Florida Extension Soil Testing Lab (ESTL), and three store-bought pH meters were also used. For pH testing, soil samples were collected from three different sites. The results of the three store-bought pH meters were compared with those of one portable scientific pH meter and the pH results from the ESTL. Results from the portable scientific pH meters and the ESTL were very similar. Results from store-bought pH meters were significantly different from those of the portable scientific pH meters and pH results from the ESTL. Store-bought soil pH meters produce inconsistent readings. Since store-bought pH meters are inaccurate, they should not be substituted for scientific pH meters or those used in the university's soil testing laboratory.

*Corresponding author. Email: gricketts@ufl.edu

Proc. Fla. State Hort. Soc. 136:207. 2023.

—Scientific Note—

Education About Harvesting Rainwater; More Likely to Install Whatever the Cost

Lisa Strange[1]* and Lynn Barber[2]

[1]*Taylor County Extension, University of Florida, IFAS, Perry, FL*

[2]*Hillsborough County Extension, University of Florida, IFAS, Seffner, FL*

Keywords. rain barrel, water harvesting, watering plants, water saving

Irrigating landscapes in economically distressed counties is significantly less attainable than in economically stable areas. Average earnings and education are higher in urban counties compared to rural ones (Parker et al., 2018) .Taylor County residents were less inclined to install a rain barrel because of financial inhibitors, and less concerns of water savings as private wells are commonly a source of water for both irrigation and drinking. The motivation for saving water is not a priority when water is plentiful and at no measurable cost. To improve their practices and impacts, the extension agent will try alternative strategies to solve problems such as offering rain barrels at a lower cost to enable lower income counties to be able to enjoy the benefits of rainwater properties, reduce stormwater runoff, and enhance their efforts of good environmental stewardship.

Materials and Methods

Educational programming in Taylor and Hillsborough Counties have differing results. Both counties' presentations include information on how to decrease potable water use in landscapes using rain barrels/cisterns. Both include demonstrations/information on how to create a rain barrel from a food grade tote, installing a spigot, reconfiguring the top of the barrel, creating overflow accommodations, use for harvested water from roof structures versus non-roof surfaces, connecting multiple barrels, painting the barrel, and maintenance needed.

Results

Hillsborough County residents attending a rainwater harvesting workshop installed the barrel provided for a $5.00 fee. Taylor County residents paid a higher fee of $45 after attending a workshop. The objective of our analysis is to compare whether living in an economically distressed county versus a non-economically distressed county when attending a workshop affects the decision whether to purchase a rain barrel.

In Taylor County seventy-five percent indicated after attending rainwater harvesting workshop that they had a better understanding of the purpose and practicality of a rain barrel and would purchase and install one. Fifty-two percent (n = 44) of participants purchased a rain barrel after the workshops. In Hillsborough County the knowledge increase post workshop was eighty-five percent, and all households received a fitted barrel.

The workshops increased both counties residents' awareness of water quality and impacts on plants. Socioeconomic status does not seem to deter the desire to learn. Information received from a water harvesting workshop helped residents to make an informed decision and they were more likely to purchase a rain barrel regardless of cost. Hillsborough County had larger audiences because of the population difference and longevity of the program.

Literature Cited

Parker K., Horowitz J.M., Brown A., Fry R., Cohn D'V., & Igielnik R. 2018. Demographic and economic trends in urban, suburban and communities. Pew Research Center. Washington, D.C. https://www.pewresearch.org/social-trends/2018/05/22/demographic-and-economic-trends-in-urban-suburban-and-rural-communities/

*Corresponding author. Email: lstrange@ufl.edu

Proc. Fla. State Hort. Soc. 136:208–212. 2023.

The Use of Polymer-Coated Urea in North Florida Field Corn Production

Jay Capasso[1]*, Kelly Morgan[2], Vivek Sharma[3], Kevin Athrearn[1], Lakesh Sharma[2], and Joel Love[1]

[1]*University of Florida, IFAS, North Florida Research and Education Center, Suwannee Valley, Live Oak, FL*

[2]*University of Florida, IFAS, Soil, Water and Ecosystem Sciences Department, Gainesville, FL*

[3]*University of Florida, IFAS, Agricultural and Biological Engineering Department, Gainesville, FL*

Keywords. basin management action plans, controlled-release fertilizer, nitrogen use efficiency

Field corn is an important crop grown in the Suwannee Valley region of North Florida. Field corn requires high nitrogen inputs to obtain optimal yields compared to other row crops commonly grown in the region, such as peanuts and cotton. Basin management action plans (BMAPs) have been implemented in this region to reduce the leaching and runoff of nutrients, such as nitrogen, into local bodies of water. In this experiment, an on-farm trial was conducted comparing the use of polymer-coated urea as a controlled-release fertilizer (CRF) to split applications of conventional fertilizer. Plant tissue and soil samples, collected at 0–15, 15–30, 30–60, and 60–90 cm depths, were sampled approximately every two weeks throughout the growing season. The study found that CRF treatment had no significant impact on corn yield, moisture content, soil nitrate concentrations below the root zone, or plant tissue nitrogen concentrations in both the 2021 and 2022 seasons. However, contrast results did reveal significant differences between treatments at specific depths and sampling dates. The conclusion of this study is that CRF treatment did not improve yield or tissue nitrogen content compared to split applications of conventional fertilizers. The maize weevil (*Sitophilus zeamais*), a major corn insect pest, likely contributed to the observed lower yields.

The Suwannee Valley is an agricultural region in North Florida with large acreage of row crops, such as field corn, peanuts, and cotton. Compared to other regional row crops, field corn requires higher nitrogen inputs to reach an optimal yield and therefore has higher potential for nitrogen leaching. The Suwannee Valley contains both the Suwannee and Santa Fe Rivers. In order to safeguard water quality and comply with basin management action plans (BMAPs) imposed for the rivers, it is necessary to improve nitrogen use efficiency (NUE) in agriculture. Controlled-release fertilizers (CRF) could potentially improve NUE. Controlled-release fertilizers are composed of soluble fertilizers surrounded by an outer coatings such as a polymer. Thicker coatings result in slower rates of nutrient release allowing CRF blends to be tailored for specific crops (Lawrencia et al., 2021; Du et al., 2008). The rate at which polymer-coated CRF releases nutrients is primarily influenced by soil temperature, with warmer soil temperatures resulting in a faster rates of nutrient release. The use of CRF reduces the amount of soluble N available at any one time compared to conventional fertilizers, minimizing the risk for N leaching from high-intensity rainfall events or irrigation (Liu et al., 2014; Simonne and Hutchinson, 2005).

There are advantages and disadvantages to CRF compared to conventional fertilizers. Along with the potential to increase NUE, CRF can reduce the number of fertilizer applications farmers need to make throughout the field corn growing season. The CRF blends used in research trials and extension demonstrations in the Suwannee Valley for field corn production are usually formulated to release over 90 days and are applied near the planting time. In contrast, it is generally beneficial to conduct multiple applications of conventional fertilizer due to the mobility of soluble N in soil and the various other loss mechanisms of applied N (Kabir et al., 2021). By reducing the frequency of fertilizer application, CRF saves growers time and money spent on inputs such as labor and fuel. It is especially beneficial for growers who lack center pivot fertigation systems or high clearance applicators, as it enables them to maintain fertilizer in the root zone even when corn height surpasses their high clearance application capabilities. The major disadvantage of CRF is the high cost compared to conventional fertilizers, which serves as the main barrier preventing the widespread adoption of this technology (Liu et al., 2021; Zhao et al., 2013).

Controlled release fertilizers, in the form of polymer-coated urea, have been investigated as a method of increasing NUE in North Florida in field corn, watermelon, and carrots. However, few trials in the region have consisted of replicated treatments, multiple seasons of experimentation, and examined the movement of nitrate below the crop root zone. This research aims to compare field corn fertilized via polymer-coated CRF to grower

Thank you to my co-authors and the grower cooperator for support conducting the trial. We also thank the Florida Department of Agriculture and Consumer Services for funding the trial during the 2021 season through the BMP Mini-grant Program.

*Corresponding author. Email: jcapasso@ufl.edu

standard practices of using dry water-soluble fertilizer in an on-farm trial. The objectives of this research were to: 1) determine if the use of CRF increases field corn yield; and 2) compare soil nitrate concentrations from various depths and plant tissue nutrient concentrations between fertilizer treatments. We hypothesized that CRF would increase field corn yield compared to the grower standard treatment. In addition, we hypothesized that the use of CRF would result in less nitrogen leaching below the root zone and result in higher average tissue N in the CRF treatment.

Materials and Methods

Two trials took place on a Columbia County, FL, farm in the same location and field during the 2021 and 2022 field corn seasons. During the 2021 season, the trial compared two fertilizer treatments: 1) CRF treatment—140 L/ha of urea ammonium nitrate and ammonium thiosulfate mixture (28–0–0–5) applied at planting, followed by a side-dress application of 684 kg·ha^{-1} of Harrell's polymer-coated urea (26–0–15) 14 days after planting; and 2) grower standard treatment—140 L·ha^{-1} of urea ammonium nitrate and ammonium thiosulfate mixture (28–0–0–5) at planting, followed by side dress application of 448 kg·ha^{-1} of ammonium nitrate and potassium chloride blend (20–0–12) at 30 and 40 days after planting. During the 2022 season, the trial also compared two fertilizer treatments; 1) CRF treatment—168 L/ha of urea ammonium nitrate, ammonium thiosulfate, and ammonium polyphosphate mixture (23–9–0) applied at planting, followed by a side-dress application of 661 kg·ha^{-1} of Harrell's polymer-coated urea (26–0–15) 7 days after planting; and 2) Grower standard treatment—168 L·ha^{-1} of urea ammonium nitrate, ammonium thiosulfate, and ammonium polyphosphate mixture (23–9–0) at planting, followed by side dress application of 448.34 kg·ha^{-1} of ammonium nitrate and potassium chloride blend (20–0–12) at 30 and 40 days after planting. The experimental design follows a non-randomized complete block design, with treatments ordered instead of randomized to facilitate the grower cooperator in managing fertilizer applications. Each treatment was replicated four times in the field, with the combination of one replicate of both treatments serving as a block (Fig. 1). Corn was planted at a seeding rate of 71,660 seeds per hectare in 2021 and 69,200 seeds per hectare in 2022 with a row spacing of 91.44 cm. The corn variety utilized during the 2021 season was pioneer 1847VYHR. The corn variety utilized during the 2022 season was pioneer 3016VYHR. The corn plot was four rows (3.05 m) wide and approximately 250 m long. The corn was irrigated based soil moisture sensor data collected from an adjacent corn field. The trial field, located in a low lying area with a shallow water table, required infrequent irrigation. In the 2021 season, multiple irrigation events occurred, but no irrigation was applied during the 2022 season.

Soil samples were collected using an augur approximately every 2 weeks throughout the growing season from 0–15, 15–30, 30–60, and 60–90 cm depths from each replicated plot. Plant tissue samples from each replicated plot were collected approximately every 2 weeks between the V5 (5 leaf collars present) and R6 (physiological maturity) growth stages. Soil and plant tissue samples were sent to Water's Agricultural Laboratories, Inc. (Camilla, GA), where soils were tested for nitrate and tissues were analyzed for nitrogen content.

The yield was determined by hand-harvesting corn and using the 1/1000 of an acre method (0.0004 ha) (Lauer 2022). Two yield measurements were randomly collected from the middle two rows along 4.45 m in each replicated plot. Corn ears were counted and weighed before and after shelling. A moisture analyzer was used to determine the moisture content of the shelled corn.

A two sample *t*-test was used to compare average yield results from both fertilizer treatments. A three-way analysis of variance (ANOVA), including time as a repeated measure in a linear model, was used to compare the main effects of treatment, depth, and time as well as their interaction effects on soil nitrate concentrations. A two-way ANOVA, including time as a repeated measure in a linear model, was used to compare the main effects of treatment and time as well as their interaction effects on corn tissue percent N. Results were determined statistically significant at a significance level of $P < 0.05$. Data were entered into Microsoft Excel and analyzed using the R 4.1.3 statistical program.

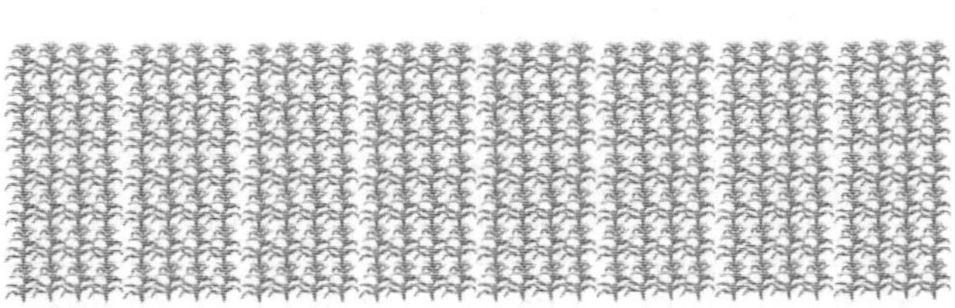

Fig. 1. Experimental design for an experiment conducted at a farm location near Lake City, FL. GS = grower standard treatment. CRF = controlled-release fertilizer.

Results

2021. The average yield result for the CRF treatment was 10038.57 kg·ha^{-1} and the average yield result for the grower standard treatment was 10612.89 kg·ha^{-1} (Fig. 2). A two-sample *t*-test was used to determine that there were no significant differences ($P = 0.63$) in yield results between treatments. The average grain moisture content in the corn at harvest was 17.32% in the CRF treatment and 18.38% in the grower standard treatment. Treatment was not found to have a significant effect ($P = 0.10$) on corn grain moisture content.

In the CRF treatment, the average soil nitrate at the 0–15, 15–30, 30–60, and 60–90 cm depths were 4.68, 1.57, 0.84, and 1.08 ppm, respectively. In the grower standard treatment, the average soil nitrate at the same depths was 4.01, 1.95, 1.55, and 1.03 ppm, respectively (Fig. 3). The three-way ANOVA found that treatment did not have a significant effect ($P = 0.3543$) on total soil nitrate concentration. However, depth ($P < 0.0001$) and time ($P < 0.0001$) were found to have significant effects on soil

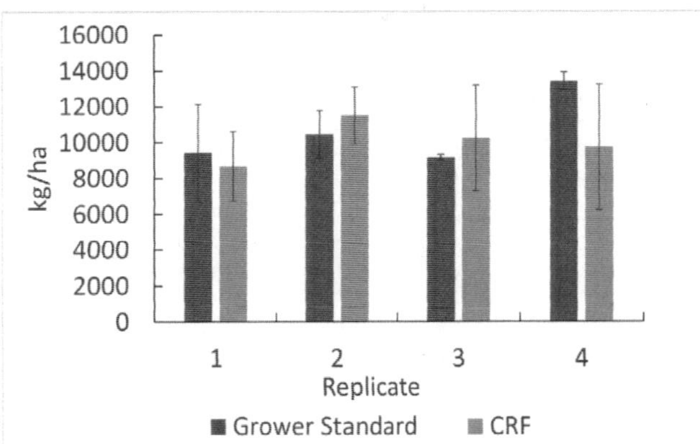

Fig. 2. Average corn yield results from the 2021 growing season collected from the controlled release fertilizer (CRF) and grower standard (GS) treatments at a farm location near Lake City, FL.

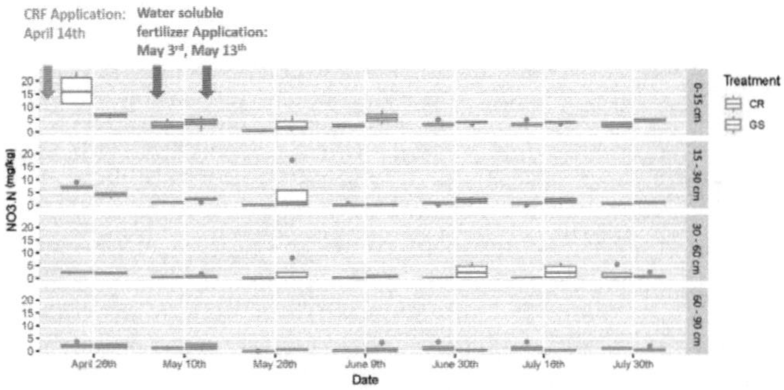

Fig. 3. Average soil nitrate results from 0–15, 15–30, 30–60, and 60–90 cm depths following applications of growers standard (GS) fertilization and controlled release fertilizer (CRF) collected during the 2021 growing season at a farm location near Lake City, FL.

nitrate concentrations. The interaction between treatment and time was significant ($P < 0.0001$); however, the interaction between treatment and depth was not significant ($P = 0.5351$). Contrast results found significant differences between treatments at 23 ($P = 0.0056$) and 53 ($P = 0.0425$) days after planting at the 0–15 and 15–30 cm depths, respectively (Table 1).

Throughout the season, the average plant tissue N concentration was 2.58% in the CRF treatment and 2.39% in the grower standard treatment (Fig. 4). The two-way ANOVA found treatment did not indicate a significant effect ($P = 0.7593$) on plant tissue percent N. However, time did have a significant effect ($P = 0.0002$) on plant tissue N but the interaction between treatment and time was not significant ($P = 0.4527$).

2022. The average yield result for the CRF treatment in 2022 was 8189.54 kg·ha^{-1} and the average yield result for the grower standard treatment was 8352.09 kg·ha^{-1} (Fig. 5). A two-sample *t*-test was used to determine that there were no significant differences in average yield measurements between treatments ($P=0.92$). The average moisture in the corn at harvest was 16.80% in the CRF treatment and 17.24% in the grower standard treatment. Again, treatment was not found to have a significant effect

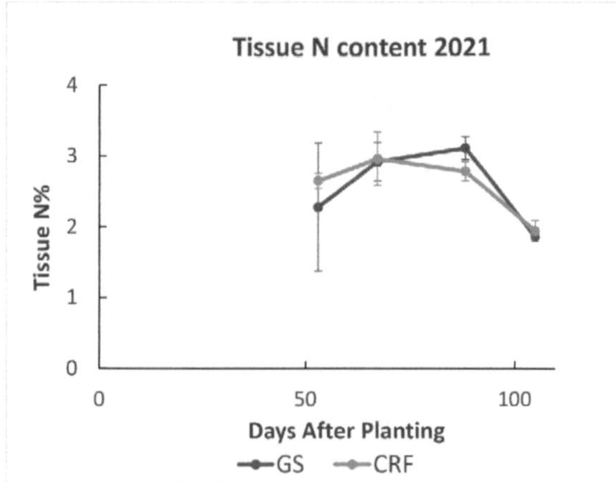

Fig. 4. Average corn plant tissue N content over each sampling event following applications of growers standard (GS) fertilization and controlled release fertilizer (CRF) during the 2021 growing season at a farm location near Lake City, FL. Error bars represent standard deviations at each sampling event.

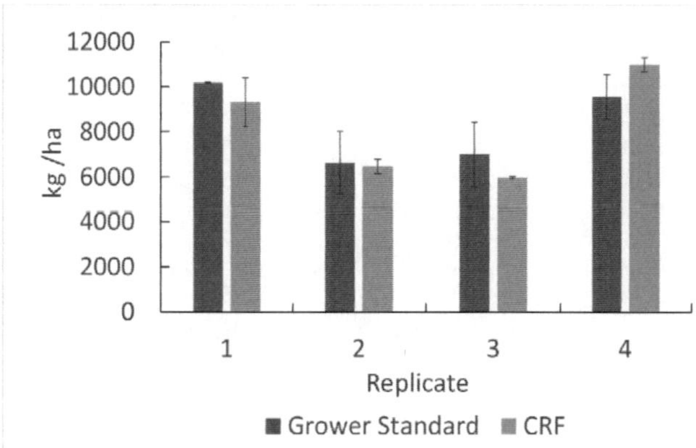

Fig. 5. Average corn yield results from the 2022 season collected from the controlled release fertilizer (CRF) and grower standard (GS) treatments at a farm location near Lake City, FL.

Table 1. Average marginal mean and contrast soil nitrate values by fertilizer treatment, soil depth, and days after planting during the 2021 season at a farm location near Lake City, FL.

Fertilizer treatment	Days after planting	Soil depth 0–15 cm	15–30 cm	30–60 cm	60–90cm
Average controlled release fertilizer (CRF) nitrate values (mg/kg)	23	16.57	6.97	2.26	2.09
	37	2.69	1.11	0.59	1.27
	53	0.38	0.24	0.05	0.01
	67	2.53	0.17	0.34	0.33
	89	3.30	0.85	0.42	1.39
	119	2.73	0.75	1.77	1.04
Average grower standard (GS) nitrate values (mg/kg)	23	6.57	3.96	2.13	1.94
	37	3.56	2.23	0.97	1.82
	53	2.89	4.98	2.32	0.60
	67	5.62	0.30	0.59	0.93
	89	3.75	1.82	2.74	0.54
	119	4.53	1.19	0.86	0.83
Contrasts (CRF – GS) $P < 0.05$ *	23	$P = 0.0056$ *	$P = 0.1201$	$P = 0.9303$	$P=0.9211$
	37	$P = 0.5590$	$P = 0.4832$	$P = 0.8041$	$P=0.7160$
	53	$P = 0.2186$	$P = 0.0425$ *	$P = 0.2023$	$P=0.7018$
	67	$P = 0.1116$	$P = 0.9303$	$P = 0.8717$	$P=0.6971$
	89	$P = 0.7682$	$P = 0.5337$	$P = 0.1950$	$P=0.5820$
	119	$P = 0.2866$	$P = 0.7705$	$P = 0.5604$	$P=0.8937$

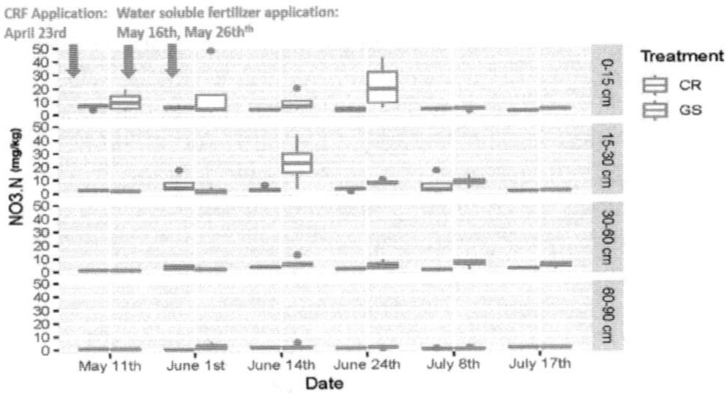

Fig. 6. Average soil nitrate results collected from 0–15, 15–30, 30–60, and 60–90 cm depths following applications of growers standard (GS) fertilization and controlled release fertilizer (CRF) during the 2022 growing season at a farm location near Lake City, FL.

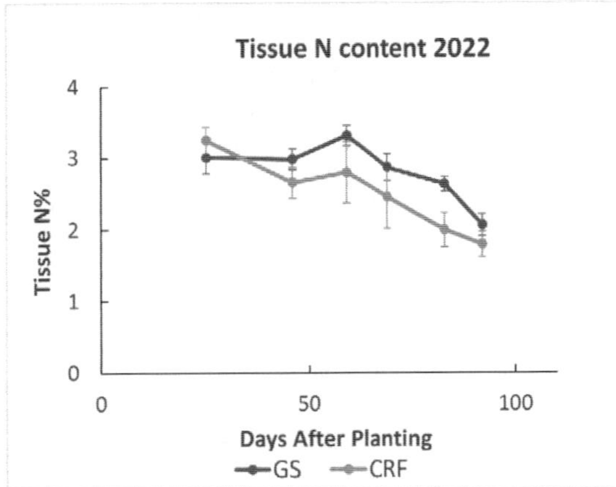

Fig. 7. Average corn plant tissue N content over each sampling event following applications of growers standard (GS) fertilization and controlled release fertilizer (CRF) during the 2022 growing season at a farm location near Lake City, FL. Error bars represent standard deviations at each sampling event.

($P = 0.394$) on corn grain moisture content. In the CRF treatment, the average soil nitrate at the 0–15, 15–30, 30–60, and 60–90 cm depths was 4.01, 3.39, 1.71, and 0.90 ppm, respectively. In the grower standard treatment, the average soil nitrate at 0–15, 15–30, 30–60, and 60–90 cm depths was 9.36, 7.20, 3.76, and 1.28 ppm, respectively (Fig. 6). The three-way ANOVA found that treatment had a significant effect ($P = 0.049$) on soil nitrate concentration. Depth ($P = 0.0017$) and time ($P = 0.0016$) were also found to have significant effects on soil nitrate concentrations. The interaction between treatment and time was significant ($P = 0.0077$), but the interaction between treatment and depth was not significant ($P = 0.1829$). The interaction between depth and time was significant ($P = 0.01$). The three way interaction between treatment, depth, and time was also significant ($P = 0.0096$). Contrast results found significant differences between treatments at 59 ($P = 0.0139$) and 69 ($P = 0.0184$) days after planting at the 15–30 and 0–15 cm depths, respectively (Table 2).

Throughout the season, the average plant tissue N concentration was 2.49% in the CRF treatment and 2.81% in the grower standard treatment (Fig. 7). Results from the two-way ANOVA indicate that treatment did not have a significant effect ($P = 0.0821$) on plant tissue percent N. Time was found to have a significant effect ($P < 0.0001$) on plant tissue percent N. The interaction between treatment and time was significant ($P = 0.0334$). The first tissue sampling event of both seasons showed a higher tissue N percentage in the CRF treatment compared to the grower standard, but most subsequent samplings found a higher N percentage in the grower standard treatment.

Discussion

These results suggest that the CRF treatment did not show significant differences in yield or corn grain moisture content compared to the grower standards treatment in the tests conducted in both 2021 and 2022 . Overall average corn moisture was lower in the 2021 and 2022 seasons in the CRF corn but not significantly different. If corn moisture had been found to be significantly lower in the CRF treatment, it could possibly benefit growers by indicating earlier maturity, enabling an earlier harvest.

Table 2. Average marginal mean and contrast soil nitrate values by fertilizer treatment, soil depth, and days after planting during the 2022 season at a farm location near Lake City, FL.

Fertilizer treatment	Days after planting	Soil depth			
		0–15 cm	15–30 cm	30–60 cm	60–90 cm
Average controlled release fertilizer (CRF) nitrate values (mg/kg)	25	6.43	2.25	0.89	1.02
	46	5.30	6.94	3.28	0.62
	59	3.70	2.68	3.20	1.60
	69	3.30	2.58	1.26	0.95
	83	3.38	5.07	0.38	0.12
	92	1.94	0.80	1.27	1.10
Average grower standard (GS) fertilizer nitrate values (mg/kg)	25	10.16	1.68	1.08	0.73
	46	14.43	1.45	1.16	2.24
	59	8.97	22.44	6.07	2.17
	69	21.14	7.30	4.10	1.25
	83	3.79	7.53	5.07	0.31
	92	3.35	1.08	3.84	1.28
Contrasts (CRF – GS) $P < 0.05$ *	25	$P = 0.4007$	$P = 0.8915$	$P = 0.9638$	$P = 0.9446$
	46	$P = 0.0963$	$P = 0.2453$	$P = 0.6172$	$P = 0.7185$
	59	$P = 0.2608$	$P = 0.0139$ *	$P = 0.5057$	$P = 0.8902$
	69	$P = 0.0184$ *	$P = 0.3036$	$P = 0.5109$	$P = 0.9422$
	83	$P = 0.9197$	$P = 0.5657$	$P = 0.3061$	$P = 0.9618$
	92	$P = 0.7364$	$P = 0.9461$	$P = 0.5487$	$P = 0.9680$

The polymer-coated CRF applied 1–2 weeks after planting had comparable results in total soil nitrate concentrations compared to split applications of side-dressed conventional fertilizer at 30 and 40 days after planting (DAP). Higher nitrate concentrations were observed in the grower standard treatment compared to the CRF treatment during the 2022 season, but only at the surface (0–15 cm) at 69 days after planting and upper soil horizons (15-30 cm) at 53 days after planting, but not at depths below the root zone (Table 2). Therefore, we speculate that similar levels of leaching occurred in each treatment, as there were no significant differences between treatments at lower soil depths below the root zone. The CRF treatment exhibited significantly higher soil nitrate concentrations only once throughout the two seasons. This occurred during the 2021 season 21 days after planting following the application of CRF fertilizer and preceding the split applications of conventional fertilizer. This highlights the importance of considering temporal dynamics when analyzing results due to the significant effects of time and depth on soil nitrate concentrations.

While the treatments did not show significant effects on plant tissue N in either year, time did have a significant effect during both seasons and the interaction between time and treatment was significant during the 2022 season. This indicates that any influence of treatment on tissue N may differ due to the timing of fertilizer application or at different stages of plant growth.

The maize weevil (*Sitophilus zeamais*), a major corn insect pest, was observed in the trial at harvest during both the 2021 and 2022 field corn seasons. Damage to corn kernels from adult females chewing holes to lay eggs and from larvae feeding from within the kernel was observed during the corn shelling and weighing process. Yield losses by maize weevil infestations have been reported between 20 to 90% in both the field and post-storage process (Abebe et al., 2009). The pest is believed to be the cause for the low yields as the corn did not reach its yield potential based on field history and applied N (personal communication with the grower).

Conclusion

The study found that the use of CRF did not cause significant differences in yield compared to the grower standards treatment in both the 2021 and 2022 seasons. The application of CRF applied 1–2 weeks after planting, resulted in similar soil nitrate concentrations as split applications of side-dressed conventional fertilizer at 30 and 40 days after planting. However, there were significant differences in soil nitrate concentrations between treatments in the 2022 season, primarily observed at the surface and upper soil horizons. These findings suggest that nitrate leaching was similar between the CRF and the grower standard treatment, with no significant differences observed at soil depths below the root zone. The presence of the maize weevil pest was observed during both seasons and resulted in damage to corn kernels. The infestation is believed to have contributed to yield losses. Future research should be conducted in locations with reduced maize weevil pest pressure, preferably under center-pivot irrigation, as most field corn in the Suwannee Valley Agricultural area is produced using center-pivot irrigation.

Literature Cited

Abebe F., Tefera T., Mugo S., Beyene Y., & Vidal S. 2009. Resistance of maize varieties to the maize weevil *Sitophilus zeamais* (Motsch.) (Coleoptera: Curculionidae). *African Journal of Biotechnology* .8(21). https://doi.org/10.4314/ajb.v8i21.66077

Davies B., Coulter J.A., & Pagliari, P.H. 2020. Timing and rate of nitrogen fertilization influence maize yield and nitrogen use efficiency. *PLoS ONE*. 15(5):e0233674. https://doi.org/10.1371/journal.pone.0233674

Diaz D.A.R., Sawyer J.E., Barker D.W., & Mallarino, A.P. 2010, Runoff nitrogen loss with simulated rainfall immediately following poultry manure application for corn production. *Soil Sci. Soc. Am. J.* 74:221-230. https://doi.org/10.2136/sssaj2009.0134

Du, C., Tang D., Zhou J., Wang H., & Shaviv A. 2008. Prediction o nitrate release from polymer-coated fertilizers using an artificial neural network model. *Biosystems Engineering* 99:478–486. https://doi.org/10.1016/j.biosystemseng.2007.12.003

Lauer J. 2002. Methods for calculating corn yield—Wisconsin corn agronomy. Available online: http://corn.agronomy.wisc.edu/AA/A033.aspx

Lawrencia D., Wong S.K., Low, D.Y.S., Goh B.H., Goh J.K., Ruktanonchai U.R., Soottitantawat A., Lee L.H., & Tang S.Y. 2021. Controlled release fertilizers: A review on coating materials and mechanism of release. *Plants* 10:238. https://doi.org/10.3390/plants10020238

Liu G., Zotarelli L., Li Y., Dinkins D., Wang Q., & Ozores-Hampton M. 2014. Controlled-release and slow-release fertilizers as nutrient management tools. Extension Publication HS1255/HS1255, University of Florida, Gainesville, FL. https://doi.org/10.32473/edis-hs1255-2014

Liu G., Simonne E.H., Li Y., Hutchinson C.M., Warren M., & Lands S. 2021. Controlled-release fertilizers for commercial potato production in Florida. Extension Publication HS941, University of Florida, Gainesville, FL. https://doi.org/10.32473/edis-hs187-2003

Kabir T., De Laporte A., Nasielski J., & Weersink A. 2021. Adjusting nitrogen rates with split applications: Modelled effects on N losses and profits across weather scenarios. *European Journal of Agronomy*. 129:126328. https://doi.org/10.1016/j.eja.2021.126328

Zhao B., Dong, S., Zhang J., & Liu P. 2013. Effects of controlled-release fertiliser on nitrogen use efficiency in summer maize. *PLOS ONE*. 8:e70569. https://doi.org/10.1371/journal.pone.0070569

Proc. Fla. State Hort. Soc. 136:213. 2023.

—Scientific Note—

Alternative Sprinkler Systems for Reduced Water Use in Florida Strawberry Production

Josue St. Fort[1], Carlene A. Chase[2]*, and Vivek Sharma[3]

[1]*School of Natural Resources and the Environment, University of Florida, IFAS, Gainesville, FL*

[2]*Horticultural Sciences Department, University of Florida, IFAS, Gainesville, FL*

[3]*Agricultural and Biological Engineering Department, University of Florida, IFAS, Gainesville, FL*

Keywords. *Fragaria* ×*ananassa*, bare-root transplants, irrigation, low-volume

Strawberry (*Fragaria* ×*ananassa*) production on raised beds covered with black plastic mulch in Florida uses a dual irrigation system: drip irrigation system and overhead sprinkler irrigation. The latter is used during bare-root transplant establishment (late September to mid-October) to cool the plastic mulch during high daytime air temperatures to prevent the desiccation of transplants while they produce new roots. Sprinkler irrigation is also used later in the season for frost and freeze protection (Poling et al., 1991). Growers typically employ an impact sprinkler system for these practices, which can result in very high-water use (Santos et al., 2012). Therefore, this study aimed to conserve water during these operations by evaluating four low-volume sprinkler systems as alternatives to the impact sprinkler system in an on-station experiment and by comparing one of the best-performing low-volume sprinklers with the grower's sprinkler system in an on-farm trial.

Findings

The Jain Mini-Revolver, Netafim Supernet, Xcel Mini-Wobbler, and Xcel Wobbler sprinklers were compared with an impact sprinkler at the Plant Science Research and Education Unit in Citra, FL, during the 2021–22 and 2022–23 seasons. The systems were arranged in a randomized complete block design with four replications and were run for eight hours per day for 11 days beginning immediately after transplanting 'Florida Brilliance' bare-root transplants. Water use was determined for the establishment and frost/freeze protection. Transplant survival, plant vigor, vegetative growth parameters, and yield were measured to evaluate the efficacy of the low-volume sprinklers. During freezing nights, the irrigation was triggered automatically at 37 °F in 2021–22 and 34° F in 2022–23 for frost/freeze protection of the crop. The percentage of damaged fruits and flowers per irrigation treatment was used to evaluate the adequacy of the water application.

One of the best-performing, low-volume sprinklers (Jain Mini-Revolver) was compared to the grower's system (Nelson Rotator) in an on-farm experiment in Plant City, FL, during the 2022–23 season. The Nelson Rotator was adopted by the grower as a lower-volume sprinkler than the standard impact sprinkler. The experimental design was a complete block with four replications with a restriction on randomization of the sprinkler systems. Water use, transplant survival, plant vigor, vegetative growth, and yield of the 'Odessa' cultivar were assessed. Freeze protection was initiated manually at 32 °F at the farm.

We found that the low-volume sprinklers reduced water use during establishment by up to 79% in the on-station trial, with no adverse effects on 'Florida Brilliance' establishment and growth in both seasons. Water use for frost/freeze protection also was significantly lower with the low-volume sprinklers. However, the Xcel Wobbler and Mini-Wobbler sprinklers will require initiation of irrigation at temperatures above 34 °F for effective frost protection. Yield was significantly higher with the Netafim Supernet system in 2021–22 than with the other low-volume sprinklers, but not statistically significantly different from the impact sprinkler. In 2022–23, only the Xcel Wobbler system resulted in a significantly lower yield than the impact sprinkler system.

A 42% reduction in water use was observed during transplant establishment without adverse effects on the growth or yield for the on-farm trial. The low-volume sprinkler system provided effective protection for the crop from freeze damage, and no injury was detected on flowers and fruits. Thus far, our results indicate that low-volume sprinklers may be viable options for water conservation in Florida strawberry production.

Suggestions for Future Direction

The on-farm trial will be repeated during 2023–24 to confirm the performance of this low-volume sprinkler during establishment and freeze protection. Since more low-volume nozzles per unit area are required, it is possible that the equipment and installation costs may be greater than for sprinkler systems currently in use. Therefore, an economic feasibility study will aid grower and policy making decisions regarding adoption.

Literature Cited

Poling E.B., Fuller H.P., & Perry K.B. 1991. Frost/freeze protection of strawberries grown on black plastic mulch. *HortScience*. 26:15–17. https://doi.org/10.21273/HORTSCI.26.1.15

Santos B.M., Salame-Donoso, T.P., & Whidden A.J. 2012. Reducing sprinkler irrigation volumes for strawberry transplant establishment in Florida. *HortTechnology*. 22:224–227. https://doi.org/10.21273/HORTTECH.22.2.224

*Corresponding author. Email: cachase@ufl.edu

Proc. Fla. State Hort. Soc. 136:214. 2023.

—Scientific Note—

Micro-forests Yield Macro Benefits: Utilizing Micro-forests in Climate Mitigation Efforts

MARGUERITE BECKFORD*

Citrus County Extension, University of Florida, IFAS, Sarasota, FL

Many rapidly urbanizing communities recognize the importance of preserving and expanding tree canopy. However, balancing competing priorities between providing greenspace for the public with land development demands continues to be a challenge. Micro-forests can provide a feasible solution to this challenge by providing the ecosystem benefits of tree canopy expansion, on relatively small parcels of greenspace. Micro-forests are a biodiverse group of densely planted native trees, based on the Miyawaki Method, and the World Economic Forum estimates that planting micro-forests can increase the worldwide capacity of communities to sequester up to 10 gigatons of carbon by the year 2050.

The University of Florida/IFAS Extension urban forestry program *Treejuvenation™ Florida* partnered with two local community volunteer groups, providing technical assistance for the Sarasota Urban ReForestation (SURF) micro-forest project. The SURF micro-forest was established with 250 trees, in a local park, on a parcel of land approximately 14,000 sq. ft. in area. To promote the micro-forest, *Treejuvenation™ Florida* launched a 'Tree Trail Tour' scavenger hunt to educate visitors on the specific ecosystem benefits of each tree.

Ninety-six percent of participants on the SURF micro-forest Tree-Trail-Tour indicated an increase in knowledge about ecosystem benefits of urban trees, and the contribution of micro-forests to mitigating climate change. With a mature native tree having a lifetime capacity to sequester more than 3,000 lbs. carbon-dioxide-equivalent of carbon, these 250 tree plantings represent a carbon sequestration potential of close to 1M pounds of carbon. Following on the success of the original micro-forest project, a second micro-forest was installed with 1000 trees on a greenbelt in a nearby community.

Installing micro-forests is a feasible solution to addressing the challenge of balancing urban communities' needs for ecosystem benefits of increased tree canopy, with limited land availability. Efforts are underway to install micro-forests in additional locations and to use them for urban forestry extension demonstration sites and program activities. Using micro-forests in urban forestry extension programs can promote awareness of the benefits of urban trees, as well as increase support for the contribution of urban afforestation efforts to climate change mitigation.

Treejuvenation™ Florida's project and resource partners include: Dr. Michael Andreu, Associate Professor, University of Florida School of Forestry, Fisheries and Geomatic Sciences, Sarasota Veterans for Common Sense; Sarasota County Parks and Natural Resources Dept. and iTree© My Tree Benefits database.

*Corresponding author. Email: mbeckford@ufl.edu

Proc. Fla. State Hort. Soc. 136:215. 2023.

Assessment of *Sargassum* spp. Management Strategies in Southeast Florida

Afeefa A. Abdool-Ghany[1,2], Trent Blare[3,4*], and Helena M. Solo-Gabriele[1]

[1]*Department of Chemical, Environmental, and Materials Engineering, College of Engineering, University of Miami, Coral Gables, FL*

[2]*Department of Biological Sciences, Institute of Environment, Florida International University, Miami, FL*

[3]*University of Florida IFAS, Tropical Research and Education Center, Homestead, FL*

[4]*International Potato Center, Ecuador Country Office, Quito, Ecuador*

—abstract—

This study evaluated the economic viability of *Sargassum* management strategies in southeast Florida. Semi-structured interviews were conducted with key informants, beach managers and potential users. To compare the strategies, a cost and market analysis was conducted to assess the value of *Sargassum* compost relative to other compost products available in the market. The findings indicate that composting *Sargassum* is an economically viable option and suggests the potential for recouping some of the costs associated with beach cleaning through the sale of compost. Furthermore, the utilization of *Sargassum* compost could offer a cost-effective alternative as a soil amendment, especially in the face of rising fertilizer costs. However, it is crucial to address concerns regarding compost quality due to contaminants such as plastics and arsenic before implementing a composting operation. This aspect necessitates thorough investigation and remediation to ensure that the resulting compost meets quality standards and is free from potentially harmful contaminants.

The full paper can be found at https://doi.org/10.1016/j.rcradv.2023.200175

*Corresponding author. Email: trent.blare@cgiar.org

Agroecology and Natural Resources Section

Proc. Fla. State Hort. Soc. 136:216–222. 2023.

Bamboo Production Practices in Florida: Water and Fertilization Requirements

Davie M. Kadyampakeni[1]*, Labake O. Agunbiade[1], Cyrus J. Januarie[1], Steven Sargent[2], Hardev Sandhu[3], Joao Vendramini[4], Hardeep Singh[5], and Lisa Hickey[6]

[1]*University of Florida, IFAS, Citrus REC, Lake Alfred, FL*

[2]*University of Florida, IFAS, Horticultural Sciences, Gainesville, FL*

[3]*University of Florida, IFAS, Everglades REC, Bell Glade, FL*

[4]*University of Florida, IFAS Cattle Range REC, Ona, FL*

[5]*West Florida REC, Jay, FL*

[6]*University of Florida, IFAS, Extension, Palmetto, FL*

Keywords. best management practices, *Dendrocalamus asper*, precision fertilization

Bamboo production is estimated at around 1,300 acres in Florida, growing by 50% annually. Currently, growers are using trial and error methods to estimate production requirements related to water and fertilization and which varieties to grow. Stahl and Sargent (2021) provided a synopsis and comprehensive review of postharvest management practices for bamboo in Florida. In this review, we further elucidate the production requirements for bamboo using lessons and literature from other regions. In the near future, ongoing research will enable further updates on the production needs and best management practices of bamboo as the industry grows.

Bamboo species are commonly found in temperate, tropical, and subtropical regions, and they are known for a wide variety of uses such as for industrial purposes, for making paper, as construction and building materials, as fresh edible shoots, and even in traditional medicine. They have high and fast growth rate and biomass production (Kleinhenz and Midmore, 2001). Bamboo plants have recently been explored for their potential use in phytoremediation. Recent studies found that several bamboo species show very high endurance in heavy metal contaminated soils with specific heavy metal absorption capacities (Yan et al., 2015; Emamverdian et al., 2018; Jiang et al., 2019). Bamboo species can be a great substitute for food crops due to its rapid growth and annual regenerative abilities. This species of plants is one of the underutilized plants for food production even though it is very rich in nutrients. The edible young shoots are rich in nutrients and bioactive compounds which are essential for metabolism and normal functioning of the organs of the human body to promote good health and well-being (Ikewuchi and Ikewuchi, 2009). Research conducted by Saini et al. (2017) showed the mineral nutrients present in fresh shoot of bamboo. Some of the mineral nutrients analyzed were potassium (K), phosphorus (P), magnesium (Mg), calcium (Ca), sulfur (S), silicon (Si), sodium (Na), and chlorine (Cl). The result showed high amount of K and P, and moderate incidence of Mg, Ca, S, and Si in the shoots of the bamboo studied. The authors further emphasized the fact that the fresh shoots of bamboo are rich in the above-mentioned mineral elements although, the intention of the authors was to address the issue of hidden hunger due to mineral deficiencies in the food consumed. With this knowledge, the availability of certain nutrients where bamboo is grown will not only improve bamboo production but also enhance the quality of the bamboo species and other parts of the bamboo consumed (fresh young shoots) and those used for traditional medicine (leaves). Morton et al. (2023) established the relationship between the availability of certain soil nutrients and human health.

The potential contribution of bamboo to sustainable natural resources management, such as biomass productivity, and its ecological benefits have only been recently recognized (Song et al. 2011; Liu et al. 2018; Srivaro et al. 2018; Dwivedi et al. 2019). Hence, there is the need to adopt new and sustainable measures for the production and development of underutilized plant species such as bamboo. To understand and explore the growth patterns of bamboo species, it is critical to explore the productivity and nutrient dynamics of the specie.

In recent years, growers have been growing bamboo in Florida especially in areas where farmers have removed or abandoned citrus groves. Bamboo can be used for a variety of uses including timber, wood and several other recreational uses. As a group of generally tall, woody grass species, bamboo in the West is an enigma in terms of its science and technology, and its history, habit and the opportunities it presents. Given its central role in the lives of so many in Asia—reports suggest that up to 2.5 billion people use it as a source of food and fibre-and its increasing importance in Africa and the Americas, it might be expected that significant amounts of research aimed at increasing productivity would have been undertaken by the scientific community, but the reality is otherwise (Midmore, 2009).

The authors are grateful for the funding from the University of Florida Institute of Food and Agricultural Sciences project # 00132311-212-2200.

*Corresponding author. Email: dkadyampakeni@ufl.edu

This paper reviews and discusses water and fertilization requirements for bamboo plants for consideration of adaptation to Florida conditions. A discussion on the processing needs was already provided by Stahl and Sargent (2021).

Fertilizer Needs and Nutrient Distribution in Bamboo Production

Bamboo farming has grown exponentially over the years. With the crisis in the citrus industry, growers in Florida are exploring alternative crops to citrus. One of the crops commonly resonating with the growers is bamboo due to its potential in the booming marketing and its ease of production in the tropical region. However, little is known in terms of nutrient management regarding Florida's sandy soils. This leaves growers with no option but to trial and error about nutrient management practices. In this review, we shed more light on nutrient management practices used in other parts of the world which could be a foundation for developing Florida site- and soil-specific practices that are based on the right time, right amount, right placement, and right timing (UC Sustainable Agriculture Research and Education Program, 2017).

Bamboo plants have an extensive root system (Tripathi & Singh, 1996) that is effective at absorbing nutrients that are available to the plant, which is why past research has shown that bamboo responds quickly to fertilization (Li et al. 1998; Shanmughavel & Francis, 2001; Kleinhenz et al. 2003; Gao et al. 2016). Bamboo growth and production are usually restricted by limited availability of nutrients in the soil; therefore, fertilizers are frequently used to increase the productivity and quality of bamboo (Zhang, et al. 2014). The need for regular fertilization is important to maintain productivity of bamboo due to high amount of nutrients removed during harvesting of edible shoots and culms (Kim et al., 2018).

A three-year study conducted in India by Shanmughavel and Francis (2001) to determine bio-productivity and nutrient cycling processes of bamboo forests. The bamboo seedlings were grown in soil applied with 3800 kg·ha^{-1} (N), 360 kg·ha^{-1} (P), 3600 kg·ha^{-1} (K), 1600 kg·ha^{-1} (Ca), and 1800 kg·ha^{-1} (Mg). There was an increase in total biomass (47.48 t·ha^{-1}), above-ground biomass with a concomitant increase of diameter at breast height (DBH) and height. In this study, the authors estimated that 94% of the N, P, Ca, and Mg applied was taken up by the bamboo species plant and retained in the non-photosynthetic biomass, and only 6% was returned to the soil. This implies that bamboo plants are heavy nutrient feeders, and this study therefore emphasizes the importance of fertilizer application to soil particularly after harvesting of bamboo shoots. Another study conducted by Kim et al. (2018) was carried out to determine the effects of regular fertilization on the nutrient distribution of biomass components (culms, branches, leaves, roots, rhizomes) in a bamboo stand regularly fertilized for three decades (Table 1). Fertilizer was applied based on the bamboo forest guidelines. Higher culm densities (3400–4732 culm/ha), and slightly higher DBH (8.37–12 cm) was recorded. There was also significant increase of Ca in culms, P and K in branches as well as N, P, and Mg in leaves of the regularly fertilized plots indicating an increased uptake of these nutrients. There was no significant increase in concentration of K even though it was regularly applied (Table 2). A previous study by Piouceau et al. (2014) also reported that there was no significant increase in K concentration in leaves with increasing fertilization. This may be attributed to rapid translocation from leaves for the new growth of new tissues. Older bamboo had higher concentrations of N, P, K concentration in leaves compared to bamboo less than a year old. Calcium concentration in above-ground bamboo components was significantly higher in older bamboo plants than less than a year-old bamboo. There was generally increased concentration of N, P, and K in the roots and rhizome of fertilized plots compared with unfertilized plots. The results from this study suggested that regular fertilization is a crucial nutrient management practice to sustain the productivity of above-ground components (shoots and culms) in bamboo plants. Therefore, the strategic use of fertilizer in the management of bamboo for shoot production may provide economic benefits to bamboo growers.

Table 1. Nutrient stocks of bamboo components in a bamboo stand focusing on potassium (K), calcium (Ca) and magnesium (Mg) (n = 3) (Adapted from Kim et al. 2018).

Bamboo component	Treatment	Nutrient stock (kg·ha^{-1})[z] K	Ca	Mg
Culms	Fertilized	510.4 (61.7)[y]	6.4 (0.7)	9.8 (1.2)
	Unfertilized	243.5 (17.5)	7.8 (0.6)	18.3 (1.3)
	P-value	0.01[z]	0.22	< 0.01
Branches	Fertilized	29.0 (2.7)	0.9 (0.1)	1.9 (0.2)
	Unfertilized	55.6 (3.3)	1.0 (0.1)	3.1 (0.2)
	P-value	< 0.01	0.56	0.01
Leaves	Fertilized	52.4 (5.3)	13.8 (1.4)	5.2 (0.5)
	Unfertilized	62.7 (4.5)	26.3 (2.6)	6.8 (0.6)
	P-value	0.22	0.01	0.10
Roots	Fertilized	14.7 (2.7)	1.0 (0.3)	1.9 (0.4)
	Unfertilized	39.8 (2.8)	6.1 (0.5)	14.9 (0.9)
	P-value	< 0.01	< 0.01	< 0.01
Rhizomes	Fertilized	16.8 (6.8)	0.8 (0.3)	1.0 (0.3)
	Unfertilized	27.9 (18.0)	2.2 (1.8)	5.6 (4.6)
	P-value	0.60	0.48	0.37

[z]1 kg·ha^{-1} = 0.892179 lbs/ac

[y]Values in parenthesis are standard error. Bold values denote significance at $P < 0.05$.

Nitrogen Availability, Effects and Requirements for Bamboo Plants

Nitrogen is a strong indication of soil fertility and a limiting component for plant growth and development. Nitrogen is mostly found in organic forms in terrestrial environments, where its availability is constrained by the chemistry and structure of those forms (Qian et al., 2021). Just like many other plants, bamboo, through its roots, takes up N from the soil and is absorbed as ions such as ammonium (NH_4^+) and nitrate (NO_3^-). Nitrogen is the most utilized macronutrient in bamboo production. Nitrogen is responsible for giving newly sprouting leaves and shoots green color and promoting growth. For bamboo plants, which are fast-growing and versatile, a fertilizer high in N may be best. Understanding the effects of N on bamboo growth is essential for optimizing cultivation practices and enhancing bamboo productivity. Numerous research studies have examined how N affects bamboo growth, concentrating on features including biomass output, height, and culm development. Nitrogen is essential for the creation of proteins and enzymes and is a major component of chlorophyll, the pigment that is responsible for photosynthesis. Greater photosynthetic activity and total plant growth are frequently linked to adequate N supply.

Yang et al. (2022) investigated how N fertilization affected the growth of a commonly grown type of bamboo called Moso bamboo (*Phyllostachys edulis*). Higher N levels were found to

Table 2. Amount of total (kg·ha^{-1}) of applied potassium (K) and leaf nutrient levels before shoot emergence. (Adapted from Treynor and Midmore, 2009).

	Site 3 Treatments						Site 2 Treatments			
	25%		100%		200%		25%		100%	
Year	Soil	Leaf	Soil	Leaf	Soil	Leaf	Soil	Leaf	Soil	Leaf
2001–02	74	–[z]	297	–	594	–	42	–	168	–
2002–003 (26 Oct 2002)[y]	72	2.05	287	2.05	574	1.92	33	1.68	132	1.69
2003–04 (7 Oct 2003)	91	1.75	363	1.59	726	1.59	92	1.43	369	1.38
2004–05 (11 Oct 2004)	94	1.73	374	1.64	748	1.65	93	1.66	369	1.37

[z]Not available.

Sampling dates for leaf K in brackets ( ).

positively influence shoot growth, resulting in enhanced biomass accumulation, diameter at breast height, and clump height. They did, however, also highlight the significance of applying nitrogen in a balanced manner to prevent any unfavorable outcomes, such as nutrient imbalances and environmental issues. Li et al., (2018) found similar trends while applying N-fertigation. On the other hand, increased nitrogen supply to bamboo leads to reduced N use efficiency in young bamboo plants (Sofi et al., 2018). This underscores the necessity of understanding the appropriate N requirements for bamboo species, considering growth, culm, and root development. In efforts to develop N application guidelines for bamboo production, various experiments have been conducted and each presented its own optimal amount of N regarding bamboo stand density (Table 3).

Zhang et al. (2014) found that this rate of N (207 kg·ha^{-1}) improved bamboo yield as compared to the control. In their study, where they compared site-specific nutrient management (SSNM) to farmer fertilizer practices (FFP), bamboo plants increased the emergence rate, diameter, and number of shoots produced if fertilizer is applied, and applied on the specific site of the clump in contrast to FFP´s broadcasting method of fertilizer application. Other literature by Kim et al. (2018) revealed that regular application of N at the rate of 244 kg·ha^{-1} N improves shoot and new culm production. Increasing N levels from 250 to 400 kg·ha^{-1} N significantly increased shoot and culm production in Zhejiang Province, China (Gao et al., 2016). Additionally, at the rate of 250 kg·ha^{-1} N the photosynthetic functions of the leaves were also promoted, i.e. quantum production and electron transport in photosystem II, and transduction of absorbed light energy in photosystem I. Therefore, for better yield production and promotion of photosynthesis, 250 kg·ha^{-1} N was concluded optimal in the study. On the other hand, increasing the rate of N fertilization does not always increase the yield for bamboo if other conditions are not met like optimal irrigation rate. Kleinhenz et al. (2003) emphasized the need for finding the right combination. In their study, 250, 375, and 500 kg·ha^{-1} N had no significant differences in yield while under low irrigation while at higher irrigation rates, 500 kg·ha^{-1} N improved shoot yield significantly. Nevertheless, others went as far as to find the 1972 kg·ha^{-1} N rate optimal for Moso bamboo with a density combination of 3000 plants/ha. This demonstrates the need to understand the variability of N rates for bamboo growth along with planting density, and the right irrigation rate.

Phosphorus Availability, Effects, Requirements and Deficiency for Bamboo Plants

Bamboo, known for its rapid growth and gigantic shape, has a root structure that helps in nutrient absorption owing to P. Phosphorus is a vital nutrient that is particularly important to bamboo growth because of its complex effects on plant development (Sun et al., 2023). Bamboo's access to P is greatly impacted by the characteristics of the soil. Phosphorous availability in the root zone is influenced by various factors, including microbial activity, soil pH, and the amount of organic matter present (Hinsinger, 2001). Thus, keeping the soil at its ideal state is essential to guaranteeing bamboo plants a steady supply of phosphorus. Phosphorus is available for plant uptake in the form of orthophosphates, $H_2PO_4^-$ and HPO_4^{2-} (Roberts & Johnston, 2015). Phosphorus participates in essential physiological processes within plants (Khan et al., 2023). It is integral to the structure of nucleic acids, adenosine triphosphate (ATP) synthesis, and other critical biomolecules. However, in bamboo growth, the role of P extends to root development, energy transfer, and overall plant vigor.

Research has demonstrated that P benefits bamboo species' root development (Yang et al., 2021). A more widespread and effective root system is linked to adequate phosphorus levels through improved root elongation and branching (Ni et al., 2021). In energy transfer activities, P is crucial, especially in ATP synthesis (Khan et al., 2023), which bamboo needs for rapid growth and shoot development. Furthermore, P supports the growth of bamboo overall by aiding in photosynthesis, which converts light energy into chemical energy (Liu et al., 2017).

Table 3. Summary of rates of nitrogen used with varying plant density.

Location	Rate used (kg·ha^{-1})[z]	Planting density (plants/ha)	Variety	Author
China (Southeast)	207	3301–3435	*Phyllostachys edulis*	Zhang et al. (2014)
China (Mainland)	3945	2700–3750	*Phyllostachys edulis*	Maoyi et al. (1991)
South Korea	244	–[y]	*Phyllostachys edulis*	Kim et al. (2018)
China (Zhejiang Province)	250–400	160–180	*Phyllostachys edulis*	Gao et al. (2016)
Australia	500	–	*Phyllostachys edulis*	Kleinhenz et al. (2003)
Chongqong	400	–	*Phyllostachys pubescens*	Li et al. (1998)

[z]1 kg·ha^{-1} = 0.892179 lb/ac.

[y]Not available.

Table 4. Summary of rates of P used in past studies.

Location	Rate used (kg·ha^{-1})^{z}	Variety	Author
China	50	*Phyllostachys edulis*	Zhang et al. (2019)
China	300	*Phyllostachys negra*	Bo et al. (2021)

^{z}1 kg·ha^{-1} = 0.892179 lb/ac.

If P is not supplied in sufficient quantity, bamboo plants develop deficiency. This deficiency has a detrimental impact on the growth of the bamboo plants. According to with Wang et al. (2020) and Sun et al. (2023), low phosphorus reduces chlorophyll content and compromises the functioning of photosystem II and CO_2 assimilation, eventually lowering bamboo leaf photosynthetic capacity. In a different study, Phosphorus deficiency impeded sucrose to starch conversion and hexose allocation, limiting cell division and elongation necessary for shoot growth (Lemoine et al. 2013).

Various rates of P have been used in different studies (Table 4). In a nutshell, various studies have indicated that the optimal rate of P fertilization for bamboo growth is between 50 and 300 kg·ha^{-1} (Bo et al., 2021; Zhang et al., 2019). Excessive application of P may have negative consequences. Sustaining availability within this goal range ensures efficiency while meeting the nutritional needs of plants.

Silicon Mineral Nutrient in Bamboo Production

Silicon (Si) is classified as a non-essential element that is beneficial for increased productivity and sustainable production (Liang et al. 2007; Kleiber 2018). Plants take up Si in the form of silicic acid, which is transported to shoots and after loss of water, it polymerizes as silica gel on the surface of leaves and stems (Ma et al., 2001). Silicon has proven to be beneficial to aid plant tolerance to biotic and abiotic stress (Luyckx et al. 2017; Lopez-Perez et al. 2018; Yan et al. 2018). Bamboo species have the potential to accumulate large amount of silica in the above-ground biomass part particularly the leaves (Motomura et al. 2004). The rate of Si translocation from root to shoot is usually above 90%, which leads to some physio-molecular responses such as improving essential nutrient uptake, plant growth and development, enzymatic activity and genetic expression (Kim et al. 2017). Bamboo being a monocotyledonous plant belonging to the family *Poaceae* tends to accumulate high amount of Si (Currie and Perry, 2007). Therefore, it is imperative to include plant-available Si to fertilizers as a management approach to protect plants from environmental stress and soil nutrient depletion to achieve sustainable productivity (Guntzer et al., 2012). Although Si distribution (Table 5) in bamboo has been described in several studies (Collin et al. 2012; Ding et al. 2008b; Li et al. 2006; Lux et al. 2003), the effects of Si on bamboo growth are poorly described (Ma and Takahashi 2002). Collin et al. (2012) assessed the interaction between Si and copper (Cu) metal concentration in bamboo, their results showed that Cu concentration was relatively low whereas Si was accumulated at a level of 82–159 mg·g^{-1} in leaf dry weight. In another study, Collin et al. (2013) investigated the effect of Si supplementation on bamboo (*Gigantocloa* sp. Malay Dwarf) growth and Cu sensitivity grown in hydroponics culture. The result showed high Si concentration in the leaves (218 mg·g^{-1}) while Si concentration in stems and roots of the bamboo averaged 65 mg·g^{-1} and 11 mg·g^{-1}, respectively and they observed a decrease in Si in the nutrient solution. There was also an increase in the concentration of Ca, Mg and iron (Fe) in the nutrient solution. Interestingly, they realized that increasing

Table 5. Average silicon concentration in different components of bamboo.

Bamboo parts	Concentration (mg·g^{-1})	References
Leaves		
Culms	48–68	
	1.2–3.5	Mitsutoshi and Takenaka (2014)
Leaves		
Stem		
Roots	218	
	65	
	11	Collin et al. (2013)
Leaves	82–159	Collin et al. (2012)
Shoots	1400	Saini et al. (2017)
Leaves	48–60	Ma and Takahashi (2002)
Culms	9.2–14.6	Umemura and Takenaka (2014).
Stems	5.7–27	Collin et al. (2012)

Si concentration in the nutrient solution did not significantly influence the growth parameters. Even though Si has often been attributed to promote plant growth in literature, there have been studies on increase in total biomass weight of some horticultural crops such as cucumber, rose (Voogt and Sonneveld 2001; Hwang et al. 2005). However, it is not well known whether the growth stimulation was caused by Si nutritional effect or alleviation of biotic and abiotic stresses (Collins et al. 2013).

Biochar Application in Bamboo

Biochar is a highly aromatic material that is rich in carbon (C) (60 to 85%) and is produced from the pyrolysis of organic plant biomass in the presence of limited oxygen (Lehmann, 2007). The use of biochar as a soil amendment has been shown to influence the physical properties of the soil such as texture, structure, porosity and available water capacity. In addition to this, studies have shown that biochar can also improve soil fertility and affect plant growth (Cayuela et al. 2014) controls N cycling and retain soil nutrients by reducing NO_2 emissions (Bai et al. 2015). Xu et al. (2022) conducted a study to determine the effect of biochar on bamboo seedling growth and to provide new insights into biochar fertilization in bamboo plantations to improve productivity. This study made use of bamboo branches as raw materials to prepare biochar. According to the authors, bamboo branches are typically considered as a waste and converting it to biochar was a way of utilizing it positively and addressing environmental and social concerns associated with it. From their result, biochar proved to be effective in increasing soil organic carbon (SOC), total nitrogen (TN), and soil water content. Surprisingly, they also discovered that biochar alone did not significantly increase plant biomass of bamboo seedlings, but the combination of biochar and mineral nutrient significantly increased the biomass of leaves. This shows that biochar helps to improve soil quality and nutrient availability, but the synergetic effect of biochar and mineral fertilizers would greatly benefit bamboo production (Xu et al. 2022).

Irrigation Management and Water Requirements for Bamboo

In Florida, there is limited information and apparently no data on the water requirement of bamboo for optimal production. Various aspects of the growth and development of bamboo have been documented for other regions such as Asia, Africa and portions of Europe including the optimal irrigation systems (Kleinhenz and Midmore, 2001; Awe and Ogedengbe, 2011; Piouceau et al. 2014;). For example, the actual evapotranspiration (ET) and the crop coefficients (K_c) for the five tropical and temperate species of three-year-old bamboo plants, i.e. *Bambusa oldhamii, Bambusa multiplex, Bambusa vulgaris, Phyllostachys aurea* and *Pseudosasa japonica*, were studied in lysimeters for a period of more than one year under a tropical climate and the average evapotranspiration (ET) rates for the bamboo species ranged from 4 to 7 mm·day^{-1} with maximum values of between 10.7 and 17.1 mm·day^{-1} during the wet season, and an average K_c of 1.1 to 1.9 in France. The ET was correlated to weather parameters, especially minimum temperatures. The differences in ET rates between the bamboo species were explained by morphological parameters, particularly the total aboveground biomass. In another report, researchers from Australia emphasized that irrigation rates would depend on the season and cultivar genus of the bamboo (Kleinhenz and Midmore, 2001). Bamboo species distribution and growth are more strongly influenced by precipitation than any other climatic variable, except temperature (Biswas, 1988). Bamboo has an estimated transpiration rate of 3300 mm, a number close to double the annual rainfall of Florida. Therefore, introduction of irrigation would improve bamboo production particularly in areas with limited rainfall amount and distribution (Kleinhenz and Midmore, 2000).

Several researchers have tried to study and quantify the effects of different irrigation approaches that will best suit bamboo management for effective productivity and economic advantage. Research conducted by Qui et al. (1992) in China, shows a dramatic increase in bamboo productivity associated with increased annual rainfall. Kleinhenz and Midmore (2002), conducted research that showed that the emergence of shoots coincided with peak rainfall seasons (January–March) in the then Northern Territory of Australia. In another study conducted by Kleinhenz et al. (2003) to estimate the effects of different rates of water and fertilizer on soil water, growth and yield components of bamboo. Their results showed that irrigation treatments had a much higher effect on bamboo production than the fertilizer treatment applied. Interestingly, they also observed that the high-irrigation treatment with high N-application produced the highest shoot yields. This shows that growth and yield of bamboo shoots was considerably affected by soil water availability much more than the fertilizer rates. In order words, bamboo did not respond to higher fertilizer rates when soil water conditions were limiting. In another field trial conducted by Treynor and Midmore, (2009) to determine the effects of irrigation management, nitrogen fertilization on shoot and culm yields of two varieties of bamboo (*Dendrocalamus asper* and *Dendrocalamus latiflorus*). From their study, they observed that supplying high-volume of irrigation to mature clumps only before and during the shoot season seems to have the same effect on shoot yield as when irrigation occurred throughout the dry season. They observed increase in shoot yield and early shoot growth that was due to higher fertilizer rates which was influenced by a good wet season. This discovery may help growers save on irrigation pumping costs. From the experiment, shooting of the bamboo species began when the upper profile reached 100 mm and 80 mm at the 50-cm profile for two different sites studied, respectively. The application of fertilizer effectively hastened the timing of shoot appearance and number of shoots especially in the wetter seasons, but the culm yield was not affected by fertilizer rate.

Concluding Summary

It is important to develop guidelines for the application of N, P, other nutrients and water management practices adapted to Florida site-specific soil characteristics for sustainable bamboo production. Eventually, studies can be conducted by drawing as a foundation the guidelines discussed from other bamboo-producing regions. They proved effective regarding planting density, bamboo variety, irrigation, and management practices. In the beginning, these studies could be done under controlled environmental conditions and later extended to the on-farm scale in order to refine and develop best water and nutrient management practices adapted for Florida soils and sub-humid climate.

Literature Cited

Awe G.O. & Ogedengbe K. 2011. Performance evaluation of bamboo (*Bambusa vulgaris* Schrad)-pipe and medi-emitter in a gravity-flow drip irrigation system. *Int. J. Agric. For.* 1(1):9–13.

Awol A. 2015. Nutrient, mineral and bioactive constituent evaluation of bamboo shoots grown in Masha area, South-West of Ethiopia. *American Scientific Research Journal for Engineering, Technology, and Sciences*. 7(1):15–25.

Bai S.H., Reverchon F., Xu C.Y., Xu Z., Blumfield T. J., Zhao H., van Zwieten L., & Wallace H.M. 2015. Wood biochar increases nitrogen retention in field settings mainly through abiotic processes. *Soil Biology and Biochemistry*. 90:232–240. https://doi.org/10.1016/j.soilbio.2015.08.007

Biswas S. 1988. Studies on bamboo distribution in north-eastern region of India. *Indian Forester*. 114(9):514–531.

Bo L., Yuzuo T., Yongsheng R., Xianwei L., & Chuan F. 2021. Effect of formula fertilization on yield and quality of black bamboo. *Soil and Fertilizer Sciences in China*. http://dx.doi.org/10.11838/sfsc.1673-6257.20067

Cayuela M.L., Van Zwieten L., Singh B.P., Jeffery S., Roig A., & Sánchez-Monedero M.A. 2014. Biochar's role in mitigating soil nitrous oxide emissions: A review and meta-analysis. *Agriculture, Ecosystems and Environment*. 191:5–16. https://doi.org/10.1016/j.agee.2013.10.009

Collin B., Doelsch E., Keller C., Panfili F., & Meunier J.D. 2013. Effects of silicon and copper on bamboo grown hydroponically. *Environmental Science and Pollution Research*. 20:6482–6495. https://doi.org/10.1007/s11356-013-1703-0

Collin B., Doelsch E., Keller C., Panfili F., & Meunier J.D. 2012. Distribution and variability of silicon, copper and zinc in different bamboo species. *Plant and Soil*. 351:377–387. https://doi.org/10.1007/s11104-011-0974-9

Currie H.A., & Perry C.C. 2007. Silica in plants: Biological, biochemical and chemical studies. *Annals of Botany*. 100(7):1383–1389. https://doi.org/10.1093/aob/mcm24 7

Ding T.P., Zhou J.X., Wan D. F., Chen Z.Y., Wang C.Y., & Zhang F. 2008. Silicon isotope fractionation in bamboo and its significance to the biogeochemical cycle of silicon. *Geochimica et Cosmochimica Acta*. 72(5):1381–1395. https://doi.org/10.1016/j.gca.2008.01.008

Dwivedi A.K., Kumar A., Baredar P., & Prakash O. 2019. Bamboo as a complementary crop to address climate change and livelihoods—Insights from India. *Forest Policy and Economics*. 102:66–74. https://doi.org/10.1016/j.forpol.2019.02.007

Emamverdian A., Ding Y., & Xie Y. 2018. Phytoremediation potential of bamboo plant in China. *Ecology, Environment and Conservation*. 24(1):530-539.

Gao P., Zuo Z., Zhang R., Qiu Y., He R., Gao R., & Gui R. 2016. Optimum nitrogen fertilization *for Phyllostachys edulis* productivity and photosynthetic response. *Agronomy Journal*. 108(1):448-458. https://doi.org/10.2134/agronj2015.0324

Guntzer F., Keller C., & Meunier J-D. 2012. Benefits of plant silicon for crops: A review. *Agron. Sustain. Dev.* 32:201–213. https://doi.org/10.1007/s1359 3-011-0039-8

Hinsinger P. 2001. Bioavailability of soil inorganic P in the rhizosphere as affected by root-induced chemical changes: a review. *Plant and Soil*. 237(2):173–195. https://doi.org/10.1023/A:1013351617532

Hwang S.J., Park H.M., & Jeong B.R. 2005. Effects of potassium silicate on the growth of miniature rose 'Pinocchio' grown on rockwool and its cut flower quality. *Horticultural Society Journal*. 74(3):242–247 https://doi.org/10.2503/jjshs.74.242

Ikewuchi J.C. & Ikewuchi C.C. 2009. Comparative study of the mineral element composition of some common Nigerian medicinal plants. *Pacific Journal of Science and Technology*. 10(1):362–366.

Jiang M., Liu S., Li,Y., Li X., Luo Z., Song H., & Chen Q. 2019. EDTA-facilitated toxic tolerance, absorption and translocation and phytoremediation of lead by dwarf bamboos *Ecotoxicology and Environmental Safety*. 170:502–512. https://doi.org/10.1016/j.ecoenv.2018.12.020

Khan,F., Siddique A.B., Shabala S., Zhou M., & Zhao C. 2023. Phosphorus plays key roles in regulating plants' physiological responses to abiotic stresses. *Plants*. 12(15). https://doi.org/10.3390/plants12152861

Kim Y.H., Khan A.L., Waqas M., & Lee I.J. 2017. Silicon regulates antioxidant activities of crop plants under abiotic-induced oxidative stress: a review. *Frontiers in Plant Science*. 8:510. https://doi.org/10.3389/fpls.2017.00510

Kim C., Baek G., Yoo B.O., Jung S., & Lee K.S. 2018. Regular fertilization effects on the nutrient distribution of bamboo components in a Moso bamboo [*Phyllostachys pubescens* (Mazel) Ohwi] stand in South Korea. *Forests*. 9(11):671. https://doi.org/10.3390/f9110671

Kleiber T. 2018. The role of silicon in plant tolerance to abiotic stress. p. 253–267. In: Hasanuzzaman M., Fujita M., Oku H., Nahar K., & Hawrylak-Nowak, B. (eds.). Plant nutrients and abiotic stress tolerance. Springer, Singapore. https://doi.org/10.1007/978-981-10-9044-8_11

Kleinhenz V., & Midmore D.J. 2000. Estimating water usage of bamboo. *Access to Asian Vegetables*. 28:1–2.

Kleinhenz V., & D.J. Midmore. 2001. Aspects of bamboo agronomy. *Advances in Agronomy*. 74:99–153. https://doi.org/10.1016/S0065-2113(01)74032-1

Kleinhenz V., & Midmore,D.J. 2002. Improved management practices for culinary bamboo shoots. *RIRDC Publication*. (02/035).

Kleinhenz V., Milne J., Walsh K.B., & Midmore D.J. 2003. A case study on the effects of irrigation and fertilization on soil water and soil nutrient status, and on growth and yield of bamboo (*Phyllostachys pubescens*) shoots. *J. Bamboo Rattan*. 2(3):281–293.

Lehmann J. 2007. A handful of carbon. *Nature*. 447(7141):143–144. https://doi.org/10.1038/447143a

Lemoine R., Camera S.L., Atanassova R., Dédaldéchamp F., Allario T., Pourtau N., Bonnemain L., Laloi M., Coutos-Thévenot P., Maurousset L., Faucher M., Girousse C., Lemonnier P., Parrilla J., & Durand M. 2013. Source-to-sink transport of sugar and regulation by environmental factors. *Frontiers in Plant Science*. 4:272. https://doi.org/10.3389/fpls.2013.00272

Li J., J. Gao, H. Gao, B. Zhang & L. Shi. 2019. Effects of nitrogen treatment on growth characteristics in Moso seedling. *Biomed. J. Sci. Tech. Res.* 15(4):11517–11523. http://dx.doi.org/10.26717/BJSTR.2019.15.002734

Li R., Werger M.J.A., During H.J., & Zhong Z. C. 1998. Carbon and nutrient dynamics in relation to growth rhythm in the giant bamboo *Phyllostachys pubescens*. *Plant and Soil*. 201:113–123 https://doi.org/10.1023/A:1004322812651

Li S.D., & Xu C.D. 1997. The history of Chinese bamboo industry nd the challenge for development in 21st century. p. 4. In: Paper summaries. International bamboo workshop, bamboo towards 21st century. Research Institute of Subtropical Forestry, Chinese Academy of Forestry Anji, Zhejiang, PR China.

Li, Z.J., Lin, P., He, J.Y., Yang Z.W., & Lin Y.M. 2006. Silicon's organic pool and biological cycle in moso bamboo community of Wuyishan Biosphere Reserve. *Journal of Zhejiang University Science B*. 7:849–857. https://doi.org/10.1631/jzus.2006.B0849

Liang Y., Sun,W., Zhu Y.G., & Christie P. 2007. Mechanisms of silicon-mediated alleviation of abiotic stresses in higher plants: A review. *Environmental Pollution*. 147(2):422–428. https://doi.org/10.1016/j.envpol.2006.06.008

Liu E., Pimpin L., Shulkin M., Kranz S., Duggan C.P., Mozaffarian D., & Fawzi W.W. 2018. Effect of zinc supplementation on growth outcomes in children under 5 years of age. *Nutrients*. 10(3):377. https://doi.org/10.3390/nu10030377

Liu C, Wang Y, Jin Y, Pan K, Zhou X, & Li N. 2017. Photoprotection regulated by phosphorus application can improve photosynthetic performance and alleviate oxidative damage in dwarf bamboo subjected to water stress. *Plant Physiol Biochem*. 118:88–97. Epub 2017 Jun 8. PMID: 28624684. https://doi.org/10.1016/j.plaphy.2017.05.022

López-Pérez M.C., Pérez-Labrada F., Ramírez-Pérez L.J., Juárez-Maldonado A., Morales-Díaz A.B., González-Morales S., García-Dávila L.R., García-Mata J., & Benavides-Mendoza A. 2018. Dynamic modeling of silicon bioavailability, uptake, transport, and accumulation: Applicability in improving the nutritional quality of tomato. *Frontiers in Plant Science*. 9:647. https://doi.org/10.3389/fpls.2018.00647

Lux A., Luxová M., Abe J., Tanimoto E., Hattori T., & Inanaga S. 2003. The dynamics of silicon deposition in the sorghum root endodermis. *New Phytologist*. 158(3):437–441. https ://doi.org/10.104 6/j.1469-8137.2003.00764 .x

Luyckx M., Hausman J.F., Lutts S., & Guerriero,G. 2017. Silicon and plants: current knowledge and technological perspectives. *Frontiers in Plant Science*. 8:411. https://doi.org/10.3389/fpls.2017.00411

Ma J.F. & Takahashi E. 2002. Soil, fertilizer, and plant silicon research in Japan. Elsevier.

Ma J.F., Miyake Y., & Takahashi E. 2001. Silicon as a beneficial element for crop plants. *Studies in Plant Science*. 8:17–39. https://doi.org/10.1016/S0928-3420(01)80006-9

Maoyi F., Jingzhong X., & Mingyu F. 1991. Fertilization studies in bamboo stands with different end uses II. High yield management method for bamboo shoot stands. *Forest Research*. 4(3):238–245. http://lykxyj.xml-journal.net/en/article/id/19910302

Midmore D. 2009. Silvicultural management of bamboo in the Philippines and Australia for shoots and timber: Proceedings of a workshop held in Los Baños, the Philippines November 22–23 2006. CQ University.

Morton C.M., Pullabhotla H., Bevis L., & Lobell D.B. 2023. Soil micronutrients linked to human health in India. *Scientific Reports*. 13(1):13591. https://doi.org/10.1038/s41598-023-39084-8

Motomura H., Fujii T., & Suzuki M. 2004. Silica deposition in relation to ageing of leaf tissues in *Sasa veitchii* (Carriere) Rehder (Poaceae: Bambusoideae). *Annals of Botany*. 93(3):235–248. https://doi.org/10.1093/aob/mch034

Ni H., Su, W., Fan S., & Chu H. 2021. Effects of intensive management pratices on rhizosphere soil properties, root growth, and nutrient uptake in Moso bamboo plantations in subtropical China. *Forest Ecology and Management*. 493:119083. https://doi.org/10.1016/j.foreco.2021.119083

Piouceau J., Bois G., Panfili F., Anastase M., Dufossé L., & Arfi V. 2014. Effects of high nutrient supply on the growth of seven bamboo species. *International Journal of Phytoremediation*. 16(10):1042–1057. https://doi.org/10.1080/15226514.2013.810583

Piouceau J., Panfili F., Bois G., Anastase M., Dufossé L., & Arfi V. 2014. Actual evapotranspiration and crop coefficients for five species of three-year-old bamboo plants under a tropical climate. *Agricultural*

Water Management. 137:15–22.

Qiu G.X., Shen Y.K., Li D.Y., Wang W., & Huang Q.M. 1992. Bamboo in subtropical eastern China. p. 159–188. In: Long S.P., Jones M.B., & Roberts M.J., (eds.). Primary productivity of grass ecosystems of the tropics and sub-tropics. Chapman and Hall, London.

Qian Z., Xiao S., Gao J., & Zhuang S. 2021. Effects of Bamboo (*Phyllostachys praecox*) cultivation on soil nitrogen fractions and mineralization. *Forests*. 12(8):1109. https://doi.org/10.3390/f12081109

Roberts T.L. & Johnston A.E. 2015. Phosphorus use efficiency and management in agriculture. *Resources, Conservation and Recycling*. 105:275-281.

Saini N., Rawat K., Bisht M.S., & Nirmala C. 2017. Qualitative and quantitative mineral element variances in shoots of two edible bamboo species after processing and storage evaluated by wavelength dispersion x-ray fluorescence spectrometry. *Int. J. Innov. Res. Sci. Engine Technol*. 6(5):8265–8270. https://doi.org/10.15680/IJIRSET.2017.0605167

Shanmughavel P. & Francis K. 2001. Bioproductivity and nutrient cycling in bamboo and acacia plantation forests. *Bioresource Technology*. 80(1):45–48. https://doi.org/10.1016/S0960-8524(01)00060-8

Sofi P., Chopan J., Islam M., Bhat G., Malik A., & Rather T. 2018. Nitrogen use efficiency, nutritive value of leaves and survival of Moso bamboo (*Phyllostachys pubescence*) seedlings under the influence of N-fertigation. *Chemical Science Review and Letters*. 7(26):673–679.

Song X., Zhou G., Jiang H., Yu S., Fu J., Li W., Wang W., Ma Z., & Peng, C. 2011. Carbon sequestration by Chinese bamboo forests and their ecological benefits: Assessment of potential, problems, and future challenges. *Environmental Reviews*. 19:418–428. https://doi.org/10.1139/a11-015

Srivaro S. 2018. Potential of three sympodial bamboo species naturally growing in Thailand for structural application. *European Journal of Wood and Wood Products*. 76(2):643–653. https://doi.org/10.1007/s00107-017-1218-3

Stahl S.L. & Sargen S.A. 2021. Preservation of postharvest quality of fresh bamboo shoots via suppression of lignification, browning, and decay: A review. *Proc. Florida State Hort. Soc*. 134:167–173.

Sun J., Li J., Koyama K., Hu D., Zhong Q., & Cheng, D. 2023. The morphology and nutrient content drive the leaf carbon capture and economic trait variations in subtropical bamboo forest. *Frontiers in Plant Science*. 14:1137487. https://doi.org/10.3389/fpls.2023.1137487

Traynor M. &Midmore D.J. 2009. Cultivated bamboo in the Northern Territory of Australia. In: Midmore D.J. (ed.). Silvicultural management of bamboo in the Philippines and Australia for shoots and timber. Australian Center for International Agricultural Research Proceedings. 129:108–123.

Tripathi S.K. & Singh K.P. 1996. Fine root dynamics in a dry tropical bambo savanna in India. p. 23–28. In: Rangelands in a sustainable biosphere. Proc. Fifth International Rangeland Congress. Salt Lake City, UT.

Umemura M. & Takenaka C. 2014. Biological cycle of silicon in moso bamboo (*Phyllostachys pubescens*) forests in central Japan. *Ecological Research*. 29:501–510. https://doi.org/10.1007/s11284-014-1150-5

UC Sustainable Agriculture Research and Education Program. 2017. Soil Nutrient Management. What is Sustainable Agriculture? University of California Division of Agriculture and Natural Resources, Davis, CA. https://sarep.ucdavis.edu/sustainable-ag/soil-nutrient-management

Voogt W. & Sonneveld C. 2001. Silicon in horticultural crops grown in soilless culture. *Studies in Plant Science*. 8:115–131. https://doi.org/10.1016/S0928-3420(01)80010-0

Wang S., Chen T., Liu E., & Liu C. 2020. Accessing the nursing behaviour of Moso bamboo (*Phyllostachys edilus*) on carbohydrates dynamics and photosystems. *Scientific Reports*. 10(1):1–12. https://doi.org/10.1038/s41598-020-57643-1

Xu Y., Yang N., Ge X., & Zhou B. 2022. Biochar combined with nitrogen alters rhizosphere soil nutrients and microbial communities and promotes growth of moso bamboo seedlings. *Forests*. 13(7):1043. https://doi.org/10.3390/f13071043

Yan G.C., Nikolic M., Ye M.J., Xiao Z.X., & Liang Y.C. 2018. Silicon acquisition and accumulation in plant and its significance for agriculture. *Journal of Integrative Agriculture*. 17(10):2138–2150. https://doi.org/10.1016/s2095-3119(18)62037-4

Yan W., Mahmood Q., Peng D., Fu W., Chen T., Wang,Y., Li S., Chen J,, & Liu D. 2015. The spatial distribution pattern of heavy metals and risk assessment of moso bamboo forest soil around lead–zinc mine in Southeastern China. *Soil and Tillage Research*. 153:120–130. https://doi.org/10.1016/j.still.2015.05.013

Yang Z., Zhou B., Ge X., Cao Y., Brunner I., Shi J., and Li M. 2021. Species-specific responses of root morphology of three co-existing tree species to nutrient patches reflect their root foraging strategies. *Frontiers in Plant Science*. 11:618222. https://doi.org/10.3389/fpls.2020.618222

Yang K., Zhu C., Zhang J., Li Z., Liu,Y., Song X., and Gao Z. 2022. Nitrogen fertilization in bamboo forest accelerates the shoot growth and alters the lignification process in shoots. I*ndustrial Crops and Products*. 187:115368. https://doi.org/10.1016/j.indcrop.2022.115368

Zhang Q., Wang, J., Wang G., Shamsi I.H., and Wang X. 2014. Site-specific nutrientmanagement in a bamboo forest in Southeast China. *Agriculture, Ecosystems & Environment*. 197:264–270. https://doi.org/10.1016/j.agee.2014.08.013

Zhang M., Zhang W., Bai S.H., Niu Y., Hu D., Ji H., & Xu Z. 2019. Minor increases in *Phyllostachys edulis* (Moso bamboo) biomass despite evident alterations of soil bacterial community structure after phosphorus fertilization alone: Based on field studies at different altitudes. *Forest Ecology and Management*. 451:117561. https://doi.org/10.1016/j.foreco.2019.117561

Zhang Q., Wang J., Wang G., Shamsi I.H., & Wang X. 2014. Site-specific nutrient management in a bamboo forest in Southeast China. *Agriculture, Ecosystems and Environment*. 197:264–270. https://doi.org/10.1016/j.agee.2014.08.013

Proc. Fla. State Hort. Soc. 136:223. 2023.

The Rest of the Story: Three Years of Research With "Natural" Herbicides and Aquatic Plants

Lyn A. Gettys*, Kyle L. Thayer, Joseph W. Sigmon, and Jennifer H. Bishop

University of Florida, IFAS, Fort Lauderdale Research and Education Center, Davie FL

—abstract—

The 2019 Florida Fish and Wildlife Conservation Commission (FWC) "pause" in herbicidal aquatic weed control spurred interest in evaluating the efficacy and selectivity of alternative products for invasive species management in aquatic systems. We investigated the activity of a range of concentrations of acetic acid, d-limonene and combinations of the two products after foliar applications to selected invasive floating and native emergent plants. Plants were co-cultured in 68L mesocosms until floating plant coverage was > 80%, then treated once with a single product or combination (plus surfactant). Four replicates were prepared for each treatment and all plants were maintained for 8 weeks after treatment. Plants were then evaluated for visual quality, destructively harvested and placed in a forced-air drying oven for 2 weeks to determine reduction in biomass compared to untreated control (UTC) plants. In most cases, native emergent plants were not seriously damaged by treatments. Floating invasive species were adequately controlled (>80% reduction in biomass) by some treatments. However, these treatments are not yet labeled for aquatic use and deployment of these treatments at scale would result in a 16-fold to 47-fold increase in material costs alone. These data reveal that alternative products may be useful for aquatic weed management but more research – including field trials – is necessary to confirm these results.

*Corresponding author. Email: lgettys@ufl.edu

Proc. Fla. State Hort. Soc. 136:224. 2023.

—Scientific Note—

Teaching Children Florida-Friendly Landscaping

Leslie N. Munroe*, Kaitlyn McCarty, and Andrea Lazzari

University of Florida, IFAS, Indian River County Extension, Vero Beach, FL

Keywords. environmental awareness, Florida-Friendly, 4-H garden club, junior Master Gardener, water quality, water quantity, youth development

We developed a project to provide the young students with the best possible gardening experience by using the "Growing Up Florida-Friendly," for all garden management discussions and decisions with the children in conjunction with the Junior Master Gardener curriculum for youth development and to: 1) increase student awareness of Florida's water quality and quantity issues; 2) introduce Florida-Friendly Landscaping to youth participants; 3) help them learn how to use the principles to manage and plan their gardening activities; and 4) encourage them to use their critical thinking skills while working in the garden.

We found that the children's awareness increased of how their actions in the yard and garden can affect the environment. The children are having a good experience gardening. They are enjoying the success of putting the right plant in the right place and lower-maintenance gardening activities. The young gardeners explain what they are doing to classmates who are not in the garden club. Their enthusiasm for the club is generating more interest in the program.

The students were allowed to plan small projects and manage them themselves using Florida-Friendly Principles. The students enjoyed making decisions and implementing their plans. The children demonstrated that they have unique aesthetic preferences.

Our students use the Florida-Friendly Landscaping principles to manage their selection and care of plants they take home. They share knowledge with members of their household. It is a huge confidence builder for them to be trusted and to be regarded as responsible.

We will continue to use this combination of curricula to administer programming with the young people. This region-specific gardening education combined with hands-on gardening experiences contributes to building environmentally conscious and responsible young people in Florida's unique ecosystem. Additional fundraising is underway to give current and future students more options for their annual projects.

*Corresponding author. Email: lnmunroe@ufl.edu

Proc. Fla. State Hort. Soc. 136:225–226. 2023.

Case Study: Using the State Florida-Friendly Landscaping™ Survey to Estimate Water Savings

Tiare Silvasy[1]*, Jennifer W. Marvin[2], C.J. Bain[2], and Esen Momol[2]

[1]*University of Florida, IFAS, Hillsborough County Extension, Seffner FL*

[2]*University of Florida, IFAS, Center of Land Use Efficiency, Gainesville FL*

Keywords. Extension education, water savings impacts, survey methods

The goals of extension education are to increase knowledge, inspire behavior change and ultimately show program impacts to make the world a better place. Extension agents who provide education about water conservation strategies and sustainable landscape practices have the opportunity to utilize the Florida-Friendly Landscaping™ (FFL) behavior change survey. This can be used as a sole data collection source or supplemental to surveys agents already send out. The state FFL office provides a comprehensive survey to track behavior change related to water conservation practices. Extension agents from all over the state are encouraged to share their educational contacts with the state office. Once a year, the state office sends a survey to the contacts received from all counties. The survey asks questions including: "Did you reduce watering in the summer?", "Did you reduce watering in the winter?", "Did you calibrate your irrigation system to deliver 1/2" to 3/4" instead of 1" or more?" To better understand how agents can use these survey results, we examine one case study using data from Orange County. In 2022, an extension agent participated in educational events and submitted class participant emails (n = 1858) to the state office. The results were collected and converted to water savings metrics using formulas based on accepted assumptions. Based on client self-reported follow-up surveys from Orange County (n = 226), the annual water savings due to adoption of FFL practices was approximately 5.5 million gallons. This is enough water to supply 63 households with water per year, water bill savings valued at $23,800, and utilities water delivery savings of $14,300. It remains imperative to document effectiveness of water conservation programs which help to protect Florida's water resources. The state FFL survey is an effective tool available for use by extension agents, faculty, specialists, and staff to quantify water savings and report measurable impacts.

Florida freshwater resources are limited. With Florida's population increasing by over 1000 people per day, we expect to run out of water by the year 2070. Water conservation is an important part of the solution, with lawn and landscapes accounting for more than 50% of household water use. Extension agents deliver educational programs to inform residents on sustainable landscaping practices, including strategies to conserve water. Documenting behavior change is important to both the agents and statewide program leaders to show water conservation impacts. These impacts from water savings are easy to quantify because they can convert to gallons and dollars saved using formulas based on accepted assumptions (Boyer & Dukes, 2015). Capturing and measuring results from water savings is an important addition to the report of accomplishments (ROA) for residential horticulture agents, Florida-Friendly Landscaping™ (FFL) agents, and Master Gardener Volunteer coordinators. Water conservation is also critical to our natural resources and deferring the costs for more expensive water treatment options such as desalinization (Copeland & Carter, 2015; Borisova et al., 2017; US EPA, 2022). Reducing water use can benefit residents financially by reducing both water and energy bills.

Materials and Methods

Surveys are excellent tools to gather data and measure results from educational programs. Extension agents have the choice to create their own survey or use the survey administered by the state FFL office or do both to track behavior change. Surveys are sent to class participants by email six months after the educational event to allow time for residents to make changes in their landscape. To use the state FFL survey, agents can upload their educational contacts email addresses in the educator's login portal on the FFL website (Fig. 1). All that is required to use the state FFL survey is the person's first name, last name, and email address. Survey questions to assess change related to watering practices include reduce watering in the summer, reduce watering in the winter, reduce irrigation from 3 days a week to 2 days per week, calibrate your irrigation system to deliver 1/2"to 3/4"instead of 1" or more, convert turfgrass area to landscaped bed with micro irrigation, and convert spray head nozzles to multi-stream nozzles.

Fig. 1. Use the educator's login on the Florida-Friendly Landscaping website to access the portal to submit email contacts to be surveyed.

*Corresponding author. Email: tsilvasy@ufl.edu

Orange County Impact Statement		
In Florida in 2022, Orange County FFL efforts resulted in a combined estimated water savings of	5,534,526	gallons of water annually, based on Extension clients' reported behavior change.
This is enough water to supply nearly	63	households with water per year [based on the average of 88,000 thousand gallons per household per year].
This water savings is valued annually at a total of	$23,854	in water bill savings for participating households [based on the average statewide/regional value of $4.30 per 1,000 gallons].
This water savings is valued annually at a total of	$14,390	in water delivery costs for utilities [based on the average cost of $2.60 per 1,000 gallons in delivery costs] statewide.

Fig. 2. Water savings results for Orange County, FL, calculated using the Florida Friendly Landscape state survey in 2022.

Component	Results	Unit	Justification
Total water savings	5,534,526	Gallons	FFL Calculator (Boyer and Dukes, 2018)
Energy savings for water utilities	19,094	kWh	The lifecycle energy-intensity of water in cities nationally is estimated to be 3,300-3,600 kWh per million gallons delivered and treated (Copeland & Carter, 2014)
$ savings for water utilities	$14,389	$	$2.60 per 1,000 gallons in delivery costs (Borisova et al., 2021)
Greenhouse gases	13.54	Metric tons CO2	7.09 × 10-4 metric tons CO2/kWh (U.S. EPA, 2022)
Alternative water supply infrastructure	$47,098	$	$8.51 per thousand gallons (desalinization) (Borisova et al., 2021)

Fig. 3. Further impacts of water savings for energy and associated costs for Florida in the year 2022.

Results and Discussion

In 2022, 1,858 educational contacts were submitted to the state FFL survey from programs in Orange County, Florida. We received 226 responses, a response rate of 12%. Gallons of water saved and water savings in dollars is reported in Fig. 2. Household water use reduction can benefit residents financially by reducing both water and energy bills. Reducing water use also decreases the energy needed for the public utilities to treat and deliver water to residential users. These energy savings can reduce fossil fuel use, and related greenhouse gas emissions. Further impacts of these water savings can be calculated for energy savings of cost savings from deferring the investment of infrastructure for alternative water supply (Fig. 3). It remains imperative to document effectiveness of water conservation programs which help to protect Florida's water resources. The state FFL survey is an effective tool available for use by extension agents, faculty, specialists, and staff to quantify water savings and report measurable impacts.

Literature Cited

Borisova T., Warner L.A.S., Searcy J., Chaudhary A.K., & Dukes M.D. 2017. Estimating benefits of residential outdoor water conservation: A step-by-step guide. Extension publication FE1009. University of Florida, Gainesville, FL. https://doi.org/10.32473/edis-fe1009-2017

Boyer M. & Dukes M.D. 2015. Estimated water savings potential of Florida-Friendly Landscaping™ activities. Extension publication AE515. University of Florida, Gainesville, FL. https://doi.org/10.32473/edis-ae515-2015

Copeland C. & Carter N.T. 2014. Energy-water nexus: The water sector's energy use. Congressional Research Service Report R43200. Retrieved December 1, 2023, from https://sgp.fas.org/crs/misc/R43200.pdf

United States Environmental Protection Agency (US EPA). 2022. Greenhouse gases equivalencies calculator. Retrieved December 1, 2023, from https://www.epa.gov/energy/greenhouse-gas-equivalencies-calculator

Agroecology and Natural Resources Section

Proc. Fla. State Hort. Soc. 136:227. 2023.

Birding: Introduce this Popular Hobby into Your Program for Success

Jim E. Davis*

University of Florida, IFAS, Extension Sumter and Hernando Counties, Bushnell, FL

Keywords. birding, natural resources, nature, wildlife

Birding enthusiasts, also known as "birders," contribute significantly to the economy, spending an estimated $40 billion annually on supplies, birding trips, and photography equipment (USFWS, 2021). This robust spending generates over 860,000 jobs, highlighting the economic importance of the birding industry. In Florida alone, birding and wildlife viewing activities generate over four billion dollars annually and create around 44,000 jobs (Visit Florida, 2021). The COVID-19 pandemic has further fueled the growth of these numbers, emphasizing the enduring appeal of birding during challenging times. To engage this audience, a series of virtual webinars were developed to educate residents on the identification and biology of Florida's birds. The webinars also included essential photography tips and field craft techniques. The objective was to measure knowledge gain, aiming for at least a 20% improvement in participants' understanding of birds found in Central Florida.

Birding, or birdwatching, involves observing and identifying birds in their natural habitats. Birders use various methods, including naked-eye observation, binoculars, spotting scopes, and listening to bird calls and songs for identification. The primary goal of birding is to appreciate and study avian diversity, encompassing different species, behaviors, and ecological roles. Birding is pursued as both a recreational hobby and a serious scientific endeavor, providing opportunities for individuals to connect with nature, explore diverse habitats, and deepen their understanding of avian ecology. This activity can take place in various settings, ranging from urban parks to forests, wetlands, coastal areas, and even backyards. The popularity of birding is a global phenomenon, attracting individuals of all ages and backgrounds who share an admiration for birds and a commitment to their protection.

Materials and Methods

Webinars were conducted on the Zoom platform, lasting approximately one hour each. Scheduled every other Friday from 1:00 pm to 2:00 pm, Question and answer sessions were facilitated, and closed captioning was provided. Recordings of all webinars were made available on YouTube. Pre and posttests, consisting of about five true/false or multiple-choice questions, were used to assess knowledge gain. Registration was required to collect attendee emails due to changes in the Zoom setup, and data collection was simplified using screenshots from cellular phones. To gauge behavior change, a follow-up survey was conducted through Qualtrics. Webinar promotions were carried out via social media and online newsletters.

Results

In the recent survey conducted during several birding workshops, participants attending "Common Birds Found Around Lakes and Rivers VII" demonstrated a significant increase in knowledge gain, with a 24% improvement (n = 19pre/n = 18post). Similarly, attendees of "Common Birds Found Around Lakes in Rivers VI" showed a substantial 41% increase (n = 23pre/n = 22post) in knowledge. "Birds and Babies" also proved to be an informative session, leading to a 23% boost in knowledge (n = 16pre/n = 16post). Moreover, participants of the "Wildlife Photography" workshop exhibited a notable 40% increase (n = 34pre/n = 29post) in their photography skills.

The survey's findings also revealed the positive impact of these workshops on participants' behavior and interests. An overwhelming 92% (n = 73) of respondents stated that they shared their newfound bird identification skills and knowledge with others. Additionally, 81% (n = 74) stated that they made efforts to provide new habitat for birds in their landscapes, contributing to local conservation initiatives. The workshops' influence extended beyond birding, with 31% (n = 71) of participants adopting birdwatching as a new hobby, and 33% (n = 68) embracing wildlife photography.

Discussion

The pilot project proved highly successful in reaching a new extension audience. As webinar attendance declined after the pandemic, a strategic shift towards in-person classes became increasingly popular. To enhance these classes further, citizen-science projects, such as bluebird box monitoring, will be incorporated. Participants will be required to pay a registration fee, allowing us to gather valuable data from the boxes installed in home landscapes.

Literature Cited

U.S. Fish and Wildlife Service (USFWS). 2021. Birding in the United States: A demographic and economic analysis. Accessed March 20, 2023. https://www.fws.gov/birds/surveys-and-data/birding-in-the-united-states.php

Visit Florida. 2021. Wildlife viewing and birding in Florida. Accessed March 20, 2023. https://www.visitflorida.com/en-us/things-to-do/outdoors-nature/wildlife-birding.html

*Corresponding author. Email: dvisshdn@ufl.edu

Proc. Fla. State Hort. Soc. 136:228. 2023.

—Scientific Note—

Arbor Day Planting Event—Teaching Youth the Importance of Native Trees

Leslie N. Munroe*, Kaitlyn McCarty, and Andrea Lazarri

University of Florida, IFAS, Indian River County Extension, Vero Beach, FL

Keywords. Arbor Day, environmental awareness, Florida-Friendly Landscaping, 4-H Garden Club, Myrtaceae, native plants, native trees, youth programming, natural resources

We developed an after-school project to involve young children in tree planting activities and increase their environmental awareness. Our objectives were to help the children identify benefits we receive from trees, to teach children how to plant trees, to explain why we are using a native species and to show children how to care for trees.

We found that children are excellent tree planters. They are also very understanding of many concepts. Our children look on the trees and other plants for signs of wildlife activity. They know the trees do more than just provide us with oxygen.

The project was a great opportunity to provide the children with a hands-on experience with nature. It helped give the children a sense of responsibility and care for the planet from an early age. The children enjoyed the physical act of planting their trees.

The children developed a stronger emotional bond with nature as reflected in their subsequent care for the plants. They added several small flowers to the area "to cheer up the stoppers and keep them company." This display of increased empathy towards the environment and living organisms was very pleasant to observe.

Since this planting event, the children have continued to mindfully manage the area with the stoppers (plants in family Myrtaceae). The children expressed concern for plant care over summer and holidays when they would not be directly involved in their maintenance.

Our children have become more confident in their management of the overall garden space. They perform tasks and express a sense of accomplishment and joy. The stoppers have been in the ground for almost a year. Every child is involved in the discussions about their care. They engage in problem-solving and critical thinking when they must select new flowers to add or whether the plants need to receive irrigation.

All students returned to the after-school program elected rejoin garden club for the new school year. They have been instrumental in the planning for the next Arbor Day event in January 2024. Their desire to continue engaging in eco-friendly activities beyond this initial event has been very helpful. The original set helped teach and encourage the newer students in the process.

Suggestions for Future Direction

Continue Arbor Day Events with the 4-H Garden Club. Incorporating these activities into the overall program will play a vital role in shaping environmentally conscious and responsible future residents. Each year after 2024 a different native tree will be highlighted based on suitability for the location.

*Corresponding author. Email: lnmunroe@ufl.edu

Agroecology and Natural Resources Section

Proc. Fla. State Hort. Soc. 136:229–230. 2023.

—Scientific Note—

Common Garden Phenology Differences in Leaf Emergence of Beautyberry (*Callicarpa americana*) Between Seedlings from Hardiness Zones 8 and 10

Anna Goebel[1]*, Dalton Tryba[1], Jennifer Meader[1], Elizabeth Mello[1], Dakota Rivers[1], Sofia Siaz[1], and Brian Bovard[2]

[1]*Florida Gulf Coast University, Department of Biological Sciences, Fort Myers, FL*

[2]*Florida Gulf Coast University, Department of Ecology and Environmental Studies, Fort Myers, FL*

Keywords. climate change, common garden, localized adaptations

American beautyberry (*Callicarpa americana*) is a shrub native to southeastern United States (U.S.), and lives across different U.S. Department of Agriculture plant hardiness zones (8–10). Native gardeners sell this species, as the purple fruits provide beauty to the landscape and food for wildlife. Although native gardeners stock plants from local sources, it is unclear how localized a source must be if there are growth differences due to localized adaptations. For example, common garden experiments (CGE) found northern Florida beautyberry transplanted to Missouri had reduced survival compared to local plants (Graves & Thomas, 2011). To test for the presence of potentially genetic and adaptive differences in phenology, we compared the timing of leaf emergence from plants grown from seeds and cuttings, collected from native plants in hardiness zone 8B (Tallahassee, FL) and 10A (Charlotte Harbor, FL) and grown in a common garden in Charlotte County, FL. The goal of this work was to assist native gardeners, landscapers, landscape architects and restoration land managers improve best practices by allowing better prediction of growth and seasonal patterns due to plant origins.

As part of a larger study to examine growth and phenology differences among American beautyberry from different growth zones in Florida, we propagated plants from both cuttings and seeds for common garden experiments. Previously collected data on cuttings propagated in a common garden found that southern populations showed a higher rate of leaf production and emergence, and flowering, beginning about one month sooner than cuttings taken from northern populations. Here we compared similar leaf-out data from plants grown from seeds. The time of the first appearance of new leaves (leaf emergence), as well as the number and size of leaves was recorded every two weeks over a 90-day period. Seeds collected from Tallahassee and Charlotte Harbor in February 2022 were germinated and grown outdoors in Solo® cups in a common garden environment in Charlotte County FL. Seedlings from northern Florida lost most leaves over the winter and new leaves emerged starting in early March (Fig. 1). Seedlings from southern Florida retained more leaves through the winter and spring, with new leaves emerging starting in January. Although we had seen a delay in leaf-out in the northern populations in plants grown from cuttings, we did not expect similar responses in seedlings. Consequently, we did not start data collection until after many southern plants had begun emergence (Day 61, Fig. 1). In both data sets, the southern population began leaf-out about a month sooner, significantly earlier than the northern population (Fig. 1), as indicated by an analysis of ANCOVA of leaf numbers per plant as a function of time and population ($P = 0.0015$, ANCOVA in SAS-JMP®, Version 17.1.0, SAS Institute Inc., Cary, N.C.). Plants from southern populations also produced significantly

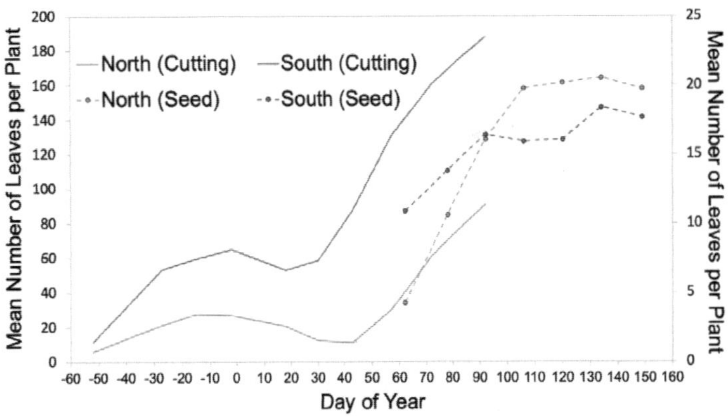

Fig. 1. Timing of leaf emergence from cuttings and seeds of American beautyberry (Callicarpa americana) grown in common gardens. Negative Day of Year values are days prior to January 1, positive numbers are days January 1 onward. The y-axes are the mean number of leaves per plant normalized for the different plant sizes (plants from cuttings were older and larger). Data from cuttings were collected in 2021 (day -55 to 95) and from seeds in 2023 (day 61-150). Plants had almost no leaves over the winter with new leaf emergence from cuttings starting about day 17 (southern population) and day 45 (northern population).

*Corresponding author. Email: agobel@fgcu.edu

more leaves earlier. These data indicate a potential for significant genetic differences among populations of American beautyberry in Florida and demonstrate the importance of using local plant material for horticultural purposes to maximize the positive ecological impacts and aesthetic appearance.

It is unclear how geographically close plants must be collected to maintain local phenotypes. Differences in hardiness zones might serve as a guide. We found phenology differences in plants from different hardiness zones approximately 400 miles apart within Florida. However, if the phenology differences have a genetic or evolutionary basis, and the hardiness zones are based on recent climate shifts, there may be a mismatch between currently defined hardiness zones and phenology differences.

The basis for phenology differences is unclear. Our findings suggest that the timing of leaf out does not appear to be set by overwintering, or at germination. We did not test if phenology differences are determined at seed set or are due to specific genetic differences. A survey of genetic differences across the full range might provide clues to localized adaptations.

Literature Cited

Graves W. & Thomas A.L. 2011. Survival and growth of (American Beautyberry) of northern and southern origin in USDA Hardiness Zones 5 and 6. *Journal of Environmental Horticulture*. 29(1):9–13.

Proc. Fla. State Hort. Soc. 136:231. 2023.

—Scientific Note—

Successes and Challenges of the Duval County Backyard Hen Permit Program

STEPHEN JENNEWEIN*

University of Florida, IFAS, Duval County Extension, Jacksonville, FL

Keywords. extension programming, poultry, urban agriculture

This scientific note aims to highlight the success and challenges of Duval County's Backyard Hen Permit Program, emphasizing its role in addressing food insecurity, promoting urban agriculture and enhancing the quality of life of the residents.

Duval County permits residents in residential zoning areas to keep backyard hens through the Backyard Hen Permit Program. Duval County Extension provides the educational resources, including a training seminar, as a prerequisite of the permit. This program educates residents on effective and responsible hen care. Since 2015, the program has led to numerous benefits, including the reduction of food waste, production of compostable manure (which helps to improve soil quality), and egg production (a source of high-quality protein).

Attendees of the Backyard Hen Seminar include a diverse range of Duval County residents. Duval County is urbanizing at a rapid pace and education on food production and agricultural appreciation is important to continued economic prosperity and public health. Many attendees have never seen any livestock, including poultry, in their lives. This class represents an introductory experience to many and a confidence building training to all. Knowledge gain was reported in 98% of survey respondents.

The Duval County Backyard Hen Permit program permits roughly 1300 households to keep some 6500 backyard hens. This leads to the production of approximately 2.2 million eggs and 273 tons of soil enriching manure annually. Many permit holders use this manure to enrich their gardens, which leads to the production of additional food. The program experienced a peak in popularity during the COVID pandemic and helped provide food security to thousands of families during the resulting supply chain disruptions. Many permit holders live in food deserts and areas with high food insecurity.

Hens can consume food waste and produce compostable manure. This reduces waste stream production and assists the county with waste management. The impact of this was acutely important during the supply chain disruptions of the pandemic as municipal waste collection became a more difficult issue to manage.

While the program has enriched the lives of many residents, challenges exist, particularly in the form of polarized opinions among residents and negative incidents that influence public sentiment. Future efforts should focus on continued collaboration with University of Florida, IFAS Extension to educate residents, address challenges, and foster a positive environment around backyard hen ownership. Additional county involvement through animal care and control and resources focused on rehoming hens are recommended. Education geared toward policy makers and homeowner associations would reduce the severity of these challenges. Municipalities considering backyard hen ordinances can draw valuable insights from Duval County's proactive approach to mitigate challenges associated with urban hen keeping.

*Corresponding author. Email: sjennewwin@ufl.edu

Author Index

H

I

J

K

L

M

N

O

P

Q

R

S

T

U

V

Thank You Sponsors

Mango.org

The Walt Disney Company

HORTSCI
Fruits and vegetables for Florida and beyond

MDPI

CRDF Citrus Research and Development Foundation

FLORIDA® TOMATOES

UF | IFAS Extension
UNIVERSITY *of* FLORIDA